T0134971

Smart Computational Strategies: Theoretical and Practical Aspects

Ashish Kumar Luhach ·
Kamarul Bin Ghazali Hawari ·
Ioan Cosmin Mihai · Pao-Ann Hsiung ·
Ravi Bhushan Mishra
Editors

Smart Computational Strategies: Theoretical and Practical Aspects

Springer

Editors
Ashish Kumar Luhach
Department of Computer Science
and Engineering
Maharishi Dayanand University
Rohtak, Haryana, India

Ioan Cosmin Mihai
Department of Police and Behavioral
Science
Alexandru Ioan Cuza Police Academy
Bucharest, Romania

Ravi Bhushan Mishra
Department of Computer Science
and Engineering
Indian Institute of Technology (BHU)
Varanasi, Uttar Pradesh, India

Kamarul Bin Ghazali Hawari
Faculty of Electrical and Electronics
Engineering
Universiti Malaysia Pahang
Pekan, Pahang, Malaysia

Pao-Ann Hsiung
National Chung Cheng University
Minxiong, Chiayi, Taiwan

ISBN 978-981-13-6297-2 ISBN 978-981-13-6295-8 (eBook)
https://doi.org/10.1007/978-981-13-6295-8

Library of Congress Control Number: 2018968531

This Springer imprint is published by the registered company Springer Nature Singapore Pte Ltd.
The registered company address is: 152 Beach Road, #21-01/04 Gateway East, Singapore 189721, Singapore

Contents

About the Editors

Dr. Ashish Kumar Luhach received his Ph.D. in Computer Science from Banasthali University, India and post-doctoral degree from Latrobe University, Australia. Since February 2018, Dr. Luhach has been working as an Associate Professor and Head of the Department of Computer Science and Engineering. He has more than a decade of teaching and research experience and has also worked with various reputed universities in an administrative capacity. He has published more than 40 research papers in journals and conference proceedings, and serves on the editorial board of various reputed journals.

Prof. Kamarul Bin Ghazali Hawari received his Ph.D. from the University Kebangsaan Malaysia (UKM) and is currently serving as a Professor and Dean of the Faculty of Electrical and Electronics Engineering, University Malaysia Pahang, Malaysia. With more than 100 research publications in reputed journals and conference proceedings to his credit, Prof. Hawari's main research interests are in Machine Vision Systems, Image Processing, Signal Processing, Intelligent Systems, Vision Control, Computer Control Systems, Thermal Imaging Analysis and Computer Engineering.

Dr. Hawari is also a member of the Board of Engineers Malaysia, Senior Member of the IEEE (SMIEEE), and Member of the IEEE Communications Society Chapter. He has edited many reputed books and received several research grants.

Dr. Ioan Cosmin Mihai is a prominent researcher in the field of cyber security, and is currently serving as an Associate Professor at the Faculty of Electronics, Telecommunications and Information Technology, University Politehnica of Bucharest, Romania and at the "Alexandru Ioan Cuza" Police Academy, Romania. Dr. Mihai is the Editor-in-Chief of various international journals, e.g. the International Journal of Information Security and Cyber Crime. The Vice-president of the Romanian Association for Information Security Assurance, Dr. Mihai has authored several books in the fields of cyber security and e-learning.

Prof. Pao-Ann Hsiung received his Ph.D. from National Taiwan University, Taipei, Taiwan, where he is currently a Professor and Dean of International Affairs. Dr. Hsiung has published more than 260 papers in top international Journals and conference proceedings, and received the 2001 ACM Taipei Chapter Kuo-Ting Li Young Researcher award for significant contributions to the Design Automation of Electronic Systems.

Dr. Hsiung is an IET Fellow and senior member of the IEEE, IEEE Computer Society and ACM. His research interests include: Smart Grids, Smart Traffic, Driver Fatigue Prediction, Landslide Prediction, Cyber-Physical System Design, Reconfigurable Computing and System Design, and System-on-Chip (SoC) Design and Verification.

Prof. Ravi Bhushan Mishra is a senior Professor at the Indian Institute of Technology (BHU), India. With more than 40 years of teaching and research experience, and more than 250 publications in reputed international journals and conference proceedings, Prof. Mishra's areas of interest include Artificial Intelligence, Multi-Agent Systems, the Semantic Web and Robotics.

Part I
Computing Methodologies

Reduce the Impact of Chronic Diseases Using MANETs—Botswana Case Study

Sreekumar Narayanan, Tazviona Gwariro Simba, Neelamegam Chandirakasan and Rajivkumar

Abstract Recent trends in advanced health systems have been evidenced to be fueled by the involvement of wireless technologies. Mobile ad hoc networks (MANETs) have an advantage as compared to cellular networks as they are exploitable for scientific purposes due to their auto transmission-enabled multi-hop networks using wireless communication to transmit data anytime, anywhere without any fixed network infrastructure. In this paper, authors assess the relevance of mobile ad hoc networks as an alternative approach to alleviate the impact of chronic diseases in Botswana health systems which are in the South-Eastern District, through the use of better communication and sensing systems for Tele-health and Tele-care Systems.

Keywords Mobile ad hoc networks · Tele-care systems · Tele-health systems · Wireless sensor networks · Routing protocols · Chronic diseases

1 Introduction

Healthy sector is a very sensitive area, where every individual is considered a major concern. There are many techniques used in [1] health sectors to prevent, alleviate, and cure diseases, but mostly the chronic diseases are of major worry due to their deadly nature of untimely or random attacks. They are imminent injuries which generally happen in case of accidents but very few people manage to tackle such

S. Narayanan (✉) · T. G. Simba · N. Chandirakasan · Rajivkumar
Botho University, Gaborone, Botswana
e-mail: sreekumar.narayanan@bothouniversity.ac.bw

T. G. Simba
e-mail: tgwariro@gmail.com

N. Chandirakasan
e-mail: neelamegam.chandirakasan@bothouniversity.ac.bw

Rajivkumar
e-mail: rajiv.kumar@bothouniversity.ac.bw

A. K. Luhach et al. (eds.), *Smart Computational Strategies: Theoretical and Practical Aspects*, https://doi.org/10.1007/978-981-13-6295-8_1

situations, mostly in the developing countries due to lack of resources and technical knowhow.

MANETs is on chronic diseases attacks and how the situation will be alleviated through the use of mobile ad hoc networks. Most of the chronic diseases are always worry due to their fatalness and untimely nature such as high blood pressure, sugar diabetes, stroke, and heart failure [2–4]. Cancer is also among chronic diseases which are in various forms like bone marrow cancer, colony cancer, blood cancer, skin cancer, etc. Thus, to cater for these diseases, close monitoring and efficient systems need to be in place all the time to alleviate their effects.

In developing countries, especially in sub-Saharan Africa like Namibia, Zambia, Angola, Zimbabwe Botswana, Malawi, Mozambique, Lesotho, Swaziland, and South Africa. There is a great need of having effective approaches that can alleviate the impact of these chronic diseases, the best way is to focus the measures more on the preventive side, where close monitoring of certain specific conditions with predetermined parameters are always kept under surveillance so that if any abnormality happens an emergency alarming is supposed to send to the concerned healthy practitioners and caregivers who are in the closest vicinity through Ad hoc network sensors to easy the situation.

If ad hoc monitoring sensors are being fully utilized in medical centers, where there are patients with different illnesses. MANETs are a new and promising technology that can be used anywhere anytime, and even in harsh conditions or underwater they are still in a position to send information. The architecture of the system is very agile because it allows mobility, it is also peer to peer and its structure and topology is self-healing. It does not have a permanent topology, and the setup is very flexible and ever changing depending on the active nodes which will be still communicating in the vicinity, which might be a result of mobility of other sensors, or merely of breakdowns due to various factors like battery failures, limited lifespan of the sensor, or node attacks.

MANETS with their flexible characteristics which make them a distinguishable communication system infrastructure in wireless system alongside cellular networks, make them a very convenient communication system in health sectors especially in situations where chronic disease attacks are a huge threat.

2 Literature Review

According to Sali Ali. Kali-Omari and Putra Sumari, 2010 says, "There are basically three main types of Ad hoc Networks which are (i) Vehicular Mobile Ad hoc Networks (VANETS), (ii) Intelligent Vehicular Ad hoc Networks (inVANETS), and (iii) Internet Based Mobile Ad hoc Networks (iMANETS)."

They are range from static, small coverage areas that are always characterized by power limitations to large-scale coverage areas which have high power and highly dynamic. There is a great diversity on routing protocols on how they should work in order to give the network best quality of service (QoS) since communication is essential at each and every second. There is a great need for complicated algorithms that will be used for the routing protocols. There is also a need for devising the shortest path from source to destination, especially if the communication is remote to remote that the nodes will be moving as a group or stationery for a certain period.

Like all wireless systems, signals are always deterred along the way as they are being propagated from source to destination. Security and privacy are very important in such systems as data can be hacked anytime by intruders. Latency is also another issue where signals are delayed along the way to reach their destination and high latency can result in loss of signals thus information. If distances between the nearest hops become longer, high chances of multipath fading can be felt and some signals also upon propagation are refracted, diffracted, reflected scattered, and might merely be absorbed when they pass through some media leading to data loss.

MANETS need durable power sources because as longer the battery life, the longer the node can support communication. Some nodes currently are now equipped with a small solar chip that will be used to charge the node's battery during day time so as to lengthen the node's lifetime by regularly replenishing battery power.

3 MANETs Used as to Activate Chronic Disease Attacks and Emergencies in Health Systems

MANETs have a great potential in improving the quality of service (QoS) in health systems, thus bearing a positive impact on alleviating chronic disease attacks and emergencies. This is made possible through the use of wireless sensors, which are used in and outside healthcare facilities to cater for individuals who are in different setups depending on the situation, especially:

(i) Inside health care facilities, [2, 3] to monitor the patient's biomedical condition, e.g., patient's blood pressure, temperature, sugar levels, heartbeat rates, etc., who will be in the Intensive Care Units, through the use of Intra-BANs and Extra-BANs. The sensor (ICU) automatically raises an alarm to alert the relevant health specialist if a certain condition becomes abnormal.

(ii) Outside healthcare facilities, put on patients can be taken care at their homes, [5] where they feel more comfortable, convenient, through the use of sensors. Patients are put on BANS, which might come as wearables in form of bangles or like wrist watches, these also will be monitoring specific biomedical condition for each patient, they can raise an alarm if the patient forgets to take his or her own medication thus reminding [2] otherwise, if the alarm rings for some time without any positive reaction the next of keen or the closest caregiver, or an ambulance is given an alert to attend the patient in time.

(iii) It is also needed to equip ambulances with sensor nodes to cater for real-time communication and monitoring [6] while ferrying the patient to the hospital, or during transference of a patient from one hospital to another. While in transit, the patient needs to be in close monitoring.

(iv) The patient can even get treatment through consultation from a remote doctor who will be overseeing [7] on the system while the patient will be on transit thus increasing the patient's chances of survival.

All this above is made possible by the implementation of Tele-care and Tele-health systems to facilitate remote monitoring, communication and easy access to patients each and every time consistently.

4 Frameworks of an Integrated Tele-care and Tele-health Systems in Health Systems

The framework for tele-health and tele-care system has been shown as in the Figs. 1 and 2, respectively.

Designing protocols for MANETs should highly consider because of their very basic nature of mobility and enabling communication in a single or multi- hop manner with ever-changing directions or paths. In order to come up with a smoothly functioning network as to enhance the quality of performance, there are lots of factors to consider [2, 7] which are (i) Loop freedom, (ii) Security, (iii) Distribution operation, (iv) Demand-based operation, (iv) Proactive operation and lastly (v) Sleep period operation which is mostly advantageous in saving battery power during idle moments where no communication will be taking place, thus lengthening battery life [2].

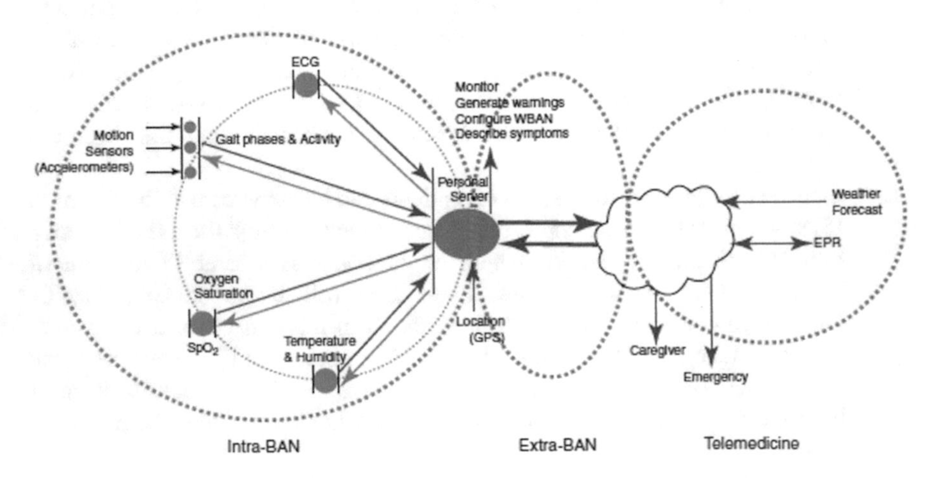

Fig. 1 Tele-health system for emergences

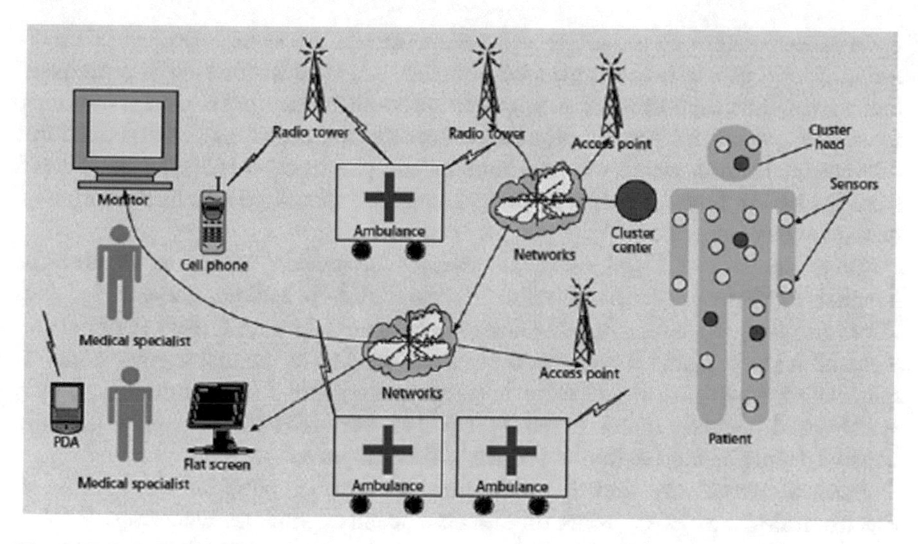

Fig. 2 Tele-care system

5 Relevance of MANETs in Tele-health Systems in Botswana

With such systems already implemented in developed countries, developing countries, like Botswana will also appreciate the proper implementation of advanced Tele-Health Systems supported MANETs. Its characteristics are better exploited in health systems in relaying monitored data or transmitted data from patient to caregiver, patient to health professional, or health professional to another health professional.

Botswana has a population of approximately two million people whereby 23% is made up of elderly population who are 50 years and above which is at great risk to the prevalent chronic diseases like sugar diabetes, high blood pressure, heart failure, asthma, hyperthermia, and hypothermia, etc. These diseases have to be always on close guard to avoid sickness as much as possible. Constant monitoring is made possible by the use of wireless sensors. These sensors will be the ones attached to each individual to monitor a specific condition on the elderly person. Also, some sensors will be working to monitor [7, 8] the patient's drug taking consistence.

It is highly advantageous to have emergence vehicles which are well equipped with proper equipment for monitoring patients and relaying data to the respective medical specialists who will be a distance from the scene as most of them will be at their hospitals. Real-time data relay is very crucial in order to help the emergence crew on how they will take care of the person who has been just involved in an emergency accident. Some people here in Botswana die on their way to the hospital in ambulances or rescue vehicles, before getting the necessary treatment at the hospital. This approach will also help in reducing fatalities in such cases.

In intensive care units (ICUs) and patient wards, the same approach of monitoring elderly people is also supposed to be implemented inside hospitals premises themselves. Intensive care units and wards should be equipped with monitoring sensors on patients for specific biomedical conditions. These sensors will be in form of wearable sensors which will still monitor the patient even if she/he has moved from the bed in times of self-relieving, the sensors should be still in a position of managing to communicate.

There are lots of hospitals in Botswana's South-Eastern, some government hospitals like Princes Marina, Extension 2 Hospital, Julia Molefe, Lobatse Hospital. There are also private hospitals in Gaborone such as Gaborone Private Hospital and Bokamoso Hospital and Ramotswa's Lutheran Hospital as the major private ones in the South-Eastern District. If these hospitals are equipped with current high tech systems and perfectly linked together, it will be very advantageous in serving the lives of people as the quality of service will be improved.

Sensors which are mostly used for monitoring patients are Oximeter, Electrocardiogram ECG Monitor, Electroencephalography, Electromyography, Graphic Hypnograms, Cardiometer, Stethoscope, Sphygmomanometer, Ophthalmoscope, Cardiotachometer, tympanometer, Auriscope, and Thermometers [2, 3]. Tricorders are also considered as the most relevant sensors in monitoring the prevalent chronic conditions and health risks in elderly people, [9] such as hypertension, stroke, respiratory diseases, and congestive heart failure.

Table 1 shows some sensors and their measured parameters and possible illness.

6 Results and Discussion

After a small scale pretest among the five hospitals in Gaborone including the private ones, in order to acquire the clarity of the already pre-validated questionnaire. It was distributed to 21 respondents including the IT Technicians who are part of hospital staff in those hospitals. The questionnaires which were responded to where 267, giving a respondents rate of 76% which was composed of 36 IT Personnel, 24 Doctors, 12 Radiologists, 35 Paramedics, 53 Other Health Specialists, 38 Physicians, and 69 Nurses. Some of the data was collected through interviews with the hospital management teams from different selected hospitals. The remaining data was collected from observations and documentation as to reduce bias of facts.

The outcome of the study showed that most of the hospitals are now implementing some of the wireless technologies in their hospital premises and Fig. 3 shows specific areas where they mostly implement wireless sensors in intensive care units and wards, and with the list is with the links to other hospitals.

Table 1 Sensors with their measured parameters

Sensor	Measured parameters	Illness	Comments
Pulse oximeter	Blood oxygen saturation SpO2, plus rate	Cardiac, respiratory	Very generic
Implantable pressure transducer	Blood pressure, intracranial pressure	Cardiac, neuronal	
Swallowable capsule(radiopill)	Gastrointestinal pressure, temperature	Gastrointestinal	
Peak flowmeter	Respiratory flow	Respiratory	Very generic
Electronic stethoscope	Heart murmurs	Cardiac	
ECG module	ECG signals (heart rate, QRS duration, ST-elevation)	Cardiac	Smart ECG exists with automatic interpretation
Electronic infrared thermometer	Body temperature	Infections, acute conditions	Very generic
EEG	Brain electrical activity	Brain function	Requires specialist interpretation of data
Ultrasound scanning	Abdominal imaging		
Electronic noise	Chemical composition of breath/fluids	Bacterial identification, drug identification	
Polysomnography	EEG, EOG, EMG, EKG, Heartbeat, breathing, body position, snoring	Sleep disorder	Integrated sensor
AMON	SpO2, BP, ECG, temperature	Cardiac	Integrated, wearable sensor
Plethysmographic ring sensor	Blood pressure, heart rate, oxygen saturation, heart rate variability	Cardiac	Integrated, wearable sensor

Figure 4 shows patient age groups who are mostly attended to and using the hospital systems.

Figure 5 shows the results of user satisfaction of MANETs.

From the results of the study, implementation of MANETs will come with a great deal in alleviating the impact of most diseases especially the chronic diseases which mostly attack elderly population which is mostly above 5 years. With the improvising of sensor networks, it becomes much easier for health specialists to monitor patients consistently. This consistent monitoring will be of much importance due to the fact that if any abnormality occurs it is attended to in time by the health specialists thus preventing the worsening of the patient's condition which might lead to loss of lives in extreme cases.

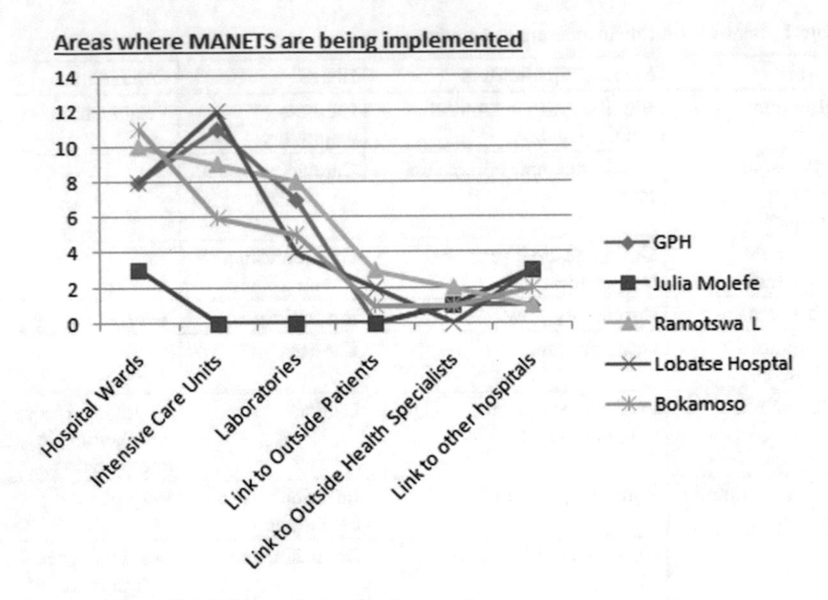

Fig. 3 Areas where MANETs are being implemented

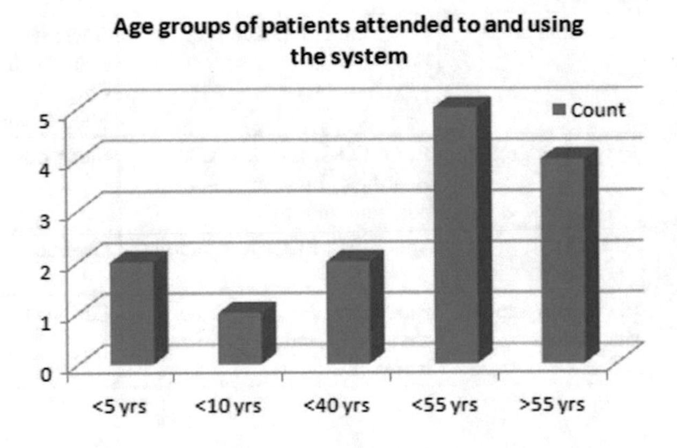

Fig. 4 Age groups of patients attended to and using the system

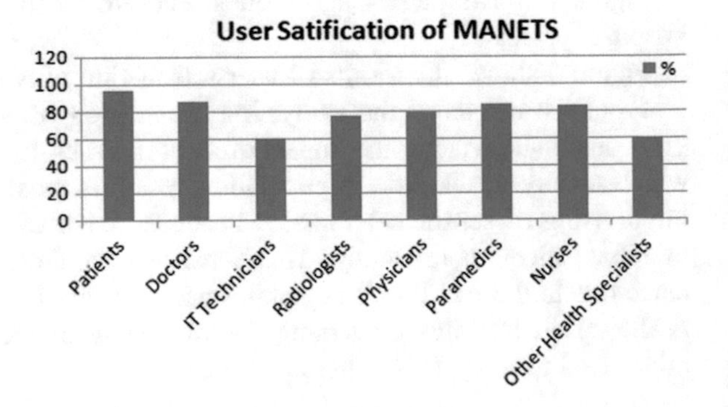

Fig. 5 User satisfaction of MANETS

7 Conclusion

As shown from the study most, MANETs are not yet fully implemented in Botswana health sectors, especially on linking with outside patients who are put on sensors in form of body area networks, and also links to outside specialists are not yet fully implemented which will make it easier to do remote Tele-Diagnosis in case of specialists scarcity. In these situations, it will serve a lot in terms of costs, time, and also sharing of expertise even if the specific health specialist is physically at a distance from the patient. The concerned patient will still get rightful medication through remote consultation and diagnosis enabled by mobile wireless sensors which will relay the required information to the specialist.

Acknowledgements The authors are thankful to the management of Botho University, Gaborone to provide their valuable support. We also grateful to the management of Makhubu Secondary Schools to conduct the research in their organization.

References

1. Andreas Schrader, D.C.: SmartAssist—Wireless Sensor Networks for Unobtrusive Health Monitoring. BMI10 (2010)
2. Bernad Fong, A.F.: Telemedicine Technologies: Information Technologies in Medicine and Telehealth. Wiley Ltd., United Kingdom (2011)
3. Lim, F.P.: The emergence of wireless sensor networks in healthcare systems. Adv. Sci. Technol. Lett. (2016)
4. Al-Omari, S.A.K., Sumari, P.: An overview of mobile adhoc networks for the existing protocols and their applications. J. Appl. Graph Theor. Wireless Ad hoc Netw. Sensor Netw. 91–110 (2010)
5. Virone, G., Wood, A., Selavo, L., Cao, Q., Fang, L., Doan, T., He, Z., Stoleru, R., Lin, S., Stankovic, J.A.: An advanced wireless sensor network for health monitoring
6. West, D.M.: Improving Health Care through. Centre for Technology Innovation (2013)
7. Ko, J., Lu, C., Srivastava, M.B., Stankovic, J.A., Terzis, A., Welsh, M.: Wireless sensor networks for healthcare (2010)
8. Hassanien, A.D.: Wearable and Implantable Wireless Sensor Network Solutions for Healthcare Monitoring. Open Access Censors (2011)
9. Kamel Boulos, M.N., Lou, R.C.: Connectivity for healthcare and well-being management: examples from six European projects. Int. J. Environ. Res. Public Health 1948–1971 (2009)

Detection and Analysis of Lung Cancer Using Radiomic Approach

Shallu, Pankaj Nanglia, Sumit Kumar and Ashish Kumar Luhach

Abstract Lung cancer is the most prevailing form of cancer and claims more lives each year in comparison of colon, prostate, and breast cancers combined. The prominent types of lung cancers are explained here along with their database details. In this work, the radiomic approach is proposed for the process of detection and analysis of lung cancer due to its impressive prognostic power. This becomes possible with the advent of computer aided detection system that not only provides a cost effective technique but also provides a noninvasive provision. The main aspect of this approach relies on its ease of implementation over the different types of databases irrespective to the category of cancer. The precise detection and prediction is possible via radiomic approach, which becomes state-of-the-art technique for lung cancer analysis. In future, the radiomic approach would be employed along with the deep learning models in conjunction with the computer-aided detection system for better diagnosis purposes.

Keywords Lung cancer · Radiomic · Segmentation · Semantic and agnostic features

1 Introduction

In the past few years, a small word "Cancer" has become a massive public health problem around the world. It is a generic term used to represent a large group of diseases that can affect any part of the body [1–4]. One key feature of the cancer is the rapid production of cells that grow beyond their usual limits, which can spread

Shallu (✉)
NITTTR, Chandigarh, India
e-mail: shallu.ece@nitttrchd.ac.in

P. Nanglia · S. Kumar
MAU, Baddi, India

A. K. Luhach
Maharshi Dayanand University, Haryana, India

A. K. Luhach et al. (eds.), *Smart Computational Strategies: Theoretical and Practical Aspects*, https://doi.org/10.1007/978-981-13-6295-8_2

Table 1 Mortality rates due to major cancer types by IARC, WHO

Types of cancer	Deaths
Lung cancer	1.69 million
Liver cancer	788,000
Colorectal cancer	774,000
Stomach cancer	754,000
Breast cancer	5710

to other organs by invading in adjoining parts of the body (metastasizing) [1]. According to cancer research agency of World Health Organization (WHO) named as International Agency for Research on Cancer (IARC) reported 8.8 million deaths in 2015, in which the most common cause of deaths was due to the cancer of Lung, Liver, Colorectal, Stomach, Breast [2, 5–7]. The mortality due to lung cancer is very high as compared to other kinds of cancer species. Statistics related to cancer mortality is depicted in Table 1.

In this paper, an approach "Radiomic" is discussed in detail, which assists in decision-making performed by Computer-Aided Diagnosis (CAD) system regarding the presence and types of cancer. Since database acts as a backbone in designing of any detection system, so a description of database used for cancer analysis, particularly for lung cancer is provided. The readers are referred to various links related to lung cancer database.

2 Lung Cancer Type and Database

Lung cancer is a type of cancer which originates from the lungs and known as "Bronchogenic Carcinomas" [3]. Lungs are the spongy organs in human body that takes oxygen and release carbon dioxide during inhaling and exhaling, respectively. The most common symptoms, which can be seen in people with early lung cancer stages are cough, rust-colored sputum, chest pain, breathing hoarseness, weight loss, loss of appetite, tiredness, weakness and frequent occurrence of infections like bronchitis and pneumonia. In addition to this, it may cause bone pain, jaundice if spread to the liver, changes in nervous system which leads to headache, dizziness, and seizures when spread to brain or spinal cord as well as lumps near the body surface if spreading to the skin or the lymph nodes. Some lung cancer may cause syndromes such as Horner syndromes, Superior Vena Cava (SVC), and Paraneoplastic syndromes [8]. Usually, most of the people ignore these symptoms because of their hectic schedules and careless behavior towards the minor health issues. Such wrong insight leads human in trouble situations, because of that sufferer diagnose with cancer in the later stages. All this happens accidently with the patients, where test become compulsory in certain cases and prompted by a medical practitioner for other medical conditions. Another major issue is the fear of patients from invasive techniques employed during the surgery and due to the fact; a large community of people is misguided that the every case of cancer passes through a painful treatment. However, proper information and awareness related to the cancer disease can avoid such circumstances. This becomes possible by means of advanced noninvasive techniques deployed for the prediction and detection of cancer with the help of computer aided diagnosis systems. Thus, in order to get a detailed information from the available datasets which are retrieved from the screening tests such as Sputum cytology, Chest X-ray and Computed Tomography (CT) scan, Positron Emission Tomography (PET) scans performed on patients of different class to examine the cancer from the procured intelligence in the structure of data sets.

2.1 *Lung Cancer Types*

Lung cancer is broadly categorized into three classes, depicted in Fig. 1:

i. Small Cell Lung Cancer (SCLC)
ii. Non-Small Cell Lung Cancer (NSLC)
iii. Lung Carcinoid Tumor.

Small Cell Lung Cancer (SCLC) or Oat Cell Carcinomas This type of cancer consists of around 10%–15% of lung cancers, also known as most aggressive cancer due to its rapid growth as compared to other types of lung cancer [3]. SCLC is strongly associated with cigarette smoking. However, the non-smokers own only 1% of this tumor. SCLC metastasizes quickly for numerous sites in the body.

Non-small Cell Lung Cancer (NSLC) It is the most common type of lung cancer and about 85% of lung cancer is NSLC. Since it is very hard to determine the exact

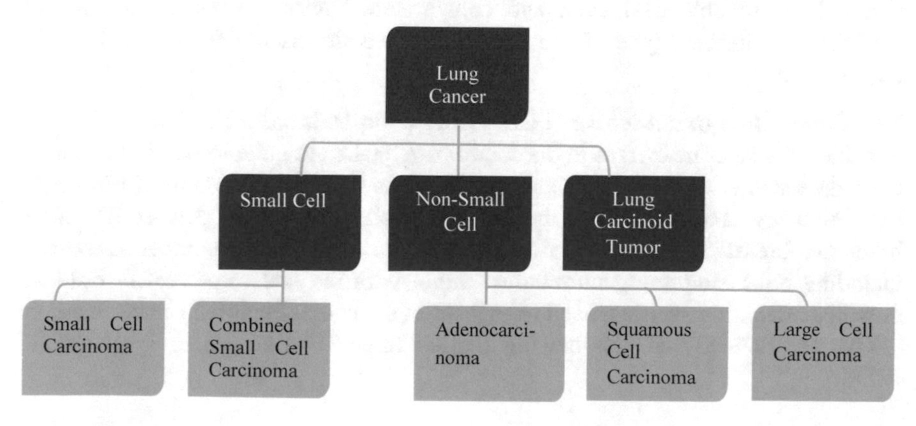

Fig. 1 Classification of lung cancer

cause for each case of cancer but still, there are many factors responsible for NSLC such as tobacco smoking, second-hand smoke, air pollution, overexposure to radon, asbestos, diesel exhaust, and inhaling of other harmful chemicals. Moreover, inherited and acquired changes in the genes also play an important role in the development of NSLC.

Lung Carcinoid Tumor These types of cancer are uncommon and tend to grow slower than other types of lung cancer. They are made up of special kinds of cells called neuroendocrine cells. Neuroendocrine cells are like neural cells in a certain way and like hormone-making endocrine cells in other ways, but do not form an actual organ like the adrenal or thyroid glands and scattered throughout the body in organs such as stomach, lungs, and intestines. In some cases, these cells undergo a certain change that leads to their abnormal growth and behavior which further form tumor cell. Neuroendocrine lung tumors are categorized into four types, viz., (i) Small cell lung cancer, (ii) Large cell neuroendocrine carcinoma, (iii) Atypical carcinoid tumor, and (iv) Typical carcinoid tumor.

The terms "small cell" and "large cell" are simply descriptive terms for the visual aspect of the cancer cells under the microscope. These characteristics of cancer cells help the doctor to determine the type of cancer, the extent of abnormality in the cells, and the location from where it originates.

2.2 *Lung Cancer Database*

The database is very important and a prerequisite condition in every study. However, the construction and collection of cases in a high-quality multimedia files are extremely difficult, time consuming, and expensive. Simultaneously, to assess the high-quality data, a wide range of skills from medical background to information technology is needed as well as a detailed ground truth is always required for reliable evaluation and comparison. Various types of datasets are available for different type of cancer and here we are discussing few of them for lung cancer:

The Cancer Imaging Archive (TCIA) Collection Database TCIA is a collection of subjects, where the images in the database of TCIA are quite interesting and have been documented in many outstanding papers as their databases contain information about the cancer type with the anatomical site at which it occurred like lung, brain, and breast. TCIA also provides information related to the number of subjects including other supporting information along with the latest updates. In order to download data, the user just need to register at the site of National Cancer Institute (NCI). The links to view the imaging data are https://wiki.nci.nih.gov/display/CIP/RIDER.

Give A Scan Database Give A Scan is the world's first patient powered open access database provided by the Lung Cancer Alliance organization for lung cancer research. This database has a collection of donated radiological imagery and metadata of people being screened for lung cancer as well as for those who have already been diagnosed and treated. In order to stimulate and accelerate lung cancer research, the database is made available in open access. http://www.giveascan.org/giveascan/browse/community/?communityId=1. These datasets have been prepared in the conventional format of Patient/Study/Series/Dataset where each individual study may include multiple scans of varying quality and anatomical orientation. The format for all stored images is "Digital Imaging and Communications in Medicine" (DICOM) which is a de facto standard in the clinical and research communities for exchanging radiological data.

Other Popular Database There are many links which are available for lung cancer databases. A few of them are listed below:

http://lungcancer.ibibiosolutions.com/;
http://www.ielcap.org/;
https://imaging.cancer.gov/programsandresources/informationsystems/lidc;
http://www.bioinformatics.org/LuGenD/;
http://anode09.isi.uu.nl/.

3 Lung Cancer Detection System

A lot of work has been carried out in the domain of lung cancer detection system and several state-of-the-art techniques are employed to solve the problem of early cancer detection. Therefore, it is necessary to understand the stages which are an integral part of a cancer detection system. The lung cancer detection system consists of mainly five stages: (i) image acquisition, (ii) preprocessing, (iii) segmentation, (iv) feature extraction, and (v) classification [9]. In order to understand the whole process, it is required to have a basic understanding of all the stages involved in the detection system. A block diagram of all the stages of the detection system is shown in Fig. 2.

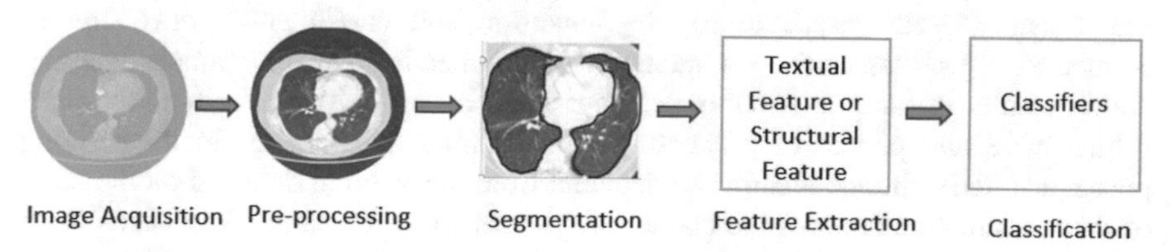

Fig. 2 Stages of cancer detection system

Preprocessing This step is performed on the acquired image to improve its quality which ultimately increases the precision and accuracy of processing algorithms usually takes place after preprocessing stage. The images are impaired by various factors such as noise, lack of contrast due to overexposed and underexposed illumination during the image acquisition process. These defects are needed to remove necessarily at the stage of preprocessing so that more information can be extracted from the image. The following techniques are used at preprocessing stage, viz.: Enhancement Filter [10, 11], Median Filtering, Adaptive Histogram Equalization [12], Smoothing filters, Noise Correction [13], Wiener filter, Auto-enhancement [14], Wavelet Transform, Anti-geometric Diffusion, Fast Fourier Transform [15], and Contrast Limited Erosion Filter [16].

Segmentation The objective of segmentation is to change the representation of an image into something that is more meaningful and easier to analyze. Segmentation of an image in lung cancer detection system involves the separation of lung nodule from the other part of the standard-of-care images. It is very critical and essential stage of the lung cancer detection system as the subsequent feature space is extracted only from the segmented volume. Therefore, the success of cancer detection system greatly depends upon the result of segmentation stage. There are several methods that are used to perform segmentation such as: region growing, thresholding, and watershed segmentation [17].

Feature Extraction Feature extraction is the most significant stage for any detection system. Several techniques and algorithms are utilized to detect and extract the features from a given image. Feature extraction stage is required when the input data is too large and having redundancy. The input data is transformed into a reduced representation set of features which at final stage used to classify an image in normal or abnormal one and creates the ground for image classification process. Average intensity, area, perimeter, eccentricity, energy, entropy, homogeneity, and correlation are some types of features that can be considered at this stage [18].

Classification or Detection The classification has a broad range of decision-theoretic methods for the image identification. The classification algorithms depend on one or more features present in the image as every feature leads to one of the different and exclusive class. The classes can either be specified as a priori through an analyst (supervised classification) or clustered automatically in prototype classes sets (unsupervised classification). Segmentation and classification have closely connected objectives as the segmentation is another form of component labeling which results in various features segment from a scene. The classification algorithms have two phases of processing, i.e., training and testing. In the training phase, distinctive image features are isolated from the training data and on the basis of these distinctive features the classes are described in which the data needs to be classified. In the testing phase, the same features are utilized to classify the image.

4 Detection of Lung Cancer Using Radiomic Approach

Radiomic is an emerging field in the medical world as it converts imaging data obtained from positron emission tomography (PET), computed tomography (CT) scan, and magnetic resonance imaging (MRI) into a high dimensional feature space using various automatically extracted data characterization algorithms [19]. In short, the radiomic approach is designed to develop a decision support tools wherein a large number of quantitative features are extracted from the standard-of-care images and these extracted features are placed in shared databases which also contains other characteristics of patients. Further, the data is subsequently mined for testing and hypothesis generation, or both which finally increase the decision power of the models.

Radiomic process can be divided into different sub processes; (i) Image acquisition, (ii) Volume of interest identification and segmentation, (iii) Feature extraction and qualification, (iv) Databases building, and (v) Classifier modeling, analysis and data sharing [20]. Pictorial representation of the whole process is depicted in Fig. 3. Out of these sub processes, feature extraction is the heart of radiomic. In practice, two types of features are extracted in radiomic: semantic and agnostic. Some of these features are represented in the Table 2. Semantic features are the part of radiologist's lexicon and used to describe the volume of interest such as shape, location or size, etc., while agnostic features are used to capture the heterogeneity in lesions through mathematically extracted quantitative descriptors [21]. Agnostic features can be divided further into three: first order, second order, or higher order statistical outputs.

First-Order Statistics It describes the individual voxels' values distribution without pertaining to spatial relationships. Histogram-based methods come under this category that reduces a volume of interest to single value of entropy (randomness), mean, median, maximum, minimum, and kutosis (flatness).

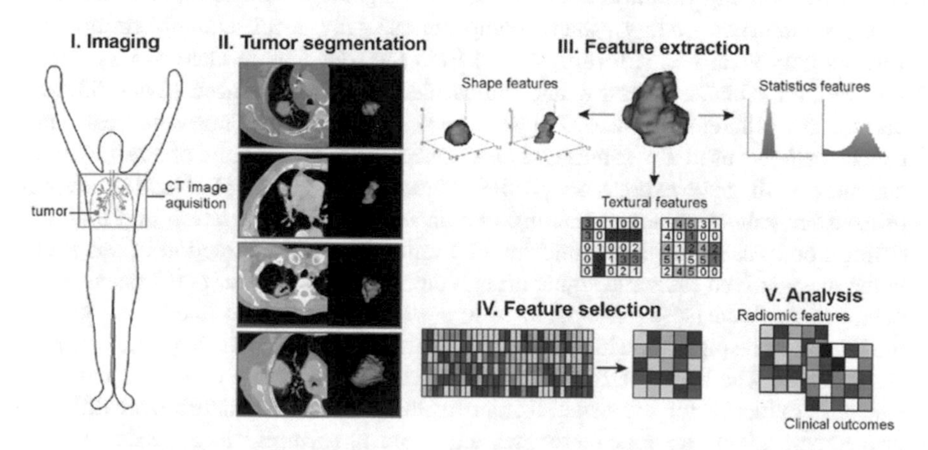

Fig. 3 Pictorial representation of radiomic workflow [18]

Table 2 Examples of radiomic features

Semantic	Agnostic
Shape	Wavelets
Location	Haralics textures
Size	Law textures
Vascularity	Histogram (Skewness, Kurtosis)
Spiculation	Laplacian transforms
Attachments or lepidics	Minkowski functionals
Necrosis	Fractal dimensions

Second Order Statistics In radiomic, "Texture" is generally used to provide a measure of intra-tumoral heterogeneity by describing the statistical interrelationship between voxels with similar or dissimilar contrast values [22–24].

Higher Order Statistics In these methods, filter grids are imposed on the image to extract repetitive or non-repetitive patterns and the number of grid elements containing voxels of specified value is computed. There are various methods to compute the value of voxel such as Wavelets, Minkowski function, and Laplacian transform. Wavelet is a filter transform in which an image is multiplied by a linear or radial "wave" matrix; while the Minkowski functional evaluate voxel patterns to detremine the intsensity above a particular threshold [25]. Laplacian transform extract an area from an image with progressively coarse texture patterns [26]. In the very first study, 182 texture features with 22 semantic features were used to describe the CT image of lung cancer which further increased up to 442 features set including wavelets and 522 features that also contain texture and fractal dimension features [19]. At present, this feature list has been expanded to 662 including Laplace transforms of Gaussian fits.

In 2014, a work by Aerts et al. [27] represents the potential of radiomic for the recognition of a general prognostic imaging phenotype existing in multiple forms of cancer by building a radiomic signature. This signature consist of best radiomic features (statistics energy, shape compactness, gray level nonuniformity, and wavelet gray level nonuniformity HLH) from the four feature classes, i.e., shape, texture, tumor image intensity, and multiscale wavelet. They used seven different datasets for different purposes. Lung1 dataset, containing 422 non-small cell lung cancer patients used for training, Lung3 dataset for association of the radiomic signature with gene expression profiles, three datasets (Lung2, H&N1, H&N2) utilized for validation and remaining two dataset, RIDER test/retest, and multiple delineation to calculate the feature stability ranks. The main attraction in this study is the lung-derived radiomic signature that demonstrated a better performance over head and neck cancer as compared to lung cancer. It happens due to the patient movement or respiration, which introduced noise in the image during acquisition of lung images. The better performance of lung-derived signature overhead and neck cancer is evidence for the generalization of this radiomic signature over different cancer types. This approach provides an ample opportunity to upgrade medical decision support system.

4.1 Challenges in Radiomics

In the past decade, radiomic research in medical world has been increased in a dramatic manner. Radiomic is enabling in diagnosis and prognosis of tumors [28, 29]. This scheme can be used to guide the selection of therapy for individual tumors when associated with genomic [30, 31] and also helps in determining the location to resect (location that is most probable to comprise important diagnostic and predictive information) [32, 33]. In spite of all these benefits of radiomic, there are still many challenges like reproducibility, big data, data sharing, and standardization that need to be handled. The biggest challenge in radiomic is sharing of data as multiple trials are needed to transmit cohort of patients and to crate datasets of sufficient size for powerful statistical analysis. Standardization is another main problem because there is no standard to report the results. These issues have become serious enough that it is necessary to provide standards for all aspects of radiomic. Recently, editors from more than 30 reputed biomedical journals have decided to impose common standards for statistical testing and to improve access to raw data [34, 35]. Many standards have been adopted by the National Institutes of Health that can also be applied across all other areas of research.

5 Conclusion

Lungs are the essential organ of our body. Thus, it is necessary to devise novel techniques that can predict cancer in its early stage and provide an effective platform in diagnosis of lung cancer. In this context, radiomic approach is rendered to meet the requirement of efficient detection and prediction of cancer types. This approach offers a great potential to accelerate precise medication and reduces the risk of infection and complications due to its non-invasive behavior. In this approach a classifier is developed to identify the cancer type by extracting the information from the standard-of-care images. Hence, radiomic has a great impact in clinical practice, which is a good alternative to histopathological technique, and provides many opportunities to improve the decision support in cancer detection. The significance of this approach is low-cost detection as this technique is able to identify general prognostic phenotype from any type of cancer. However, reproducibility and sharing of big data are major challenges due to requirement of multiple trials to crate datasets of adequate size for powerful statistical analysis. Therefore, there is a need to develop some techniques in a way that can overcome such challenges.

Radiogenomic is another approach in which radiomic data is correlated with genomic pattern. Radiogenomic can provide information even when radiomic data is not significantly related to gene expression. It also suggests about the status of mutation. Radiomic and radiogenomic both are non-invasive and low cost approaches. Currently, the field of radiomic research is concentrating on improvement of classifier models by applying different deep learning frameworks to serve the most accurate diagnoses and for better care of the patient.

References

1. Firnimo, M., Morais, A.H., et al.: Computer-aided detection system for lung cancer in computed tomography scans: review and future prospects. In: Biomedical Engineering, pp. 1–16 (2014)
2. World Health Organization, http://www.who.int/mediacentre/factsheets/fs297/en/
3. An Information Resource on Lung Cancer Testing for Pathologists and Oncologists, https://Www.Verywell.Com/Non-Small-Cell-Lung-Cancer-2249281
4. American Cancer Society, https://www.cancer.org/cancer/lung-cancer.html
5. Shallu., Mehra, R.: Breast cancer histology images classification: Training from scratch or transfer learning? ICT Express **4**(4), 247–254 (2018)
6. Gupta, S., Kumar, S.: Variational level set formulation and filtering techniques on CT images. Int. J. Eng. Sci. Technol. (IJEST) **4**(7), 3509–3513 (2012) (ISSN: 0975-5462)
7. Shallu., Mehra, R.: Automatic Magnification Independent Classification of Breast Cancer Tissue in Histological Images Using Deep Convolutional Neural Network. In: Advanced Informatics for Computing Research. ICAICR 2018. Commun. Comput. Inform. Sci. **955**, 772–781 (2019)
8. Symptoma Better Diagnosis, https://www.symptoma.com/en/ddx/lumbar-spinal-cord-tumor+superior-vena-cava-syndrome+horners-syndrome
9. Agarwal, R., Shankhadhar, A., Sagar, R.K.: Detection of lung cancer using content based medical image retrieval. In: IEEE Fifth International Conference on Advanced Computing and Communication Technologies, pp. 48–52 (2015)
10. Liu, Y., Yang, J., Zhao, D., Liu, J.: A method of pulmonary nodule detection utilizing multiple support vector machines. In: International Conference on Computer Application and System Modeling (ICCASM), vol. 10, pp. 118–121 (2010)
11. Messay, T., Hardie, R.C., Rogers, S.K.: A new computationally efficient CAD system for pulmonary nodule detection in CT imagery. J. Med Image Anal. **14**(3), 390–406 (2010)
12. Ashwin, S., Kumar, S.A., Ramesh, J., Gunavathi, K.: Efficient and reliable lung nodule detection using a neural network based computer aided diagnosis system. In: International Conference on Emerging Trends in Electrical Engineering and Energy Management (ICETEEEM), pp. 135–142 (2012)
13. Arimura, H., Magome, T., Yamashita, Y., Yamamoto, D.: Computer-aided diagnosis systems for brain diseases. In: Magnetic resonance images, Algorithms, vol. 2, no. 3, pp. 925–952 (2009)
14. Shao, H., Cao, L., Liu, Y.: A detection approach for solitary pulmonary nodules based on ct images. In: International Conference on Computer Science and Network Technology (ICCSNT), pp. 1253–1257 (2012)
15. Ye, X., Lin, X., Dehmeshki, J., Slabaugh, G., Beddoe, G.: Shape-based computer-aided detection of lung nodules in thoracic ct images. Biomed. Eng IEEE Trans. **56**(7), 1810–1820 (2009)
16. Teramoto, A., Fujita, H.: Fast lung nodule detection in chest ct images using cylindrical nodule-enhancement filter. Int. J. Comput. Assist. Radiol. Surg. **8**(2), 193–205 (2013)
17. Chaudhary, A., Singh, S.S.: Lung cancer detection on CT images by using image processing. In: IEEE International Conference on Computing Sciences, pp. 142–146 (2012)
18. Naresh, P., Shettar, Dr. R.: Early detection of lung cancer using neural network technique. Int. J. Eng. Res. App. **4**(8), 78–83 (2014)

19. Pickup, L., Talwar, A., Stalin, S., et al.: Lung nodule classification using learnt texture features on a single patient population. In: Radiological Society of North America Scientific Assembly and Annual Meeting Program. Radiological Society of North America, SSM06, Oak Brook, Ill (2015)
20. Kumar, V., Gu, Y., Basu, S., et al.: Radiomics: the process and the challenges. MagnReson Imaging **30**(9), 1234–1248 (2012)
21. Lambin, P., Rios-Velazquez, E., Leijenaar, R., et al.: Radiomics: extracting more information from medical images using advanced feature analysis. Eur. J. Cancer **48**(4), 441–446 (2012)
22. Davnall, F., Yip, C.S., Ljungqvist, G., et al.: Assessment of tumor heterogeneity: an emerging imaging tool for clinical practice? Insights Imaging **3**(6), 573–589 (2012)
23. O'Connor, J.P., Rose, C.J., Waterton, J.C., Carano, R.A., Parker, G.J., Jackson, A.: Imaging intratumor heterogeneity: role in therapy response, resistance, and clinical outcome. Clin. Cancer Res. **21**(2), 249–257 (2015)
24. Depeursinge, A., Al-Kadi, Omar, S., Mitchell, Ross J.: Biomedical Texture Analysis: Fundamentals, Tools and Challenges. Elsevier. ISBN 9780128121337. (2017-10-01)
25. Larkin, T.J., Canuto, H.C., Kettunen, M.I., et al.: Analysis of image heterogeneity using 2D Minkowski functionals detects tumor responses to treatment. MagnReson. Med. **71**(1), 402–410 (2014)
26. Grossmann, P., Grove, O., El-Hachem, N., et al.: Identification of molecular phenotypes in lung cancer by integrating radiomics and genomics. SciTrans. Med. (in press)
27. Aerts, H.J., Velazquez, E.R., Leijenaar, R.T., et al.: Decoding tumour phenotype by noninvasive imaging using a quantitative radiomics approach. Nat Commun. (2014)
28. Wibmer, A., Hricak, H., Gondo, T., et al.: Haralick texture analysis of prostate MRI: utility for differentiating non-cancerous prostate from prostate cancer and differentiating prostate cancers with different Gleason scores. EurRadiol **25**(10), 2840–2850 (2015)
29. Segal, E., Sirlin, C.B., Ooi, C., et al.: Decoding global gene expression programs in liver cancer by noninvasive imaging. Nat. Biotechnol. **25**(6), 675–680
30. Grove, O., Berglund, A.E., Schabath, M.B., et al.: Quantitative computed tomographic descriptors associate tumor shape complexity and intratumor heterogeneity with prognosis in lung adenocarcinoma. PLoS ONE **10**(3), 1–14 (2015)
31. Kuo, M.D., Gollub, J., Sirlin, C.B., Ooi, C., Chen, X.: Radiogenomic analysis to identify imaging phenotypes associated with drug response gene expression programs in hepatocellular carcinoma. J VascInterv. Radiol. **18**(7), 821–831 (2007)
32. Teruel, J.R.H.M., Heldahl, M.G., Goa, P.E., et al.: Dynamic contrast-enhanced MRI texture analysis for pretreatment prediction of clinical and pathological response to neoadjuvant chemotherapy in patients with locally advanced breast cancer. NMR Biomed. **27**(8), 887–896 (2014)
33. Klaeser, B., Wiskirchen, J., Wartenberg, J., et al.: PET/CT-guided biopsies of metabolically active bone lesions: applications and clinical impact. Eur. J. Nucl. Med. Mol. Imaging **37**(11), 2027–2036 (2010)
34. Tatli, S., Gerbaudo, V.H., Mamede, M., Tuncali, K., Shyn, P.B., Silverman, S.G.: Abdominal masses sampled at PET/CT-guided percutaneous biopsy: initial experience with registration of prior PET/CT images. Radiology **256**(1), 305–311 (2010)
35. Begley, C.G., Ellis, L.M.: Drug development: raise standards for preclinical cancer research. Nature **483**(7391), 531–533 (2012)

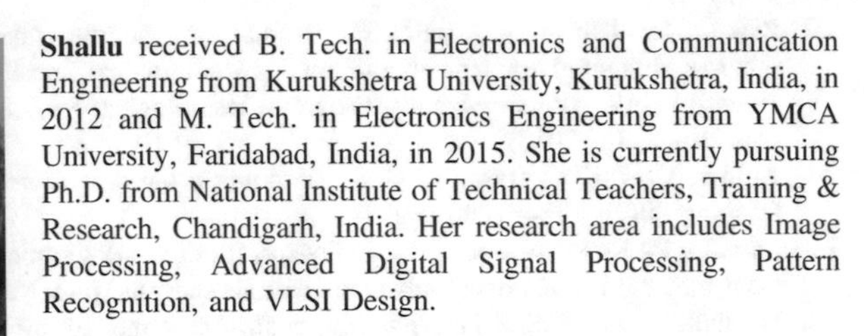

Shallu received B. Tech. in Electronics and Communication Engineering from Kurukshetra University, Kurukshetra, India, in 2012 and M. Tech. in Electronics Engineering from YMCA University, Faridabad, India, in 2015. She is currently pursuing Ph.D. from National Institute of Technical Teachers, Training & Research, Chandigarh, India. Her research area includes Image Processing, Advanced Digital Signal Processing, Pattern Recognition, and VLSI Design.

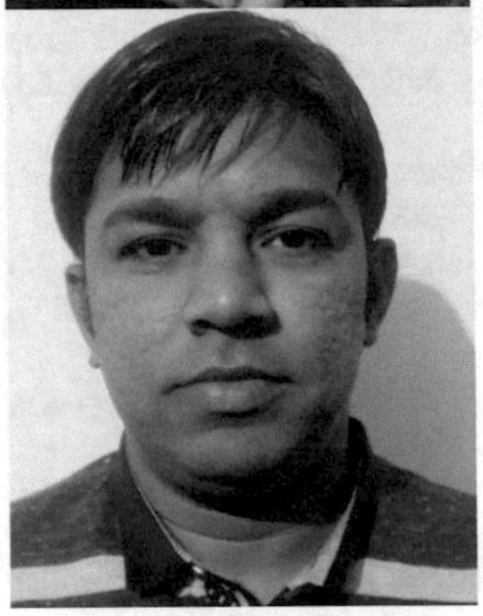

Pankaj Nanglia received B. Tech. in Electronics and Communication Engineering from Kurukshetra University, Kurukshetra, India, in 2007 and M. Tech in Electronics and Communication Engineering from Deenbandhu Chhotu Ram University of Science and Technology Sonipat, India, in 2011. He is currently pursuing Ph.D. from the Department of Electronics Engineering from Maharaja Agrasen University, Solan, Himachal Pradesh, and also working as an Assistant Professor in Department of Electronics and Communication Engineering at Maharaja Agrasen Institute of Technology, Maharaja Agrasen University, Baddi, Solan, Himachal Pradesh, India. His research area includes Image Processing, Digital Signal Processing, and communication engineering.

Sumit Kumar received B. Tech. in Electronics and Communication Engineering from Kurukshetra University, Kurukshetra, India, in 2007 and M. Tech. in Nanoscience & technology from Guru Jambheshwar University of Science and Technology, Hisar, India, in 2011. He is currently pursuing Ph.D. from the Department of Electronics Engineering at Indian Institute of Technology (ISM), Dhanbad, India and also working as an Assistant Professor in the Department of Electronics and Communication Engineering at Maharaja Agrasen Institute of Technology, Maharaja Agrasen University, Baddi, Solan, Himachal Pradesh, India. His research interests include material science, optoelectronic devices, signal processing, and nanoelectronics.

LSB Steganographic Approaches for Hiding Acoustic Data in Video by Using Chaotic and XORing Technique

Namrata Singh and Jayati Bhardwaj

Abstract Maintaining secrecy of data in the digital media is much of the worked domain and requires more of it in the present. Steganography provides the secrecy provision by hiding the different types of into some of the cover objects available. This paper proposed two schemes of video steganography for hiding a secret audio into a cover video file by embedding audio bits into the extracted frames of the video through LSB embedding technique. Embedding techniques mentioned are namely sequential and random audio bits embedding. The security of the secret data is incorporated by the concept of chaotic sequence and XORing bit patterns before embedding the audio bits into the frames. The main aim of the two techniques proposed is to find out the suitable technique that fits into the required scaling of parameters for a good steganographic technique.

Keywords Chaotic sequence · Human auditory system (HAS) · Least significant bit (LSB) · XOR

1 Introduction

Audio steganography is another form of steganographic approach that ensures privacy and secrecy in the modern times. Hiding copyrighted information, government official secrets and convert channels secret communications could be facilitated by the audio–video steganography [1]. Any kind of noise persisting in the audio is susceptible to the HAS (Human Auditory System) and hence could be easily detected. So the task lies here is to come up with an unidentifiable scheme that could meet steganographic technique requirement for hiding acoustic data. This paper represents hiding scheme of an audio into a cover video at least significant

N. Singh · J. Bhardwaj (✉)
AKTU, Lucknow, India
e-mail: jayatibhardwaj2@gmail.com

N. Singh
e-mail: nam2817210@gmail.com

A. K. Luhach et al. (eds.), *Smart Computational Strategies: Theoretical and Practical Aspects*, https://doi.org/10.1007/978-981-13-6295-8_3

bits position. The quality parameters that ensure the quality of the steganographic approaches are PSNR (Peak Signal Noise Ratio), Mean Square Error (MSE), NAE (Normalized Absolute Error), and Bit Error Rate (BRE). After embedding bits into the cover video, the values of PSNR should go high while MSE should go down, i.e., PSNR and MSE should hold inverse relation.

The cryptography provides security to the digital information flowing in the network by converting it into different forms. The security approach used here is chaotic sequence. This chaotic sequence is used in cryptographic environment in order to provide randomness to the information. These sequences are in-decomposable, normalized elements, unpredictable, and create random values between the boundary ranges of [0, 1]. XORing among the random bits of secret data and the LSBs of the frames preserve the data and it could be reconstructed successfully after the extraction. Formula for the chaotic sequence is stated as follows:

$$X_n + 1 = \mu * X_n * (1 - X_n) \tag{1}$$

All the parametric measures that ensure the quality of steganography could be defined as listed in the following equations:

$$\mathrm{SNR} = 10\log_{10}\left(P_{\mathrm{signal}}/P_{\mathrm{noise}}\right) \tag{2}$$

$$\mathrm{PSNR} = 10\log_{10}(255/\mathrm{MSE})^2. \tag{3}$$

$$\mathrm{MSE} = \frac{1}{MN}\sum_{x=1}^{M}\sum_{y=1}^{N}\left(x_{j.k} - x'_{j,k}\right)^2 \tag{4}$$

$$\mathrm{NAE} = \frac{\sum_{j=1}^{M}\sum_{k=1}^{N}\left|x_{j.k} - x'_{j,k}\right|}{\sum_{j=1}^{M}\sum_{k=1}^{N}\left|x'_{j,k}\right|} \tag{5}$$

$$\mathrm{BER} = \frac{\text{Total no. of bit errors}}{\text{Total number of bits transferred}} \tag{6}$$

Video steganography allows any kind of secret data to be inserted into the frames through a large number of redundant bits. These redundant bits replacements allow minimum distortion in the stream. Hence, it could be used as a better cover medium than all of the other available cover mediums like text, image, etc., and also maintains the privacy by being hidden from the intruders. For hiding a 2 MB size audio in 26 MB of video cover [2], the audio is sampled into 8 kHz sample frequency which results in 12,800 samples of audio. These samples are of 8-bit integer whose values lies in the range of −255 to 255. For storing the sign component the 8 bit sampled are stored in the form of 16 bit that are further converted into binary form and stored in single row array. So, this helps to determine how many audio bits are to be stored in a particular frame of video.

The paper is organized in the following manner: Sect. 1 defines the problem domain introduction, Sect. 2 presents the literature review, Sect. 3 defines the proposed schemes for the two approaches of video steganography. Section 4 concludes the paper.

2 Literature Review

Paper [3] by Cvejic and Sepanen presents an audio steganography technique by using a two steps algorithm. The proposed new LSB technique proved to be more robust and of high perceptual quality as compared to traditional LSB method. Two steps proposed LSB method push the limits of data insertion in audio from the fourth LSB layer to sixth LSB layer. The first step contains embedding the secret bits into the *i*th layer of LSB in cover audio. The second step involves shaping the impulse noise so as to change the white noise in the audio. The bit error rate is quite lower along with a challenging steganalysis of the proposed new LSB technique. Enhanced audio steganography (EAS) is proposed in paper [4] by Sridevi et al. combines steganography and cryptography by using audio. This two method scheme is implemented by using a powerful encryption method (for cryptography) and modified LSB (for steganography). The EAS provides a quite good and secure data hiding method. However, it limits the message size to 500. Main advantages of the proposed system are—the size of the cover remains same even after encoding and the negligible sound distortions observed, hence could be ignored. The EAS method mentioned in the paper proved to be robust and could be applied on any type of audio file format. A high PSNR and low MSE valued LSB technique is mentioned in [2] where an audio is hidden in the video file. The random number generator is used to find out the sequence of hiding audio samples into the frames of the cover video. Redundant bits of frames are utilized for the LSB replacement. The resultant embedded video resembles the original video, i.e., high imperceptibility. Paper [5] by Chen et al. proposed an audio signal embedding scheme that hides an audio into quantized DCT coefficients of video frames. Audio bits are converted into 0 or 1 when a nonzero quantized DCT coefficient as audio host. At the decoder side, hybrid signals so obtained extract both the audio and video signals without any loss. The proposed approach performance is quite good as compared to all other coding techniques. The original video and audio are recovered without any compromise.

A text embedding scheme in an audio is implemented by Jayaram et al. in [6]. This scheme allows large amount of secret text to be inserted into the cover audio and proves its advantages over different disadvantages namely echo hiding, spread Spectrum and parity coding. The proposed approach results are good with high security and imperceptibility. The approach is simple and leads to no change in the original audio as observed by HAS.

A secret video is hidden in another video cover by using a sequential encoding scheme by Yadav et al. in [7]. The secret video is divided into 8-bit binary valued frames in the form of individual components. XOR technique is used for the encryption purpose. XORing of the secret key and 8-bit binary values of each frame component is done. Basic LSB technique along with the unvarying pattern of BGRRGBGR is used for the security purpose.

Sorokin et al. [8] developed a synchronous method for coding audio into a video to produce a hybrid signal which is further encoded. This method allows synchronous coding transmission, storage, and playback of the video along with its related audio signals. Embedding is carried out in the quantized DCT domain. The total parity method is proposed in the paper which allows lossless data insertion without hindering with the video stream.

A colored histogram-based data hiding in video technique is proposed in paper [9]. Data is embedded into the video by dividing the pixels of each frame into two parts—right part embedded bits are counted on to the left part. The scheme mentioned here is error free, allows large data storage and also provided authentication to conform to the integrity of cover video upon extraction. The Histogram Constant Value threshold is compared with each video frame and suitable pixels are selected for the embedding purpose. Higher HCV possesses great perceptibility hence less threshold is desirable enough for the successful steganography. A comparative implementation of lossy type sequential and random byte technique is mentioned in [10]. The cover object taken is video as an advantage over the image cover disadvantages. The results show high performance of random byte technique with lesser encryption and decryption time being taken by the LSB. The resultant video shows no difference with the original video and the data is recovered successfully.

3 Proposed Work

The work comprises two different LSB techniques: Sequential and Random LSB Hiding Technique. Each of the two approaches comprises of embedding and extraction algorithms. The concept of metamorphic cryptography is being used in the proposed approaches as each approach withholds the combination of steganography and cryptography. The LSB embedding steganographic technique is used along with the chaotic sequence concept which is used for the encryption here. This encryption of the secret data makes it more secure by making it unbreakable by the intruder. Sections 3.1 and 3.2 defines the newly proposed technique for the video–audio steganography.

3.1 Sequential LSB Hiding Technique

In this technique, the audio bits are inculcated sequentially at the least significant bit position of each frame of the cover video, after checking out for proper space requirement. The whole embedding and extraction process flowcharts are shown in

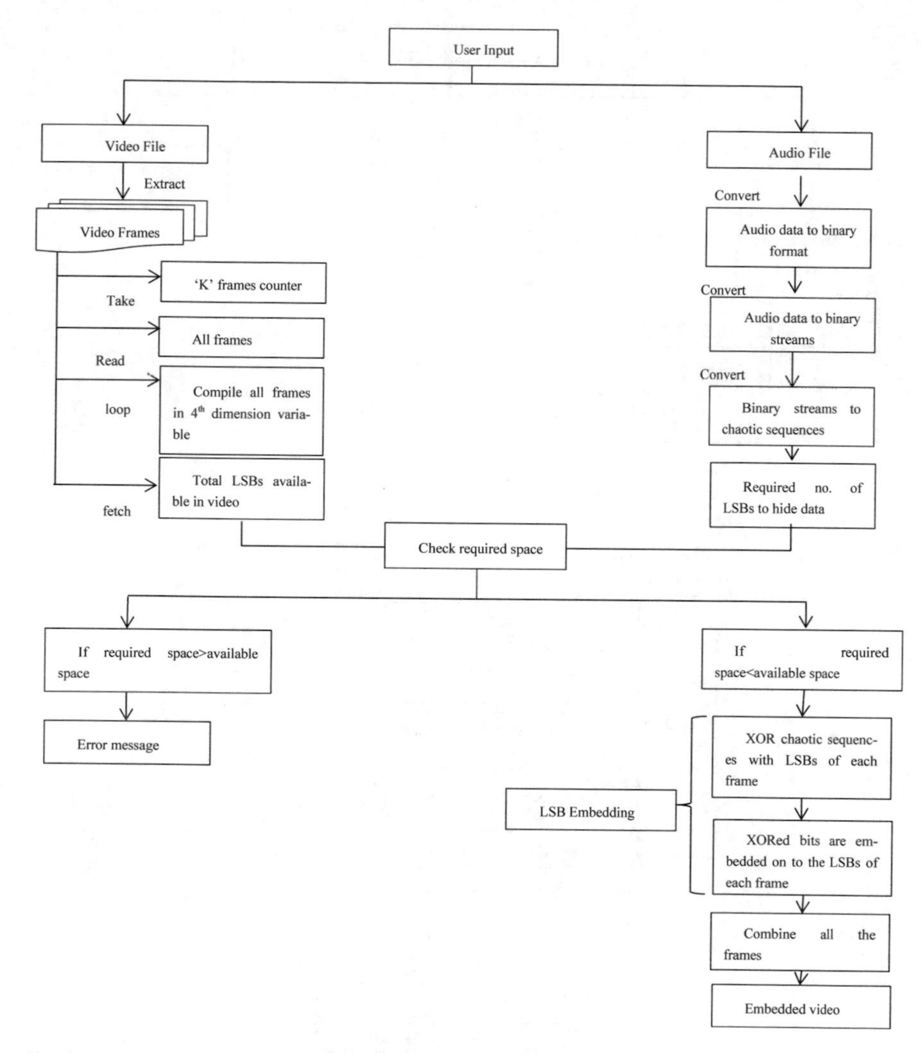

Fig. 1 Sequential LSB embedding technique

Figs. 1 and 2 respectively. Steps for the embedding scheme are defined in depth in Sect. 3.3 (Fig. 5). The video steganographic steps could be summarized as follows:

- Take a video and an audio where video size should be greater than audio size.
- Divide video into frames and check out for the number of LSBs available.
- Convert audio to its intermediate and binary form, further apply encryption formula of chaotic.
- Check for the space requirement whether LSBs available > total no. of audio bits to be inserted in each frame.

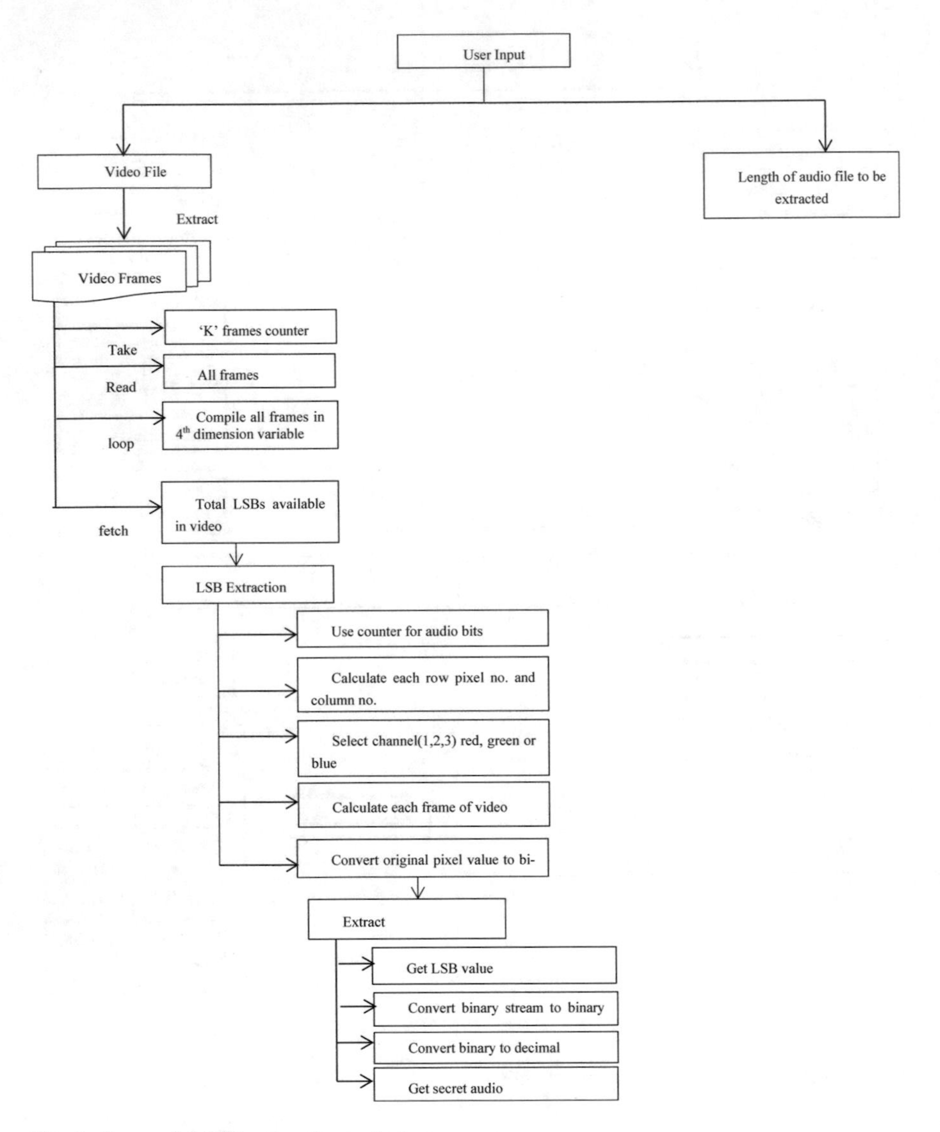

Fig. 2 Sequential LSB extraction technique

- On meeting space requirement, do XOR among the frame LSBs and the resultant chaotic audio bits.
- Replace the original LSBs of each frame with the XORed bits.
- Combine all the frames to get the video.

The extraction process of the sequential LSB technique is just the reverse of embedding process as shown in Fig. 2. Extract LSBs out of frames and reverse XOR them to get both the audio bits and original LSB bits. Convert the binary data to real audio file.

3.2 Random LSB Hiding Technique

In this technique, the secret audio bits are embedded into the LSBs of the random frames of the video. Each random frame is first checked out for the space requirement as per the required perceptibility quotients. The whole embedding and extraction process flowcharts are shown in Figs. 3 and 4. Steps for the randomized

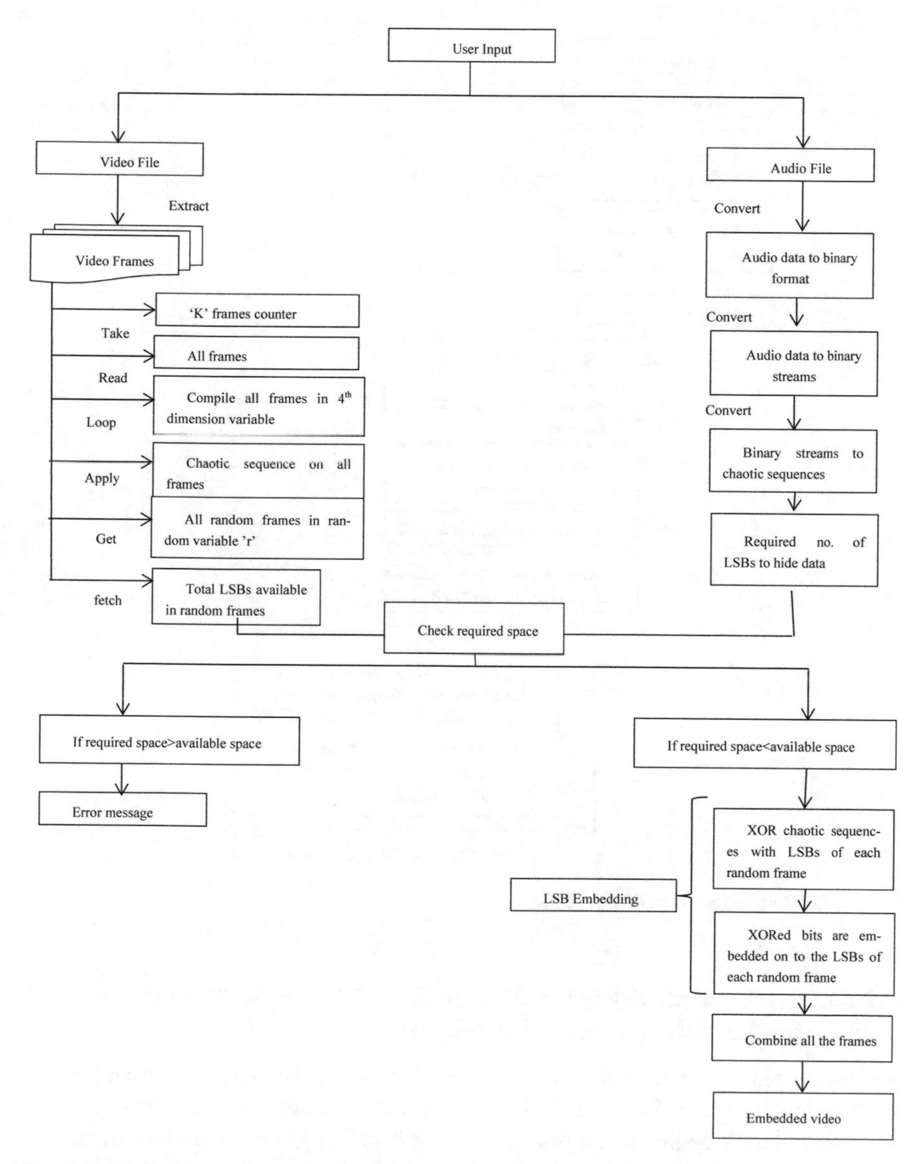

Fig. 3 Random LSB embedding technique

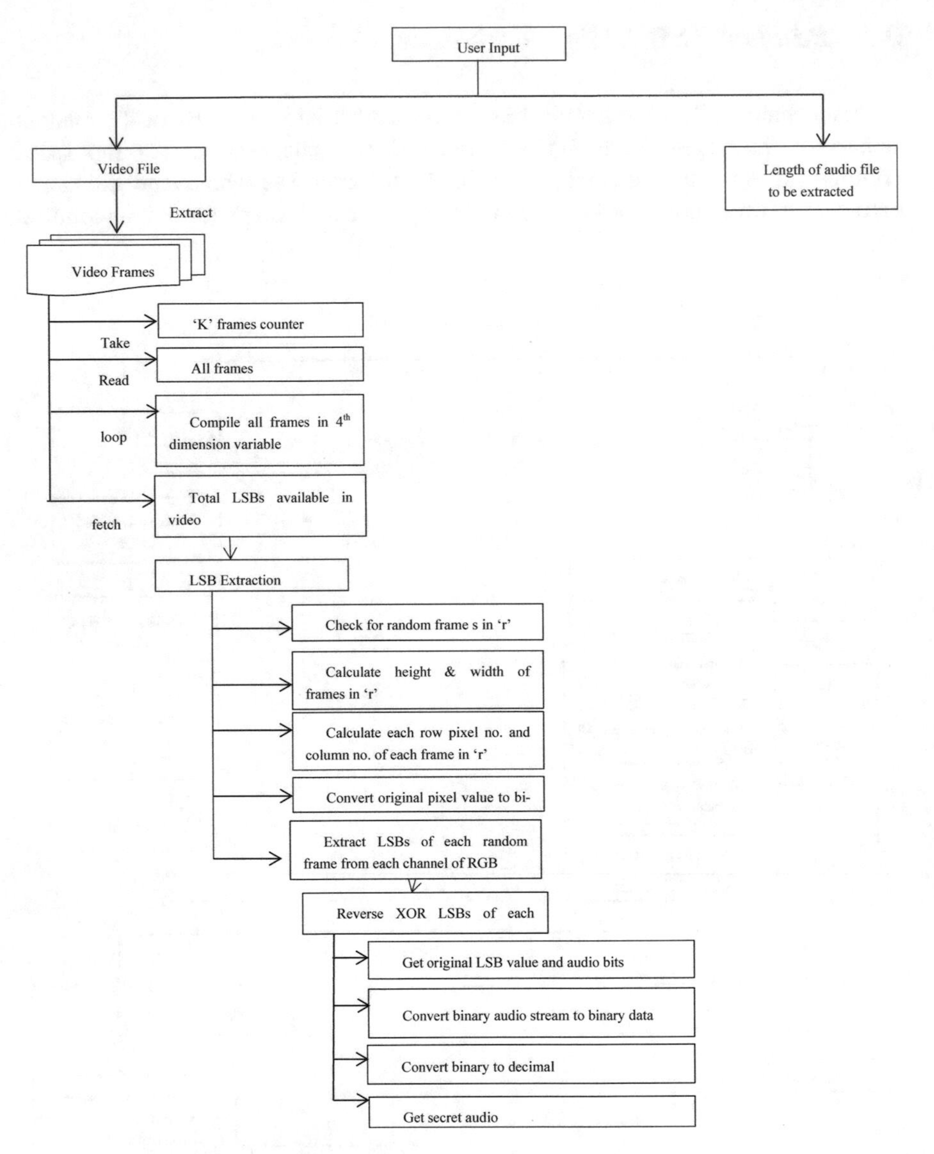

Fig. 4 Random LSB extraction technique

embedding scheme are defined in depth in Sect. 3.3 (Fig. 5). The video steganographic steps could be summarized as follows:

- Take a video and an audio where video size should be greater than audio size.
- Divide video into frames and check out for the number of LSBs available.
- Apply chaotic sequence on frames in order to select group to random frames for hiding purpose.

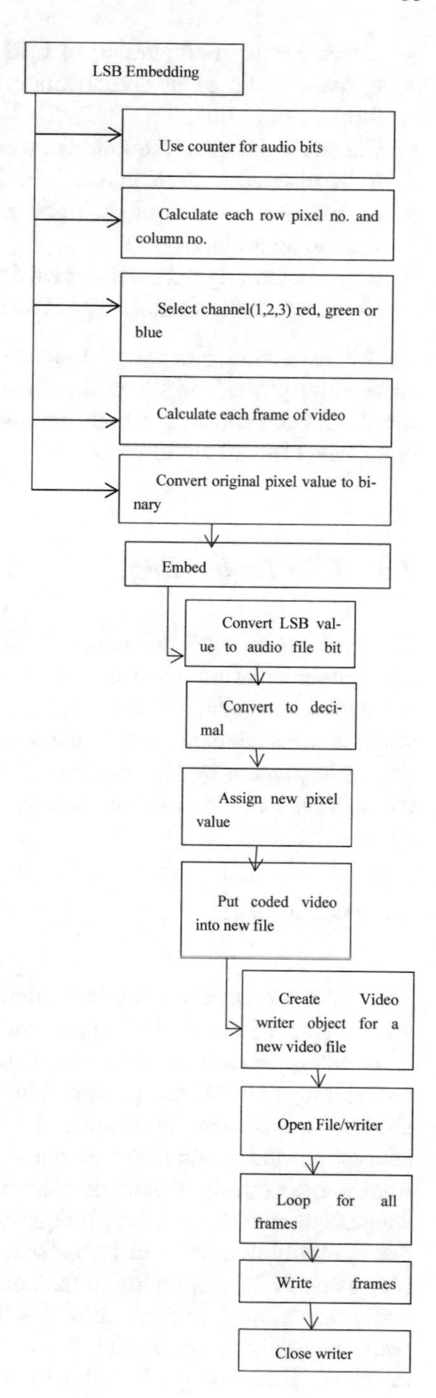

Fig. 5 LSB embedding in sequential technique

- Check for the total number of LSBs available for the audio hiding.
- Convert audio to its intermediate and binary form, further apply encryption formula of chaotic.
- Check for the space requirement whether LSBs available > total no. of audio bits to be inserted in each frame.
- On meeting space requirement do XOR among the frame LSBs and the resultant chaotic audio bits.
- Replace the original LSBs of each random frame with the XORed bits.
- Combine all the frames to get the video.

The extraction process of the random LSB technique is just the reverse of embedding process as shown in Fig. 4. Extract LSBs out of random frames and reverse XOR them to get both the audio bits and original LSB bits. Convert the binary data to real audio file.

3.3 LSB Embedding

The methods for LSB embedding vary in both sequential and random approaches. The flowchart of the processes is shown below in the Figs. 5 and 6. In Fig. 5 of sequential, the audio bits are inserted sequentially in each frame of the cover video while in the random technique, the secret bits are inserted into the random frames as shown in process by Fig. 6. Hence, both the techniques differ significantly in the methodology of video steganography

4 Conclusion

The paper presented the implementation approaches for hiding a secret audio in a video using LSB. The two approaches proposed here are sequential and random LSB hiding techniques. The embedding and extraction processes for each of the methodology are being represented in the form of algorithmic framework in Sect. 3 above. The concept of chaotic has been adopted with the aim of providing encryption and randomness to the secret acoustic data and for the selection of frames, respectively. Comparing the two of techniques on the framework grounds suggests that the random technique would be complex and would lead to high perceptibility as compared to the sequential one which is simple and would have high rate of imperceptibility in the embedded video. This is so because embedding audio data chunks into restricted frames might distort the originality of video and could be easily recognized by the HAS. Hence, randomizing methodology is less desirable. Future scope includes implementation of the schemes along with their result analysis on parametric grounds.

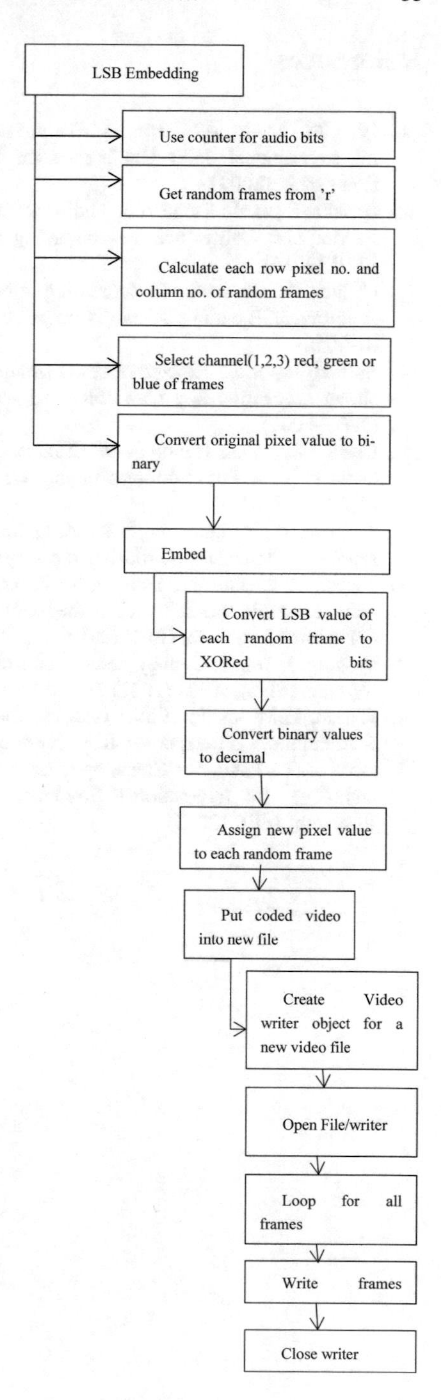

Fig. 6 LSB embedding in random technique

References

1. Qiao, M., Sumg, A.H., Liu, Q.: Feature mining and intelligent computing MP3 steganalysis. In: International Joint Conference on Bioinformatics, Systems Biology and Intelligent Computing (2009)
2. Gosalia, S., et.al.: Embedding audio inside a digital video using LSB steganography. In: 2016 International Conference on Computing for Sustainable Global Development (INDIAcom), IEEE (2016)
3. Cvejic, N., Sepanen, T.: Increasing robustness of LSB audio steganography by reduced distortion LSB coding. J. Univ. Comput. Sci. (JUCS) (2005) https://doi.org/10.3217/jucs-011-01-0056
4. Sridevi, R., et al.: Efficient method of audio steganography by modified LSB algorithm and strong encryption key with enhanced security. J. Theor. Appl. Inf. Technol. JATIT **5**(6), (2005–2009)
5. Chen, W., Li, J., Gabbouj, M., Takala, J.: Lossless audio hiding method for synchronous audio-video coding. In: International Conference on Acoustics, Speech and Signal Processing (ICASSP) (2011)
6. Jayaram, P., Ranganatha, H.R., Anupama, H.S.: Information hiding using audio steganography—a survey. Int. J. Multimedia Appl. (IJMA) **3** (2011)
7. Yadav, P., Mishra, N., Sharma, S.: A secure video steganography with encryption based on LSB technique. In: 2013 IEEE International Conference on Computational Intelligence and Computing Research, IEEE (2013)
8. Sorokin, H., et.al.: Coding method for embedded audio in video stream. In: Signal Processing Systems (SIPs), IEEE (2011)
9. Kelash, H.M., et.al.: Hiding data in video sequences using steganographic algorithms. In: International Conference on ICT Conference (ICTC), IEEE (2013)
10. Bole, A.T., Patel, R.: Steganography over video file using random byte hiding and LSB technique. In: International Conference on Computational Intelligence and Computing Research, IEEE (2012)

Machine Learning Configurations for Enhanced Human Protein Function Prediction Accuracy

Amritpal Singh, Sunny Sharma, Gurvinder Singh and Rajinder Singh

Abstract Molecular class prediction of a protein is highly relevant for conducting research in domains of disease-detection and drug discovery process. Numerous approaches are incorporated to increase the accuracy of Human protein Function (HPF) prediction task, but it is highly challenging due to wide and versatile nature of this domain. This research is focused on sequence derived attributes/features (SDF) approach for HPF prediction and critically analyzed with the WEKA data analysis tool. New SDFs were identified and included in the training dataset from the Human protein reference database, enhanced as in number of sequences and the related features for deriving the relation with various protein classes. A range of Machine Learning approaches were analyzed for prediction effectiveness and a comprehensive comparison is carried out to achieve higher classification accuracy. The Machine Learning approach is also analyzed for its limitation on application of broad spectrum data domain and remedies for the limitation were also explored by changing the configuration of data sets and prediction classes.

Keywords Bagging · Bayes Net · C5 · Decision tree · HPF · IBK · J48 · Logistic approach · PART · Random forest · SDF · Weka

1 Introduction

Protein classification is a vast domain with enormous amount of data available for research and analysis yet the knowledge about its correct perception is very limited. On the other hand Machine learning (ML) provides promising answers to not-so-clearly defined areas of research. Thus, it's a powerful tool to explore the possibilities of the enhancement of the current understanding of protein.

A. Singh · S. Sharma (✉) · G. Singh · R. Singh
Guru Nanak Dev University, Amritsar, India
e-mail: sunnysharma05@yahoo.co.in

A. Singh
e-mail: amritsinghmand@gmail.com

A. K. Luhach et al. (eds.), *Smart Computational Strategies: Theoretical and Practical Aspects*, https://doi.org/10.1007/978-981-13-6295-8_4

Decision tree [1, 2] based prediction approach of machine learning is very clear and reliable for protein classification. Being a white-box approach it clearly illustrates the sequence of computations involved at each and every stage. This plus point enables its usage by computational experts even without much knowledge of the concerned domain. Similarly, a domain expert is empowered for examining the toute followed by an expert of computation. So the gap between technical knowledge and domain expertise. Nodes and edges indicates various utilities at the different stages of computations in a Decision tree [3]. A decision tree neatly depicts the results required or outputs of various possibilities of outcome. It clearly defines the problem structure and its interpretations in a hierarchical way which is much easier to comprehend. As the model has a unique ability of considering different initial parameters and reaching a goal [4, 5]. However, recent advancements suggests that the prediction of Protein-Function is a domain where ML faces some challenges. A thorough collection of almost 65 papers in 'A Survey of Computational Methods for Protein Function Prediction' helped arrive on this aspect of ML applicability on HPF prediction [6]. So identifying the challenges and the possible solutions to overcome such situations is also the key focus of the study.

2 Protein and Protein Function Information

Proteins are the large and complex building blocks for all the life forms on this planet. They play a defining role in functionality and the structuring of organs and tissues of the body of an organism and also perform regulatory functions in it. Several diminutive components form protein. These diminutives are referred as Amino-Acid (AA) groupings. This arrangement helps in various tasks of a life cell. Proteins can be assigned to various categories based on functions, these groups for each protein are listed here as: Transport proteins, Enzymes, Hormones, Motor proteins, Immunoglobulin or Antibodies, Receptors, Storage proteins, Structural proteins, Signaling proteins. There are various diminutive units, 20 in number, listed as: Alanine, Arginine, Asparagine, Aspartic acid, Cysteine, Glumatic acid, Glycine, Glutamine, Histidine, Leucine, Lysine, Isoleucine, Methionine, Serine, Phenylalanine, Proline, Threonine, Tyrosine, Tryptophan and Valine. In the prime form the protein sequence it can be characterized as a string of Twenty AAs and they are chained to make a protein [7]. Amino part (–NH2) and the carboxylic part (–COOH) of the nearby AA is joined to generate a bond (peptide) as:

$$\begin{array}{ccccc} -\mathrm{NH2} & + & -\mathrm{COOH} & == & -\mathrm{CONH} \\ \mathrm{A-group} & & \mathrm{C-group} & & \text{Resultant bond} \\ (\mathrm{AA1}) & & (\mathrm{AA2}) & & (\text{Peptide}) \end{array}$$

Basically AA forms a Peptide bond with the other AA and an extensive grouping of AA's is formed. A series of such peptide bonds is known as a polypeptide. The chain of AA's describes the unique 3D structure and the exact function of each protein [8].

2.1 Protein Features Extraction from Sequence

Protein sequence has various features which are helpful to predict protein class. These features play a considerable role in the development of Human protein class prediction classifiers. These vital features can be extracted from a given protein sequence of amino-acids using various web based bio-informatics tools. A combination of various online tools is required to extract a set of useful features (Table 1).

Extraction of features through web based tools is a tedious task as there are various Feature Extraction tools available but none of them compute all the features on a single platform rather various tools are used for extracting different features [9] so a Sequence Derived Features Extraction Server (SDFES) was developed for this task by us in MATLAB. It was further tested for accurate extraction by comparing its results with web tools and we were able to perform accurate feature extraction with this server. SDFES follows an approach of extracting various sequence based features and they are integrated using Associative rule mining [10]. The associative rule mining plays a significant role to find the physico-chemical sequence extracted features from protein sequence. The various sequence extracted or derived features like total number of AAs, Individual number of AAs with their molecular percentage, Extinction Coefficient, Number of Positively and Negatively charged residues, molecular weight, Absorbance, Isoelectric point, Computation of Isoelectric point/Molecular weight, Aliphatic Index, Gravy, Instability Index, Volume, some residues properties Polar, NonPolar, Tiny, Small, Aliphatic, Acidic,

Table 1 Feature extraction tools

Server	Detailed information
TMHMM Server v.2.0 [23]	It predict the transmembrane helices in proteins based on Hidden Markov Model (HMM)
SignalP4.1 Server [24]	Using amino acid sequence it can identify the occurrence and place of signal-peptide-cleavage sites in various organisms like eukaryotes etc. It also predicts the signal/non-signal peptide prediction using various ANN
NetNGlyc1.0 Server [25]	It finds out the *N*-Glycosylation sites using ANN that scrutinizes the sequence environment of Asn, Xaa, Ser/Thr, sequins inside the human protein
ProtParam [26]	It finds out PI, instability/aliphatic index GRAVY, etc.
PSORT [27]	Localization spots in cells of protein are forecast by PSORT
PROFEAT [28]	PROFEAT computes AA-Composition, Di-peptide composition, Auto-correlation Descriptor, Composition-Transition-Distribution (CTD), Quasi Sequence Order (QSO) Descriptors, Pseudo-AA-Composition (PAAC), Amphiphilic Pseudo-AA-Composition (APAAC), Topological Descriptors at Atomic Level, Total-AA-Properties (TAAP), Network Descriptors

Aromatic, Basic, Charged, Total number of Atoms With C, H, N, O, S detail, information of codons, Nucleotide density and Nominal mass, Monoisotopic mass plots, are extracted from the protein sequence on a single platform using associative rule mining by union of all the algorithms together. Human protein is combination of 20 amino acids and single protein contains thousands of amino acids in a sequence so to extract single feature from the protein sequence itself is a difficult task. To extract all the above mentioned features together from protein sequence is a herculean task. But associative rule mining gives the power to solve this complex task because each calculation has few common parts which are called the intersection. The union and intersection operations are used to do handle complexity. Using this technique all the features are integrated and important results are obtained on a single platform. This shows the power of rule mining and demonstrates how we can integrate all these features in a single platform and also describes that how rule mining is helpful for machine learning. The proposed way is very easy and straightforward to use. The only manual effort to extract features is just to provide the protein-sequences; the rest of the work of extracting features is done by the server. The results were compared and verified through the output of previously available online tools. The proposed approach is easy, fast and reliable as compared to the use of other online tools. The significant contribution of this server besides providing one stop feature evaluation is the inclusion of 5 more features for class prediction which are not available in other tools to facilitate the drug discovery and its use procedures.

3 Literature Survey

Many of the protein classification methods based on Sequences, Genomics context, Phylogenomics, Structure, Protein–Protein Interaction, Gene expression, Data Integration etc., since they use the features extraction methodology for protein function prediction or protein classification; Survey shows that feature extraction form protein sequence using vector space integration methods of data Integration for protein classification plays vital role in Bioinformatics.

Jensen et al. focused research on developing a sequence based method which identified and added up the critical features for the protein function assignment task to respective classes, enzyme classification and show the benefits for protein function prediction through a linear sequence of amino acids over protein structure. They identified important attributes/features related with post-translational-modifications; also include simple features like length, composition of the polypeptide chain as well as isoelectric point [11]. Cai et al. used five physiochemical features, from a linear AA sequence, such as NVM volume, surface tension, charge and polarity using the SVM-Prot method for functional classification and showcase its importance of achieving accuracy slightly more than 71% on data set having proteins of various plants [12]. Friedberg expressed that as diversification and enhancement in the volume of pure sequence and structure related data is growing, which leads unequal enlargement in the number of uncharacterized gene products. Recently well-known methods for gene as well as for protein annotation, like homology based transformation, they are annotating fewer data and in some cases they are increasing existing incorrectly-annotated data. The author mentioned that the Contextual and Subjective definition of protein function is cumbersome in nature and expressed views for quality of function predictions [13]. Lobley et al. predict the protein function with IDRs (intrinsically disordered regions) in human protein sequence on the basis of length and position dependencies. Sequence based features were used like length, molecular weight, charge, hydrophobicity, transmembrane residue, pest region peptide and disordered related features for protein function prediction through machine learning approach [14]. Singh et al. describes how the protein related data is increasing day by day and suggested to solve the problem related to human protein function prediction and said express the need of machine learning algorithms for drug discovery. The author use Decision tree induction approach through C4.5 algorithm for the selection of best attribute for protein function prediction and presents the accuracy of 72% for human protein function prediction in contrast to an accuracy of 44% of existing prediction methodology [15]. Singh et al. describe unsupervised learning and cluster analysis approach prediction of human protein classes. The database, extracted from the human protein reference database (HPRD) was used and five AA sequences were taken for each molecular class, then sequence derived features were grabbed for each protein sequence with the help of web based tools. A clustering technique predicts the class of the query sequence [16]. Wass et al. used sequence based features; protein–protein-interaction features as well as gene expression based features and incorporate CombFunc method for protein function prediction. CombFunc was also evaluated for the predictions of gene ontology molecular function on the data set of 6686 proteins. The Uni-Prot-GOA annotation's taken out for the proteins and achieved 0.71% Precision and 0.64% recall using CombFunc [17]. Ofer et al. used the feature extracted technique and proposed biological interpretability features to predict localization-structure classes and its distinctive functional properties [18]. Gong et al. proposed GOFDR function (Sequence based method). It predicts the gene ontology or the features of the query sequence as well as features of similar sequence obtained from the database [19]. Lavezzo et al. proposed BLAST and HMMER3 tool for query processing as well as for database sequence retrieval. They extracted gene ontology terms from the retrieved sequences and followed the sequence categorization information [20]. Das et al. reviewed structure as well as a sequence function prediction method using CATH-Gene-3D and FunFHMMer classification databases. Here the categorization of query item is done by CATH-Gene3D and rendered to classification data house representing structural part of the queried sequence. The server FunFHMMer matches the sequencing with one present in CATH-Gene3D with incorporation of HMM and after that by prediction of the sequence queried [21].

4 Research Methodology

The Methodology used in the research process from data set collection to results and analysis with trial learning and testing can be expressed as follow.

1. *Collection of datasets from HPRD*: Data sets are readily available from Human Protein Reference Database (HPRD) and can be simply downloaded in .txt format.
2. *Features extraction from human protein sequence*: The results are obtained after processing of data set on the Sequence Derived Feature extraction server (SDFES), capable of extracting 34 features in order to facilitate the task of feature extraction which is quite hectic to perform on various web tools to get the feature sets. The Human protein AA-sequence is provided as input to the server which predicts or extract following features: total number of amino acids, Individual number of amino acids with their molecular percentage, Extinction Coefficient, Number of Positively and Negatively charged residues, molecular weight, Absorbance, Isoelectric point, Computation of Isoelectric point/ Molecular weight, Aliphatic Index, Gravy, Instability Index, Volume, some residues properties Polar, NonPolar, Tiny, Small, Aliphatic, Acidic, Aromatic, Basic, Charged, Total number of Atoms With C, H, N, O, S detail, information of codons, Plot a graph of Nucleotide density and Nominal mass, Monoisotopic mass features.
   These features play a vital role in prediction of protein function using machine learning.
3. *Preparing experimental dataset for WEKA*: The input is provided of human protein amino acid sequence in the capital letter format was passed to **SDFES** and the results obtained as feature parameters were passed as input to WEKA for further analysis. The server is adapted to provide results in an excel sheet which is easily comprehended by WEKA.
4. *Application of various classifiers*: WEKA [5] comes with the enhanced capability of dealing with huge databases which other popular data analytics tools lack. WEKA is a workhorse containing a blend of tools and calculus for information processing and its detailed representation with a user friendly GUI which provide ease of use for a range of computational capabilities. It is highly suitable for a wide range of Research and development activities.

Factors favoring use of WEKA in this critical analysis are:

- Free access, as it is an open source product.
- Highly versatile and equally portable.
- Extensive coverage of information acquisition and handling and visual modeling capabilities.
- User friendly GUI [22].

Various Classification techniques based on Decision Tree, Rule Mining, Lazy, Bayes Network, Meta and Functions are implemented on the data set. Various classification algorithms like Random Forest, J48, PART, BayesNet, Logistic Approach, IBK and Bagging are applied to the data set.

5. *Repeating experiment by varying attributes*: Experiment is performed by varying the attributes like number of features, sequences and classes for checking the impact of the changes on machine-learning approach and then analyzing it to provide better insight into its working.

A standard data set is derived from HPRD [4] and already well-established machine learning techniques are used in the classification process. This approach ensures any boosting or optimizing effects of enhanced or improved versions of these algorithms do not create any assessment issue and the results obtained are easily justified. However the approach is equally applicable to other versions of technique as well, as it is more data centric than algorithmic.

5 Results and Discussions

The data set used for analysis is extracted from HPRD [4] and three experiments carried out have variably 20, 24 and 34 features; these are extracted from 58 and 70 sequences respectively and is taken as input for analysis through Weka for 10 and 12 classes. Improved data set supplied to WEKA with incorporation of *Five* new features (significantly: solubility, mean and absorbance) indicated some useful consideration for improving ML applications in protein classification domain. The contribution of each sequence derived feature with its maximum and minimum range in protein class prediction is displayed in WEKA dashboard. Various Classification techniques based on Decision Tree, Rule Mining, Lazy, Bayes Network, Meta and Functions are implemented on the data set. Various classification algorithms like Random Forest, J48, PART, BayesNet, Logistic Approach, IBK and Bagging are applied to the data set, among them Random Forest outperform all of them with an achievement of overall accuracy 70.69% and classification accuracy of 97.14% for specific protein classes as shown in Fig. 1.

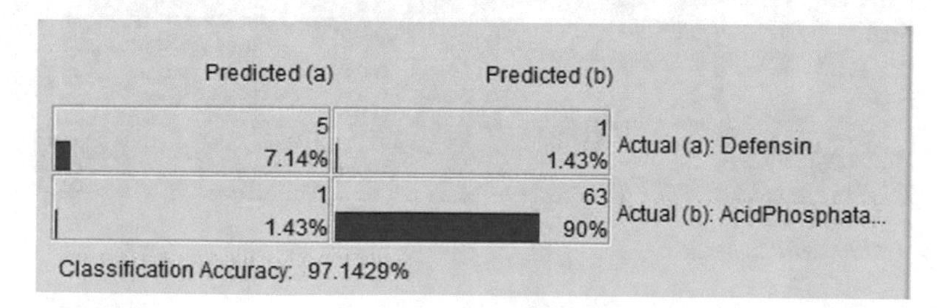

Fig. 1 Protein classification accuracy

The 10 fold cross validation is done for protein class prediction, i.e. classification of instances of random forest method is shown in run results with the detailed summary, which show the mean absolute error of 0.10% and the root mean squared error of 0.21% and describe the detail of correctly and incorrectly classified instances. It also showcases complexity improvement of −28.746 bits/instance. The detailed accuracy achievement by protein classes is indicated by a weighted average of the true positive rate, false positive rate, precision, recall-value, F-Measure, MCC and Area under ROC and PRC, which express the true classification of protein classes. The cost matrix for minimizing cost describe a gain of 8.97 at a cost of 2.0 with a good true positive rate for protein class prediction on 70 sequences is shown in Fig. 2, it describes the classification accuracy achieved for 'defensin' class at this setting (97.14%). The threshold curve for Defensin protein class for sample size and true positive rate is appropriate. The defensin class has a true positive rate of 0.833 along with the precision and recall value of 0.833 which is quite good (for 24 features, 70 sequences and 12 class experiment) and these values are generated in run information of WEKA for each experiment.

Technique wise accuracy analysis indicates, Random forest algorithm outperforms other Machine learning algorithms. Configuration wise accuracy analysis indicates that increased features increase the overall accuracy of all the ML algorithms (Fig. 3).

Accuracy comparison of various Machine Learning techniques with feature, sequence and class variations shows some obvious relations of increased overall classification accuracy with an increase in the attributes. But the interesting observation of these configurations is the impact on individual class accuracies.

The overall classification accuracy is rising, but individual class accuracy increases in some cases and decreases in the other. The details of each experiment can be better understood by observing the confusion matrix of each setting. Table 2 shows the increase and decrease in individual class accuracy with changes in features, sequence and class variations.

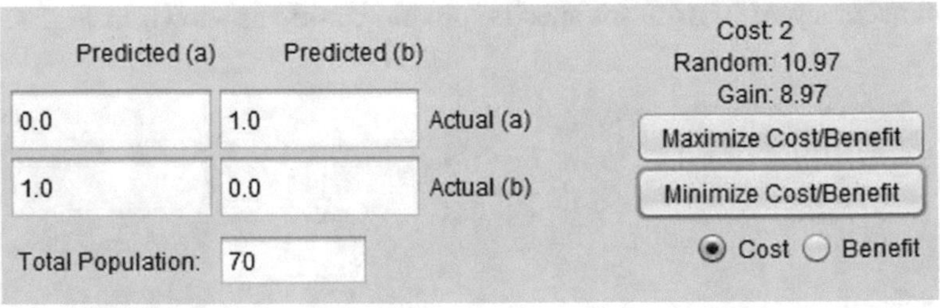

Fig. 2 Cost/benefit analysis

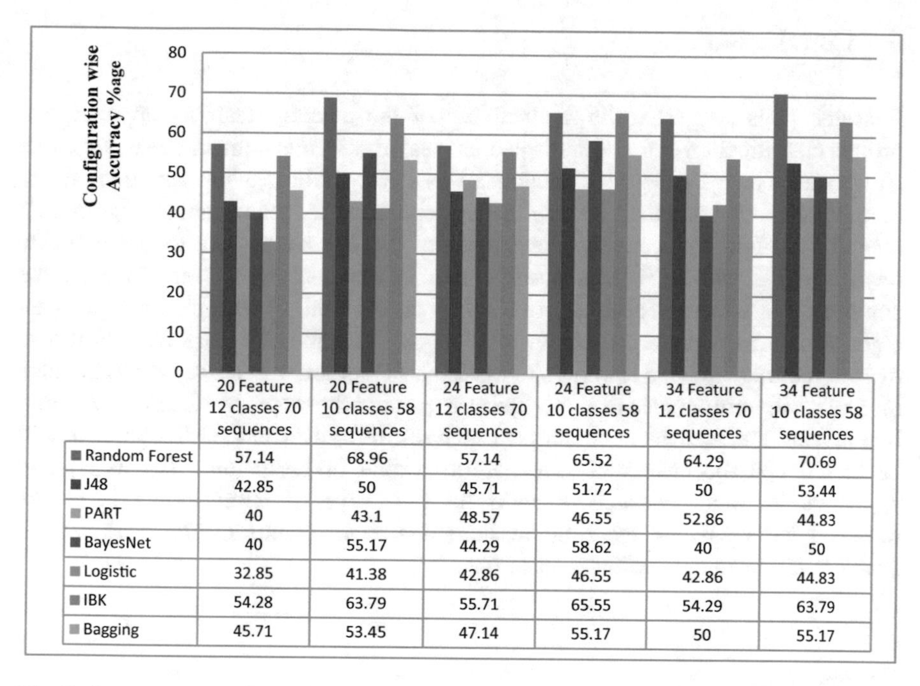

	20 Feature 12 classes 70 sequences	20 Feature 10 classes 58 sequences	24 Feature 12 classes 70 sequences	24 Feature 10 classes 58 sequences	34 Feature 12 classes 70 sequences	34 Feature 10 classes 58 sequences
Random Forest	57.14	68.96	57.14	65.52	64.29	70.69
J48	42.85	50	45.71	51.72	50	53.44
PART	40	43.1	48.57	46.55	52.86	44.83
BayesNet	40	55.17	44.29	58.62	40	50
Logistic	32.85	41.38	42.86	46.55	42.86	44.83
IBK	54.28	63.79	55.71	65.55	54.29	63.79
Bagging	45.71	53.45	47.14	55.17	50	55.17

Fig. 3 Accuracy comparison of various machine learning techniques: with feature, sequence and class variations

Table 2 Confusion matrix with 20 and 24 feature variable settings

<table>
<tr><th>CLASSES</th><th>20 features 12 classes and 70 sequences</th><th>20 features 10 classes and 58 sequences</th><th>24 features, 12 classes and 70 sequences</th><th>24 features 10 classes and 58 sequences</th></tr>
<tr><td></td><td>a b c d e f g h i j k l</td><td>a b c d e f g h i j</td><td>a b c d e f g h i j k l</td><td>a b c d e f g h i j</td></tr>
<tr><td>Defensin</td><td>5 – 1 – – – – – – – – – – | a</td><td>5 – – – 1 – – – – – – | a</td><td>5 – 1 – – – – – – – – – – | a</td><td>5 – 1 – – – – – – – – | a</td></tr>
<tr><td>AcidPhosphatase</td><td>– 3 – – – – – – 1 2 – – – | b</td><td>– 4 – – – – – 2 – – – | b</td><td>– 3 – – – – – – 1 2 – – – | b</td><td>– 3 – – – 1 2 – – – | b</td></tr>
<tr><td>VoltageGatedChannel</td><td>– 1 3 – – – 2 – – – – – – | c</td><td>– – 3 – 2 – 1 – – – | c</td><td>– – 3 – – – 2 – 1 – – – | c</td><td>– – 3 – 2 – 1 – – – | c</td></tr>
<tr><td>DNARepairProtein</td><td>– – – 2 1 – – 2 – 1 – – | d</td><td>– – – – – – – – – – – –</td><td>– – – 1 1 1 – 1 – 2 – – | d</td><td>– – – – – – – – – – – –</td></tr>
<tr><td>Decarboxylase</td><td>– – – – 4 – 1 1 – – – – | e</td><td>– – – – – – – – – – – –</td><td>– 1 – – 1 1 1 2 – – – – | e</td><td>– – – – – – – – – – – –</td></tr>
<tr><td>HeatShockProtein</td><td>– – – – – – 4 – 1 – – – – | f</td><td>– – – 4 – 1 – – – – | d</td><td>– – – – – – 4 – 1 – – – – | f</td><td>– – – 4 – 1 – – – – | d</td></tr>
<tr><td>Aminopeptidase</td><td>– – 1 – – – 4 – – – – 1 | g</td><td>– – 1 – 5 – – – – – – | e</td><td>– – 1 – – – 4 – – – – 1 | g</td><td>– – 1 – 4 – – – – 1 | e</td></tr>
<tr><td>G-Protein</td><td>– – – – 2 – – 2 – 2 – – | h</td><td>– – – – – – 4 – 2 – – | f</td><td>– – – – 2 – – 3 – 1 – – | h</td><td>– – – – – – 4 – 2 – – | f</td></tr>
<tr><td>WaterChannel</td><td>– 1 – – – – – – – 5 – – – | i</td><td>– 1 – – – – 5 – – – | g</td><td>– 1 – – – – – – – 5 – – – | i</td><td>– 1 – – – – 5 – – – | g</td></tr>
<tr><td>NucleotidylTransferase</td><td>– – 1 – 2 – – 1 – 2 – – | j</td><td>– – 1 – – 3 – 2 – – | h</td><td>– 1 – 1 1 – – 1 – 2 – – | j</td><td>– – 1 – – 2 – 3 – – | h</td></tr>
<tr><td>BCellAntigenReceptor</td><td>– 1 – – – – – – – – – 4 – | k</td><td>– – – – – – – – – 5 – | i</td><td>– – – – – – – – – – – 5 – | k</td><td>– – – – – – – – – 5 – | i</td></tr>
<tr><td>CellSurfaceReceptor</td><td>1 – – – – – – – – – 1 2 2 | l</td><td>1 – – – – 1 – 1 – 3 | j</td><td>1 – – – – – – – – – 1 – 4 | l</td><td>1 – – – – 1 – 1 1 2 | j</td></tr>
</table>

Note: The values in bold are signifies the increase and decrease in individual class accuracy with changes in features, sequence and class variations.

6 Conclusion

Research gaps suggested the applicability of the machine learning approach for protein classification and also indicated its weakness in this domain due to very vast and versatile data set. So this critical analysis clearly indicates how formulation and incorporation of 4 and then 10 new features enhanced machine learning algorithm's overall classification accuracy. For 'defensin' class accuracy rose to a remarkable level of 97% with 90% true positive rate in the confusion matrix against the combined classification accuracy of 70.69% on the data set with the random forest algorithm. For finding a particular protein class different setting can be used whose overall accuracy may be lower, but specific class prediction accuracy is high. It also highlights the importance of doing these steps at early stages of Machine Learning implementation, else the upcoming research results built with the ML approach will be biased and the error will propagate to further investigations. This is equally applicable in other research domains for the scope of improvement in results obtained from ML by working on individual components of the classification problem rather than tackling it all at once.

References

1. Wei-Feng, H., Na, G., Yan, Y., Ji-Yang, L., Ji-Hong, Y.: Decision trees combined with feature selection for the rational synthesis of aluminophosphate AlPO4-5. Natl. Nat. Sci. Found. China **27**(9), 2111–2117 (2011)
2. Information on See5/C5.0. http://rulequest.com/see5-info.html (3/3/2017)
3. Arditi, D., Pulket, T.: Predicting the outcome of construction litigation using boosted decision trees. J. Comput. Civ. Eng. **19**(4), 387–393 (2005)
4. Human Protein Reference Database. http://www.hprd.org/ (1/9/16)
5. Weka Machine Learning. https://en.wikipedia.org/wiki/Weka_machine_learning (2/3/17)
6. Amarda, S., Barbará, D., Molloy, K.: A survey of computational methods for protein function prediction. In: Big Data Analytics in Genomics, pp. 225–298. Springer International Publishing (2016)
7. Bergeron, B.: Bioinformatics Computing. Pearson Education, Delhi (2002)
8. Krane, D., Raymer, M.: Fundamental Concepts of Bioinformatics. Benjamin Cummings, San Francisco, California (2003)
9. Sharma, S., Singh, A., Singh, R.: Enhancing usability of See5 (incorporating C5 algorithm) for prediction of HPF from SDF. Int. J. Comput. Technol. **3**(4) (2016)
10. Han, J., Kamber, M.: Data Mining Concepts and Techniques, pp. 279–322. Morgan Kaufmann Publishers, USA (2003)
11. Jensen, L., Gupta, R., Blom, N., Devos, D., Tamames, J., Kesmir, C., Nielsen, H., Stærfeldt, H., Rapacki, K., Workman, C., Andersen, C., Knudsen, S., Krogh, A., Valencia, A., Brunak, S.: Prediction of human protein function from post-translational modifications and localization features. J. Mol. Biol. **319**(5), 1257–1265 (2002)
12. Cai, C.Z., Han, L.Y., Ji, Z.L., Chen, X., Chen, Y.Z.: SVM-Prot: web-based support vector machine software for functional classification of a protein from its primary sequence. Nucleic Acids Res. **31**(13) (2003)
13. Friedberg, I.: Automated protein function prediction—the genomic challenge. Brief. Bioinform. **7**(3), 225–242 (2006)

14. Lobley, A., Swindells, M.B., Orengo, C.A., Jones, D.T.: Inferring function using patterns of native disorder in proteins. PLoS Comput. Biol. **3**(8), e162 (2007)
15. Singh, M., Wadhwa, P.K., Sandhu, P.S.: Human protein function prediction using decision tree induction. Int. J. Comput. Sci. Netw. Secur. **7**(4), 92–98 (2007)
16. Singh, M., Singh, G.: Cluster analysis technique based on bipartite graph for human protein class prediction. Int. J. Comput. Appl. (0975–8887) **20**(3), 22–27 (2011)
17. Wass, M., Barton, G., Sternberg, M.: CombFunc: predicting protein function using heterogeneous data sources. Nucleic Acids Res. **40**(W1), W466–W470 (2012)
18. Ofer, D., Linial, M.: ProFET: feature engineering captures high-level protein functions. Bioinformatics **31**(21), 3429–3436 (2015)
19. Gong, Q., Ning, W., Tian, W.: GoFDR: a sequence alignment based method for predicting protein functions. Methods **93**, 3–14 (2016)
20. Lavezzo, E., Falda, M., Fontana, P., Bianco, L., Toppo, S.: Enhancing protein function prediction with taxonomic constraints—the Argot2.5 web server. Methods **93**, 15–23 (2016)
21. Das, S., Orengo, C.: Protein function annotation using protein domain family resources. Methods **93**, 24–34 (2016)
22. Weka 3—Data Mining with Open Source Machine Learning Software in Java. https://www.cs.waikato.ac.nz/ml/weka/ (2/3/17)
23. TMHMM Server, v. 2.0. http://www.cbs.dtu.dk/services/TMHMM/ (4/8/ 2017)
24. SignalP 4.1 Server. http://www.cbs.dtu.dk/services/SignalP/ (9/6/17)
25. NetNGlyc 1.0 Server. http://www.cbs.dtu.dk/services/NetNGlyc/ (7/4/17)
26. ExPASy: SIB Bioinformatics Resource Portal—ProtParam. https://web.expasy.org/protparam/ (7/6/17)
27. PSORT WWW Server. http://psort.hgc.jp/ (9/7/17)
28. PROFEAT 2015 HOME. http://bidd2.nus.edu.sg/cgi-bin/prof2015/prof_home.cgi (7/5/17)

Automatic Human Gender Identification Using Palmprint

Shivanand S. Gornale, Abhijit Patil, Mallikarjun Hangarge and Rajmohan Pardesi

Abstract Automatic human gender identification can help in a developing number of applications related to human–computer interaction (HCI), human–robot interaction and surveillances technologies. Besides, it can also assist in human face identification by reducing the issue of comparing to half of the database. Several biometrics have been used to identify the human gender, but no significant achievements have been reported in the literature. In this study, we have taken palmprint biometrics, because it contains sufficient significant discriminating information like ridges, wrinkles, and principal lines. Based on it, we are going to propose an algorithm for automatic human gender identification. It involves three steps: extraction of ROI, features computation, and classification. Gabor wavelets are employed to extract the palmprint features as they are potential in capturing discriminating textural properties of the underlying image. Its performance is evaluated with simple *K*NN classifier on publicly available CASIA palmprint Database. The results obtained are quite encouraging with average accuracy of 97.90% with 10 cross validation.

Keywords Biometrics · Gender identification · Gabor wavelets · Palmprints · KNN classifier

S. S. Gornale · A. Patil (✉)
Department of Computer Science, Rani Channamma University, Belagavi, India
e-mail: abhijitpatil05@gmail.com

S. S. Gornale
e-mail: shivanand_gornale@yahoo.com

M. Hangarge · R. Pardesi
Department of Computer Science, KASCC, Bidar, India
e-mail: mhangarge@yahoo.co.in

R. Pardesi
e-mail: madhurrajmohan1@gmail.com

A. K. Luhach et al. (eds.), *Smart Computational Strategies: Theoretical and Practical Aspects*, https://doi.org/10.1007/978-981-13-6295-8_5

1 Introduction

Gender identification was first perceived as an issue in psychophysical studies with insights on the efforts of understanding human visual processing and identifying key features employed to categorize the gender of an individual [1]. In earlier research, it has seen that there exist greater differences between male and female characteristics which can be further used to improve the performances of recognition applications in surveillance and computer vision [2, 3]. Automatic human gender identification using palmprint will be among the next most popular biometric technology, especially in forensic applications, thanks to its uniqueness and strength. In fact, palmprint data can easily be collected with low-cost devices and minimal cooperation from subjects. Moreover, several palmprint properties can contemplate to identify person gender information. Palmprints are collections of two distinguishable properties, in medical terminology that are known as palmar friction ridges and palmar flexion creases. These structures are permanent, unique, and immutable too for an individual [4]. In this study, we propose a system for automatic gender identification using palmprint. This system would be useful to enhance the accuracy of other biometric identification systems [5]. The related work is reviewed and reported in Sect. 2. The proposed methodology is discussed in Sect. 3. In Sect. 4, experimental results are described and concluding remarks are reported in Sect. 5.

2 Related Work

Identification of gender using palmprint does not have much literature. However, a few studies are reported on gender identification using hand geometry, hand shape with face was used in [6] for the biometric verification system. Geometric properties like boundary and Fourier descriptors on Zernike moments are reported in [7]. This work takes advantages of fingers and palm region separately by dividing them into six different parts (five fingers + palm) Eigenspace distance with score-level fusion with LDA has been used to identify human gender.

In [8] the authors have designed a methodology which uses geometric properties of palm such as length, aspect ratio, and width with polynomial smooth SVM (PSSVM) classifier.

The topic of Gender identification is of considerable interest in Dermatoglyphic's for medico-legal investigation. The authors in [9] found variations in different designated areas from palmprint among men and women. They showed that palm ridge density is higher in women than men. Additional to these claims are that the right and left palmprints of men does not have noteworthy distinction, but they exist in those of women. Another interesting study was reported in [10] based on the North Indian community. They have studied palmprints ridge density to infer gender in forensic investigation. These researchers have reported their claim with ROC analysis.

From the quick review made in the aforementioned paragraph, we can notice that feature extraction techniques used are of two types local and global [11]. Palm acquisition is an issue that is considered, as a few researchers suggest, contact-free [4, 12–16] and a few suggest peg aimed palm print acquisition [4]. Contact-free palm acquisition causes some disfigurements like translation, rotation, scale variation, and light variations.

In this work, we took the advantages of global features based on Gabor Wavelets as we deal with contact-free palm in our study. Our goal is to develop a generic system that can differentiate between male and female efficiently based on palmprint. From the literature, we noted that previous studies were accomplished on limited datasets with conventional tools and techniques.

3 Proposed Methodology

In this paper, we have proposed three steps namely preprocessing which involves extraction of ROI from the palmprint images. Feature computation deals with textural analysis using Gabor Wavelets and *KNN* algorithm for classification (Fig. 1).

3.1 Preprocessing

Our aim in preprocessing is to extract the ROI; to do this; First, the input image convolves with Gaussian low-pass filter; and then binarized using Otsu method [17]. Then image centroid is detected to search for two keypoints, first key point is the gap between forefinger and middle finger, and the second keypoint is the gap between the ring finger and the little finger, to set up coordinates. To determine the palmprint coordinate system, the tangent of previously located two keypoints is

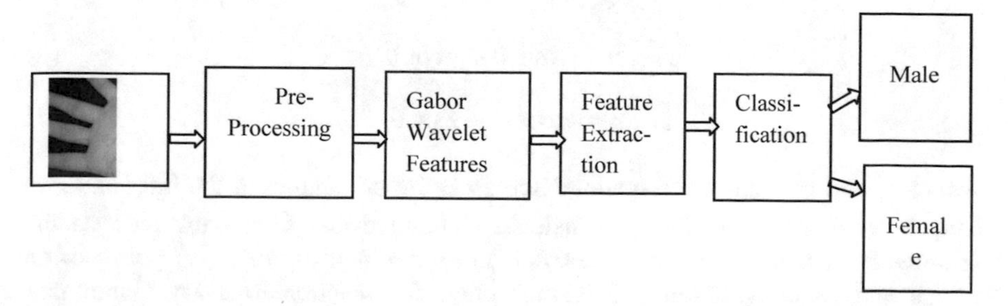

Fig. 1 Diagrammatic representation of proposed method

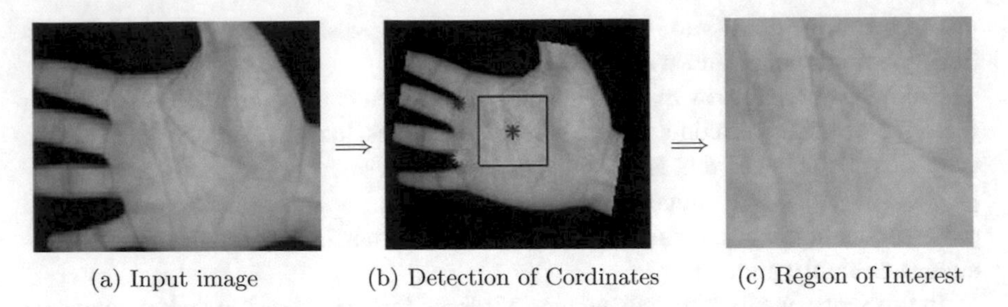

(a) Input image (b) Detection of Cordinates (c) Region of Interest

Fig. 2 An example showing ROI extraction

computed. The line joining these two keypoints is considered as *y*-axis, to locate the origin of coordinate system, based on the middle of those keypoints through which line passes and should be perpendicular to *y*-axis. After getting the coordinate system, the sub-image from coordinates is considered as ROI and these ROIs are considers for further experiments. More details on ROI extraction are given in [8]. The process of ROI extraction can be understood from Fig. 2.

3.2 Feature Extraction

In practice, it has been observed that principal lines do not contribute enough to high precision by alikeness among various palms. Although ridges play a vital role in palmprint-based gender identifications it is as yet a troublesome task to extract them precisely. Being motivated from this phenomenon, we took advantage of global texture information of palmprint ROIs. We considered Gabor wavelets due to their biologically inspired nature in information representation, as they mimic human visual system [4, 9, 16, 18]. Gabor wavelets are a popular tool in computer vision and image processing in many applications [19], such as in content-based image retrieval, texture analysis, character recognition, biometrics, etc.

$$G(x, y\colon f, \theta) = \left\{ \exp\left[\frac{1}{2}\frac{(m)^2}{\sigma_x^2} + \frac{(n)^2}{\sigma_y^2} \right] \cos 2\pi f(m) \right\} \quad (1)$$

$$m = x \sin\theta + y \cos\theta \quad (2)$$

$$n = x \cos\theta - y \sin\theta, \quad (3)$$

where σ_x^2 and σ_y^2 control the spatial extent, θ is the orientation of the filter and F is frequency of the filter. Two-dimensional Gabor wavelet filter bank extracts the features from input palm image Let $F(x, y)$ be the intensity of (x, y) points of an image and its convolution with Gabor filter, the reaction to every Gabor part representation is the complex function of real part $(G(x, y\colon f, \theta))$ and an imaginary part. In our work, feature computation is accomplished using Gabor Wavelets on ROI extracted images, feature vector is created by calculating mean squared energy and mean amplitude from each filtered ROI image. The ROI image is convoluted with Gabor wavelets considering six different orientations from (0°, 30°, 60°… 150°) and five different frequencies $f = \pi/2i$, (i = 1, 2, …, 5). The ROI extracted images is convoluted via the FFT (Fast Fourier Transformation) with the Gabor Wavelets filter bank to grab a significant amount of deviation from palmprint images. Gabor filter bank computes 30 Gabor square energy and 30 Gabor mean amplitude energy from ROI images. The obtained feature vector is of 60 features obtained each from male and female palmprint images. Further, these features are used to train and test with the classifier. Sample palmprint ROIs convolved with Gabor filters are shown in Fig. 3.

3.3 KNN *Classifier*

As per the literature, it is perceived that no other authors have tried working along with ROI and Principal Lines, generally [1, 4, 6–10, 12–20] for palmprint-based gender identification. In this paper we have employed traditional ridge-related analysis in our work: we have utilized basic classifier technique, i.e., *KNN*

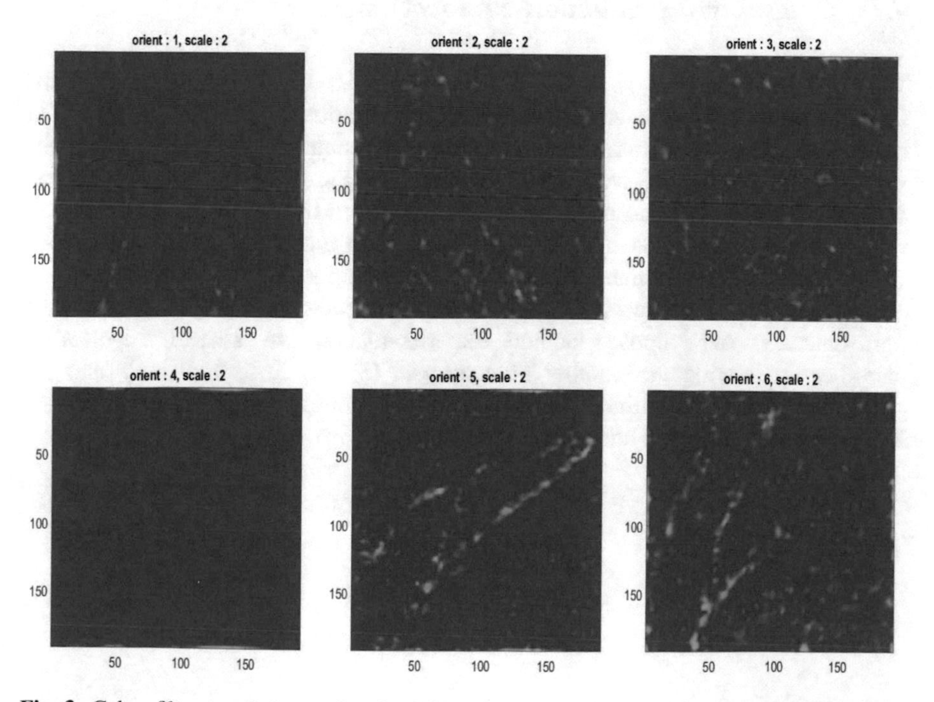

Fig. 3 Gabor filters applied on palmprint ROIs with fixed scale and different orientations

classifiers for Gender Identification. Basically, *KNN* shows the training data *M*, then finds the minimum distance *d* between training sample *X* and testing pattern *N* using

$$d_{\text{euli}}(M,N) = \sqrt{(M - N_i)^{\mathrm{T}}(M - N_j)} \tag{4}$$

$$d_{\text{city}}(M,N) = \sum_{j=1}^{n} \left(\left|M_j - N_j\right|\right) \tag{5}$$

$$d_{\cos}(M,N)_{st} = 1 - \frac{M_s N_t'}{\left(\sqrt{(M_s N_t')}\sqrt{(M_s N_t')}\right)} \tag{6}$$

$$d_{\text{corr}}(M,N)_{st} = \left(1 - \frac{(M_s N_t)(M_s N_t)'}{\sqrt{(M_s N_t')(M_s N_t'')'}\sqrt{(M_s N_t')(M_s N_t')'}}\right) \tag{7}$$

4 Experimental Analysis

4.1 Dataset and Evaluation Protocol

We have used publicly available CASIA Palmprint Database collected by the Chinese Academy of Science Institute of Automation (CASIA) [20, 21]. The database consists of palmprints captured from 312 subjects out of which 274 were male volunteers and 76 were female volunteers. From each subject, palmprint images from both left and right palms were collected by giving them prior directions. To evaluate our method, we have considered a subset of 4207 palmprints, from which 1129 are female palmprints and 3078 are male ones (Fig. 4).

In our experimental tests, we use 10-fold cross validation [22] to evaluate the performance of our proposed method. Experimentations are carried out with *KNN* classifier by altering the number of neighbors ($K = 1, 3, 5, 7, 9, 11$), and the performance of the algorithm is found optimal $K = 3$ respectively. Precision (*P*) and Recall (*R*) are computed and are represented as below

$$\text{Precision } (P) = \frac{T_{\mathrm{p}}}{T_{\mathrm{p}} + F_{\mathrm{p}}} * 100 \tag{8}$$

$$\text{Recall } (R) = \frac{T_{\mathrm{p}}}{T_{\mathrm{p}} + F_{\mathrm{n}}} * 100 \tag{9}$$

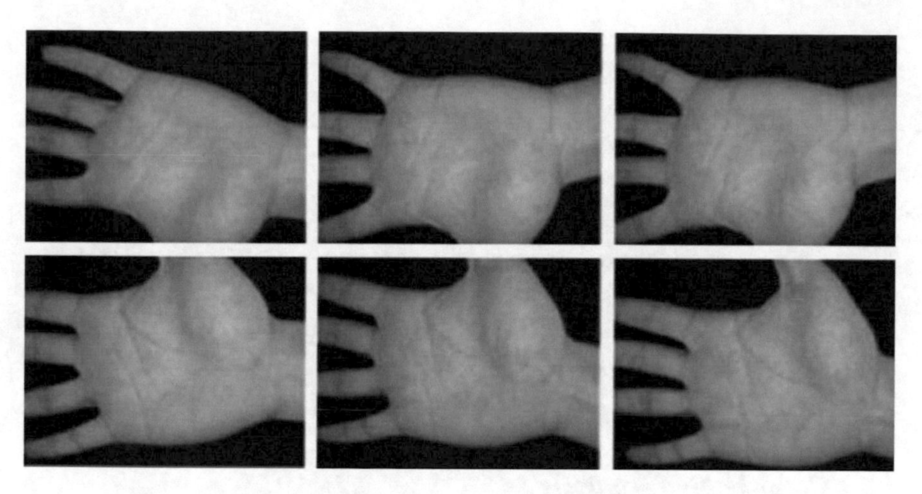

Fig. 4 Sample of CASIA database

$$\text{Accuracy }(A) = \frac{T_\text{p} + T_\text{n}}{T_\text{p} + T_\text{n} + F_\text{p} + F_\text{n}} * 100, \tag{10}$$

where T_p is True Positive, T_n is True Negative; F_p is False Positive and F_n False Negative.

4.2 Result Analysis

To realize the effectiveness of our proposed approach, we have worked out extensive experiments with *K*-nearest neighbor classifier. We also considered the different distance measures with varying *k*-values.

We observed that *KNN* outperformed when the value of $K = 1$. Highest accuracy of 97.9% is obtained with Correlation distance and lowest accuracy of 97.4% is obtained by cosine distance. City-block distance and Euclidean Distance have performed less than the Correlation distance and have yielded accurate result of 97.5 and 97.6%, respectively. Detailed results of exhaustive experiments are demonstrated in Table 1.

Table 1 Detail result analysis

*K*Value	City blocks			Cosine			Euclidean			Correlation		
	P	*R*	*A*	*P*	*R*	*A*	*P*	*R*	*A*	*P*	*R*	*A*
1	98.05	98.56	97.5	97.0	98.68	97.4	98.11	98.53	97.6	98.18	98.88	97.9
3	96.97	96.91	95.5	96.5	97.40	95.6	97.01	96.97	95.6	97.10	97.67	96.2
5	96.26	95.08	93.6	95.3	95.91	93.6	96.16	95.17	93.6	95.77	96.15	94.1
7	95.32	93.76	91.9	94.41	94.50	91.9	95.80	93.94	92.4	95.19	95.00	92.8
9	95.19	91.47	90	93.95	92.014	89.6	95.19	91.84	90.3	94.47	92.84	90.6
11	95.09	88.91	87.7	93.50	89.90	87.6	95.15	89.59	88.4	93.63	90.71	88.3

Table 2 Comapritve analysis

S. No.	Author	Features	Database	Classification method	Accuracy
1	Amayeh et al. [10]	Geometric features, boundary descriptors, Zernike moments and Fourier descriptor	20 males and 20 females palm prints	Score level fusion and LDA	98%
2	Ming et al. [6]	Geometric features like length, width and aspect ratio	180 palm print images of 30 voluntaries	PSSVM	85%
3	Present methodology	Gabor wavelets	CASIA database	*KNN*	97.90%

4.3 Comparative Analysis

In this section, we have compared our work with similar works found in the literature. In [7] the authors used Geometric features, boundary and Fourier descriptors on Zernike moments of hand geometry and achieved 98% result on dataset of 40 palmprints. Whereas in [8] the authors used very basic geometric properties such as length, aspect ratio, and width with polynomial smooth—SVM and got 85% accuracy in gender prediction. The drawback of the reported method in [7] is that it is not suitable for low resolution and far distance images captured using touch-free method, as they need touch-based palm acquisition. On the contrary, our proposed method is suitable for both the approaches: touch-based and touch-free. Our method outperformed with Gabor wavelets filter bank based texture features and very basic *KNN* classifier on large dataset consisting of 4207 ROIs of palmprints, which yielded the accuracy of 97.90%. In addition to this, our method is also immune to noise and low-resolution effects (Table 2).

5 Conclusion

In this paper, we have attempted a human gender identification problem using palmprints. Gabor Wavelets are used to extract discriminating texture features from palmprints. These features have exhibited significant performance that too with a simple *KNN* classifier. Exhaustive experimentations have been done on a relatively larger dataset of 4207 palmprint images with different distance measures and *K*-values to validate the performance of the algorithm. This analysis reveals that the proposed method has a remarkable performance with a simple classifier compared with the other methods. In future, we concentrate on the evaluation of (1) Different texture descriptors (2) sophisticated ROI extraction techniques and feature selection techniques (3) Supervised and Unsupervised classifiers to devise a best generic algorithm.

Acknowledgements The Portions of the research in this paper uses the CASIA Palmprint Database collected by the Chinese Academy of Science Institute of Automation (CASIA). Authors thanks to Chinese Academy of Science Institute of Automation for providing the database for conducting the experiment.

References

1. Gornale, S., Patil, A., Veersheety, C.: Fingerprint based gender identification using DWT and gabor filters. Int. J. Comput. Appl. **152**(4), 34–37 (2016)
2. Wu, X., Zhang, D., Wang, K.: Fisher palms based palmprint recognition. PR Lett. 2829–2838 (2003)
3. Jing, X.-Y., Tang, Y.-Y., Zhang, D.: A Fourier—LDA approaches for image recognition. Pattern Recogn. 453–457 (2005)
4. Charfi, N., Trichili, H., Alimi, A.M., Solaiman, B.: Bimodal biometrics system based on SIFT descriptor of hand images. In: IEEE International Conference on System Man and Cybernetics (SMC), pp. 4141–4145 (2014)
5. Gornale, S.: Fingerprint based gender classification for biometric security: a state-of-the-art technique. Int. J. Res. Sci. Technol. Eng. Math. **4**(3), 39–49. ISSN (Print): 2328-3491, ISSN (Online):2328-3580, ISSN (CD-ROM): 2328-3629, Dec-2014–Feb-2015
6. Wu, M., Yuan, Y.: Gender classification based on geometrical features of palmprint images. SW J. **2014** (2014). Article Id: 734564, 7 p
7. Bebis, G., Reno, NV(US), Amayeh, G., Reno, NV(US): Hand-based gender classification. Patent no. US 8,655,084 B2 DOP, 18 Feb 2014
8. Zhang, D., Kong, W.K., You, J., Wong, M.: Online palmprint identification. IEEE Trans. Pattern Anal. Mach. Intell. **25**, 1041–1050 (2003). 10.11.09/TPAMI 2003.1227981
9. Golfarelli, M., Maio, D., Maltoni, D.: On the error-reject trade-off in biometric verification systems. IEEE Trans. Pattern Anal. Mach. Intell. **19**(7), 786–796 (1997)
10. Amayeh, G., Bebis, G., Nicolescu, M.: Gender classification from hand shapes. In: 2008 IEEE Society Conference on CVPR Workshop, AK, pp. 1–7 (2008). 10.11.09/CVPRW
11. Zhang, D., Zuo, W., Yue, F.: A comparative analysis of palmprint recognition algorithm. ACM Comput. Survey (2012)
12. Wu, X., Zhang, D., Wang, K.: Palm line extraction and matching for personal authentication. IEEE Trans. Syst. Man Cyber. A Syst. Hum. 36(5), 975–987 (2006)
13. Kumar, A., Zhang, D.: Personal recognition using Hand shape. IEEE Trans. Image Process. **15**, 2454–2461 (2006)
14. Sun, Z., Tan, T., Wang, Y., Li, S.: Ordinal palmprint representation for personal identification. In: International Conference on Computer Vision and Pattern Recognition, pp. 279–284 (2005)
15. Zhang, D., Guo, Z., Lu, G., Zhang, L., Liu, Y., Zuo, W.: Online joint palmprint and palm-vein verification. Expert Syst. Appl. **38**(3), 2621–2631 (2011)
16. Kanchan, T., Krishan, K., Aparna, K.R., Shredhar, S: Is there a sex difference in palmprint ridge density? Med. Sci Law (2013). https://doi.org/10.1258/msl..2012.011092
17. Otsu, N.: A threshold selection method from gray-level histograms. IEEE Trans Man Cybern. **9**(1), 62–66 (1979)

18. Krishan, K., Kanchan, T., Sharma, R., Pathania, A.: Variability of palmprint ridge density in north indian population and its use in inference of sex in forensic examination. HOMO J. Comp. Hum. Biol. **65**(6), 476–488 (2014)
19. Ashbaugh, D.R.: Quantitative Friction Ridge Analysis: An Introduction to Basic Advanced Ridgeology. CRC Press (1999)
20. Sun, Z., Tan, T., Wang, Y., Li, S.Z.: Ordinal palmprint representation for personal identification. In: Proceeding International Conference on CVPR, vol. 1, pp. 279–284, Orlando, USA (2005)
21. http://biometrics.idealtest.org/
22. Gornale, S., Basavanna, M., Kruthi, R.: Gender classification using fingerprints based on support vector machine(SVM) with 10-cross validation technique. Int. J. Sci. Eng. Res. **6**(7) (2015)

Seasonality as a Parrondian Game in the Superior Orbit

Anju Yadav, Ketan Jha and Vivek K. Verma

Abstract In ecological modeling, switching strategy is used to represent seasonality, i.e., different environmental conditions. Switching strategies is also known as Parrondo's paradox, where the alternation of two losing games combined in deterministic and random manner yields a winning game, i.e., in game theory "lose + lose = win", in ecological systems "undesirable + undesirable = desirable" and in a dynamical system "chaos + chaos = order". The logistic map $f(x) = rx(1-x), x \in [0, 1]$shows different forms by taking $0 < x < 1$ and $0 < r \leq 4$, which helps in studying population of nonoverlapping generation such as insects. Rani in 2002, introduced the superior orbit "Mann iteration" and studied various dynamical systems and proves that they are more stable than if iterated in Picard orbit. In general, the logistic map is chaotic for values $r > 3.57$, but after iterating it in superior orbit it is extended to 21. But still chaotic range exists, i.e., in ecological system some undesirable behaviors are observed. In this paper, we consider a winter season and summer season in superior orbit and both will derive the population towards the chaos or undesirable individually and also the case of extinction and chaotic, but after applying switching strategy to them they are ordered or desirable "undesirable + undesirable = desirable". And also, we have considered a four-season model, to study either migration or immigration. Further, we have studied a noisy switching strategy for the above case and shown that desirable oscillatory behavior still prevails.

Keywords Logistic map · Superior orbit (SO) · Dynamical system · Parrondo's paradox

---

A. Yadav (✉) · V. K. Verma
Manipal University Jaipur, Jaipur, India
e-mail: anju.anju.yadav@gmail.com

K. Jha
Central University of Rajasthan, Ajmer, India
e-mail: jha.ketan555@gmail.com

A. K. Luhach et al. (eds.), *Smart Computational Strategies: Theoretical and Practical Aspects*, https://doi.org/10.1007/978-981-13-6295-8_6

1 Introduction

In 1996, a Spanish physicist named as Juan Parrondo's gave a paradox in the game theory. According to this paradox, two losing games (i.e., negative gains) are observed as winning game (i.e., positive gains) when played alternatively in either random or deterministic manner [1]. The idea can be explained as "lose + lose = win" in game theory or "chaos + chaos = order" in dynamical systems. To study the detail of Parrondo's paradox, one may refer to [5, 6, 10, 11].

In nature, various types of interactions are available and due to a unique dynamic, systems do not evolve. In such cases Parrondo's paradox could play a role in performing the interaction between different systems. Parrondo's paradox has various applications in economics, physics, biology, stock market. Refer to [6, 11, 15].

In 2005, Parrondo's paradox is applied to the combination of two nonlinear quadratic maps to control the chaos by Almedia et al. [1]. Further, it is applied on various forms of dynamical systems in many different ways. One may refer to [2, 7, 17, 20].

As we know, the logistic map $rx(1-x), x \in [0,1]$ [3, 4, 9] is a population model that has relevance in modeling ecological systems [16]. Logistic map helps for studying the population growth of species such as insect. Further, in [8, 12] alternate dynamics, i.e., Parrondo's paradox a switching strategy can also be used for the seasonality. For example, cold winter and hot summer season drives the population toward the chaos individually, i.e., undesirable behaviors whereas after performing switching on these undesirable behaviors they yield a desirable behavior.

In general, as we know that whenever chaos hits in a system then system would be unstable so overcome this problem is mandatory. The study of logistic map has been done by Rani and Agarwal via SO in 2005, they increase the stability realm of logistic map up to $r > 21$ [13]. But, still chaotic situation exists. Further, in 2016 Anju et al. have developed new superior logistic model that is used to control the chaotic situation using switching strategies [16, 19]. Researchers also gave their contribution towards SO applications in fractals and chaos, one may refer to [13, 18].

In this paper, Parronodian game is studied with superior logistic model to control the undesirable condition in ecology, i.e., the two-undesirable conditions after switching yield a desirable condition "undesirable + undesirable = desirable". In Sect. 2, the concepts or equations are taken into our account for the study. In Sect. 3, we have shown the results obtained for different seasons by applying switching strategy. In Sect. 4, we have concluded by important remarks.

2 Alternate Superior Logistic Map

Definition 2.1 Alternate Dynamics Assume two discrete dynamics $B_1\colon x_{n+1} = x_n^2 + c_1 0$, $B_2\colon x_{n+1} = x_n^2 + c_2$ and their combination for alternation is defined as

$$(B_1B_2)\colon \begin{cases} x_{n+1} = x_n^2 + c_1 & \text{when } n \text{ is odd,} \\ x_{n+1} = x_n^2 + c_2 & \text{when } n \text{ is even,} \end{cases} \quad (1)$$

where x, c, c_1, $c_2 \in R$ [1].

The above equations are in Peano-Picard iteration that is based on one-step feedback machine, denoted by formula $x_{n+1} = f(x_n)$, where f is any function of x.

Definition 2.2 Superior iterates (SO) Assume A be a subset of real numbers and $f\colon A \to A$. For $x_0 \in A$, construct a sequence $\{x_n\}$ in the following manner:

$$\begin{aligned} x_1 &= \beta_1 f(x_0) + (1-\beta_1)x_0, \\ x_2 &= \beta_2 f(x_1) + (1-\beta_2)x_1 \ldots, \\ x_n &= \beta_n f(x_{n-1}) + (1-\beta_n)x_{n-1}, \end{aligned} \quad (2)$$

where $0 < \beta_n \leq 1$ and $\{\beta_n\}$ is convergent away from 0 [13, 14].

The above sequence $\{x_n\}$ is constructed via SO that is early proposed by Mann. At $\beta = 1$ SO reduces to Picard iterates [13–15]. In this paper, we shall consider $\beta_n = \beta$ in successive approximation method.

Let us consider two logistic maps $f_1(x) = r_1 x(1-x)$ and $f_2(x) = r_2 x(1-x)$, after applying SO to $f_1(x)f_2(x)$ it becomes

$$f_1(x)f_2(x)\colon \begin{cases} x_{n+1} = \beta r_1 x_n(1-x_n) + (1-\beta)x_n & \text{where } n \text{ is odd;} \\ x_{n+1} = \beta r_2 x_n(1-x_n) + (1-\beta)x_n & \text{where } n \text{ is even,} \end{cases} \quad (3)$$

where x, r_1, $r_2 \in R$. Equation 3 is repeated n times in superior iterations to obtain the sequence of x_n, where initial value of x is x_0 and $0 < \beta \leq 1$: $x_1 = \beta(r_1 x_0(1-x_0)) + (1-\beta)x_0$,

$$\begin{aligned} x_2 = {} & \beta[r_2(\beta(r_1x_0(1-x_0)) + (1-\beta)x_0)(1-(\beta(r_1x_0(1-x_0)) + (1-\beta)x_0))] \\ & + (1-\beta)(\beta(r_1x_0(1-x_0)) + (1-\beta)x_0), \\ x_3 = {} & \beta[r_1[\beta[r_2(\beta(r_1x_0(1-x_0)) + (1-\beta)x_0)(1-(\beta(r_1x_0(1-x_0)) + (1-\beta)x_0))] \\ & + (1-\beta)(\beta(r_1x_0(1-x_0)) + (1-\beta)x_0] \\ & (1-[\beta[r_2(\beta(r_1x_0(1-x_0)) + (1-\beta)x_0)(1-(\beta(r_1x_0(1-x_0)) + (1-\beta)x_0))] \\ & + (1-\beta)(\beta(r_1x_0(1-x_0)) + (1-\beta)x_0)] + (1-\beta) \\ & \times [\beta[r_2(\beta(r_1x_0(1-x_0)) + (1-\beta)x_0)(1-(\beta(r_1x_0(1-x_0)) + (1-\beta)x_0))] \\ & + (1-\beta)(\beta(r_1x_0(1-x_0)) + (1-\beta)x_0]] \end{aligned}$$

3 Switching Strategy in the Superior Logistic Map

We have studied the switching strategy in superior logistic model to analyse the population growth and found the following results.

3.1 *Seasonality in Superior Logistic Map*

In this section, initially to study the population growth or to analyze the population we have plotted the bifurcation diagram of superior logistic map for different β values, i.e., a population growth model (see Fig. 1). To mark the orbits stability of superior logistic model, we have plotted bifurcation diagrams of superior logistic map at $\beta = 0.85, 0.8, 0.5$ (see Fig. 1). It is observed that the population is more stable than the logistic model as we move towards $\beta = 0$, the convergence range of r in superior logistic map is increased than that of logistic map.

Further from Fig. 1, we have also observed that for $r < 1$, still the population is at extinction as in the logistic model. But after applying alternation, i.e., a switching strategy to superior logistic model it is showing stable oscillations for $r < 1$,

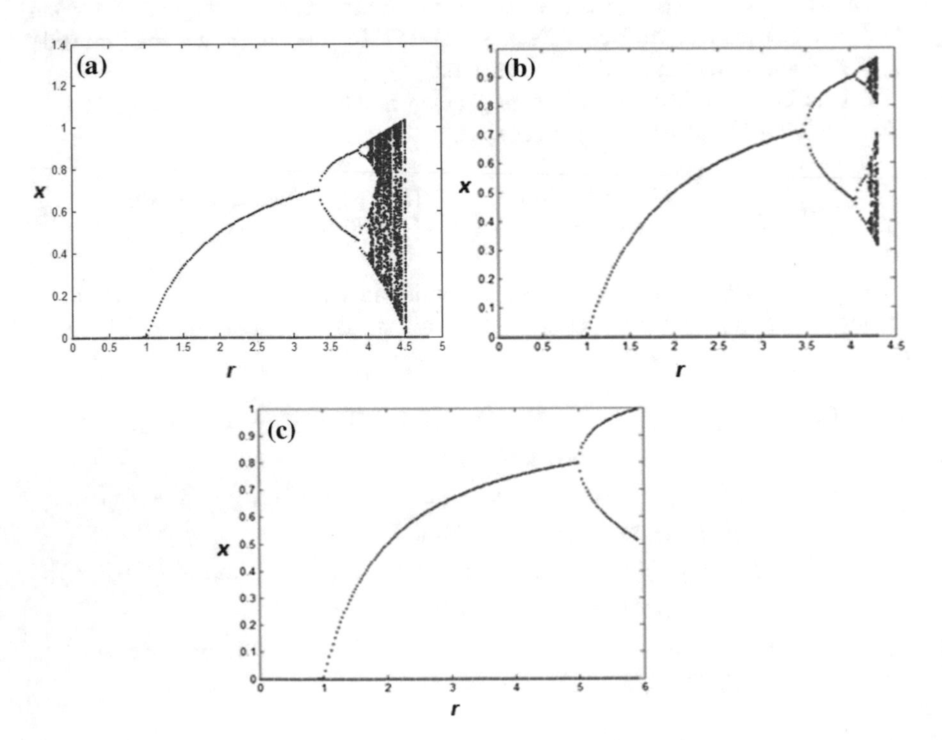

Fig. 1 Bifurcation diagram. **a** At $\beta = 0.85$, **b** at $\beta = 0.8$ and **c** at $\beta = 0.5$ of superior logistic map

see Fig. 2. From Fig. 2, take one example, $r = 4.3$, $r = 0.5$ of superior logistic map individually both conditions are harsh, where one will derive to chaotic oscillations and other will derive to the extinction. After applying the switching approach, it lead to stable oscillations or to a desirable behavior, i.e., "undesirable + undesirable = desirable".

Even by applying switching strategy, i.e., alternation we found different examples of "chaos + chaos = order" or "undesirable + undesirable = desirable" in superior logistic model (see Fig. 2). In relevance with ecology, let us take one example of two harsh conditions in which one is extreme winter and another is extreme summer, individually both will drive to chaotic trajectories, but after applying switching strategy they lead to a stable oscillation. For example, assume two chaotic parameter values $r_1 = 4.4955$, $r_2 = 4.6155$ at $\beta = 0.75$, then by implementing a switching strategy named as Parrondo's paradox, we obtain 6-periodic orbit, i.e., "$\text{chaos}_1 + \text{chaos}_2 = \text{order}$" (see Fig. 3).

3.2 Switching in Superior Logistic Model with Random Choices

Since we are considering the superior logistic map for the analysis of seasonality in ecology to show its relevance. And as we know that every year conditions of the same season will not remain same for any particular day. For example, the temperature in summer and winter on the same day may not remain same for the next year. To solve this problem, we have considered switching strategy with values of r_1 parameter is 4–4.3 at $\beta = 0.9$.

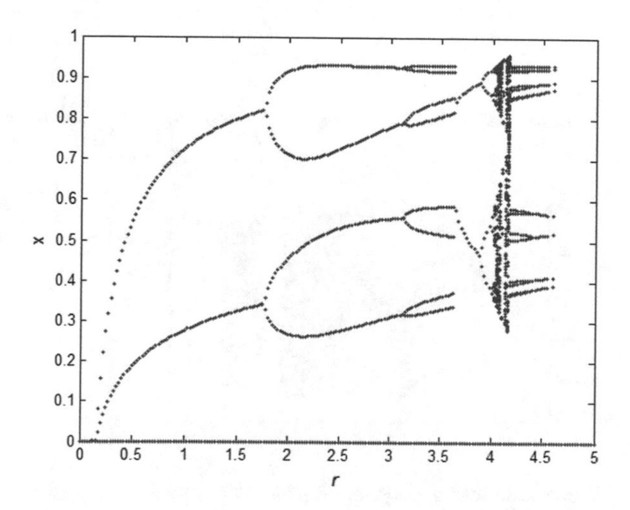

Fig. 2 Alternate bifurcation diagram at ($x_0 = 0.15$, $\beta = 0.85$, $r_2 = 4.3$)

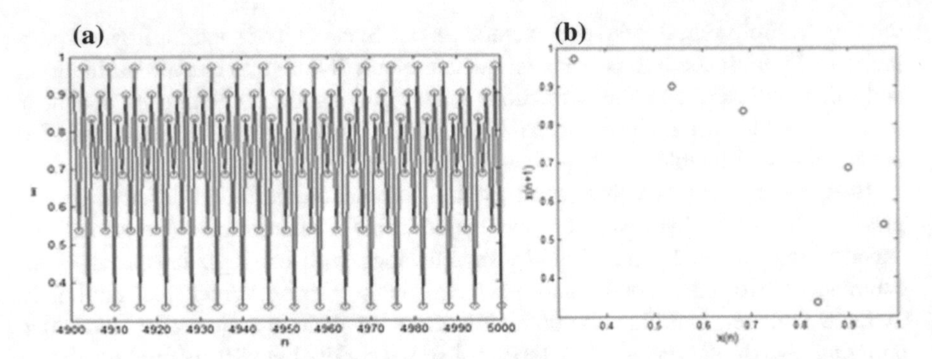

Fig. 3 6-periodic stable orbit ($\text{chaos}_1 + \text{chaos}_2 = \text{order}$) at $\beta = 0.75$, $r_1 = 4.4955$, $r_2 = 4.6155$. **a** Time series, **b** first return map in superior logistic map

$$x_{n+1}\colon \begin{cases} \beta r_1(\text{Random})x_n(1-x_n) + (1-\beta)x_n & \text{where } n \text{ is odd;} \\ \beta r_2 x_n(1-x_n) + (1-\beta)x_n & \text{where } n \text{ is even,} \end{cases}$$

where x, r_1 Random (4–4.3), $r_2 \in R$ and $\beta = 0.9$.

From Fig. 4 we have also observed that for $r_2 < 1$ still we have stable oscillations as well as near the parameter value $r_2 = 3.8$. Earlier there are deterministic oscillations seen for $r < 1$. Even the extinction range is decreased from $r_2 = 0.26$–0.18 after applying switching strategy in superior logistic model rather than logistic model (see Fig. 4b). This shows that the system becomes more stable and more population can survive for the diverse conditions.

In ecology, there are some cases in which the optimal conditions will occur for the rare event, let us say probability less than h. Then the superior logistic model can be defined as follows:

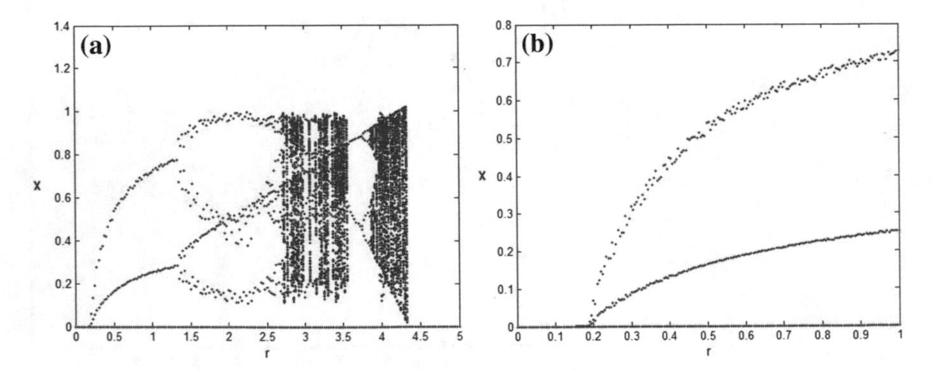

Fig. 4 Bifurcation diagram for parameter r_2 with r_1 as random values between value 4.0 and 4.3 at $\beta = 0.9$

$$x_{n+1}\colon \begin{cases} \beta r_1 x_n(1-x_n)+(1-\beta)x_n & \text{if Random}[0,1] > h \\ \beta 4 x_n(1-x_n)+(1-\beta)x_n & \text{Otherwise,} \end{cases}$$

where $h < 5\%$, $\beta = 0.9$, $r \in R$.

For example, we will consider the switching for the rare event, i.e., "viral infection" which will sometimes derive population to the extinction $r_1 < 1$. In this case, we have performed switching for the rare case, so the system will lead to extinction but due to the random values "kicks" the population move towards oscillation with noise (see Fig. 5).

3.3 Four-Season Superior Logistic Model

Finally, seasonality effects on the population can be studied by an approach, i.e., alternate dynamics. To see the application of the logistic function we can extend it to four-season logistic model and also by considering harvesting, migration and immigration. It can be represented as population by either incremental or reduction at every fourth iteration by a factor T.

Let us take constant $P = 4$, i.e., the cycle of population is four units of times, $T = 1$ represents no action case, $T < 1$ as harvesting or $T > 1$ for immigration, and four different seasons are described by a various value of the parameter r in the superior logistic model. The logistic map with parameter r for four seasons (fall, winter, spring, and summer) is as follows:

$$x_{n+1}\colon \begin{cases} TF(x_n, r_{\mathrm{F}}) & \text{if mod } [n,P]=0 \\ F(x_n, r_{\mathrm{W}}) & \text{if mod } [n,P]=1 \\ F(x_n, r_{\mathrm{Sp}}) & \text{if mod } [n,P]=2 \\ F(x_n, r_{\mathrm{Su}}) & \text{if mod } [n,\mathrm{P}]=3 \end{cases}$$

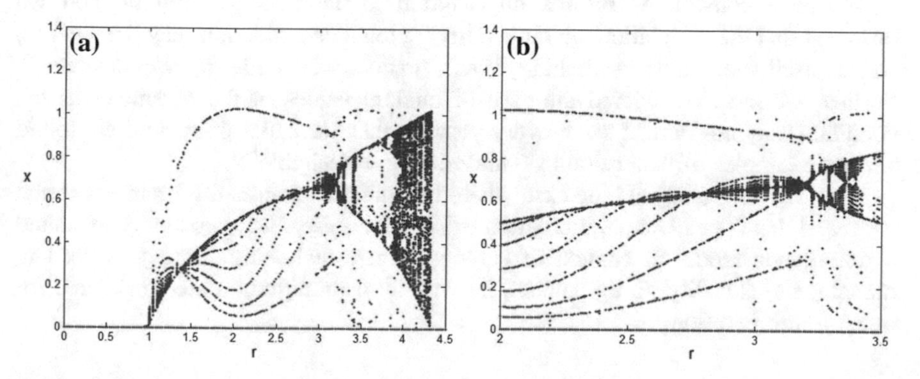

Fig. 5 Bifurcation diagram with a noise induced oscillation in the oscillations

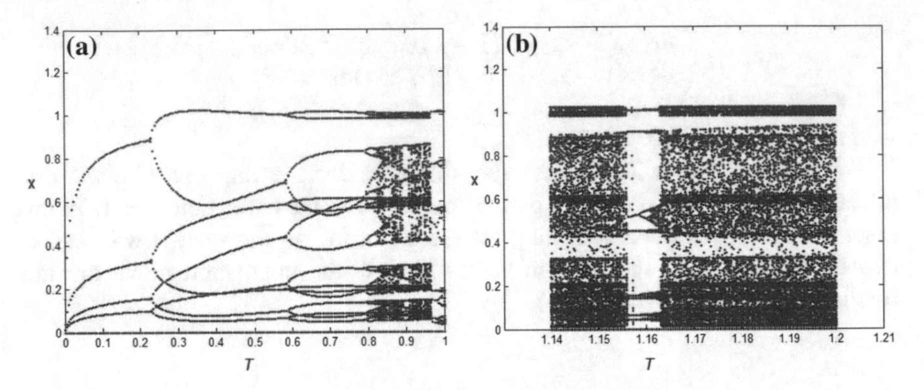

Fig. 6 Bifurcation diagram for a four-season model for harvesting and immigration

In four-season superior logistic model, we can set the T parameter values for harvesting or migration and immigration. For example, the parameter r_F value ranges between 1 and 3.3 where population will remain same, for winter season r_W we will consider the case of extinction, i.e., 0.69, for the spring and summer we will consider the chaotic trajectories for $r_{Sp} = 4$ and, $r_{Su} = 4.33$, respectively. In Fig. 6, we have observed both the cases harvesting $T < 1$ as harvesting or $T > 1$ for immigration.

4 Conclusion

In this paper, seasonality is modeled as a switching among various environmental conditions in context of ecological application. Also, we assumed the superior logistic map is a population growth model with two season which derives the harsh conditions individually but after switching it leads to a desirable or stable behavior, i.e., "undesirable + undesirable = desirable".

We have constructed bifurcation diagram to analyses the system, and we observed that the population which is deriving towards extinction may derives to a stable oscillation after switching, i.e., "extinction + undesirable = desirable". Further, we have considered the case of random values of the parameter as the condition may not remain same every year for that particular time. And we found various examples of "undesirable + undesirable = desirable".

Even we have analyzed the case where the infection occurs for a rare event and we found that the still stable oscillation exists with noise. Next, we have considered a four-season model in context of ecology, where switching strategy is used to model the seasonality. In this model, harvesting or immigration are considered for every fourth iteration.

References

1. Almedia, J., Peralta-Salas, D., Romera, M.: Can two chaotic systems give rise to order? Physica D **200**(1–2), 124–132 (2005)
2. Boyarsky, A., Gora, P., Islam, M.S.: Randomly chosen chaotic maps can give rise to nearly ordered behavior. Physica D **210**, 284–294 (2005)
3. Chauhan, Y., Rana, R., Negi, A.: Complex and inverse complex dynamics of fractal using ishikawa iteration. Int. J. Comput. App. **9**(2), 9–16 (2010)
4. Chug, R., Rani, M., Ashish: On the convergence of logistic map in Noor orbit. Int. J. Comp. App. **43**(18), 1–4 (2012)
5. Harmer, G P., Abbott, D.: Parrondo's paradox. Stat. Sci. **14**(2), 206–213. MR1722065 Zbl 1059.60503 (1999)
6. Harmer, G.P., Abbott, D., Taylor, P.G.: The paradox of Parrondo's games. Proc. R. Soc. London A Math. Phys. Eng. Sci. **456**, 247–259. Zbl 1054.91514 MR1811319 (2000)
7. Klic, A., Pokorny, P.: On dynamical systems generated by two alternating vector fields. Int. J. Bifurcat. Chaos. **6**(11), 2015–2030. MR1430981 (1996)
8. Levinsohn, E.A., Mendoza, S.A., Peacock-Lopez, E.: Switching induced complex dynamics in an extended logistic map. Choas. Soliton. Fract. **45**(4), 426–432 (2012)
9. Maier, M.P.S., Peacock-Lopez, E.: Switching induced oscillations in the logistic map. Phys. Lett. A. **374**(8), 1028–1032. Zbl 1236.92071 (2010)
10. Parrondo, J.M.R., Dinis, L.: Brownian motion and gambling: from Ratchets to paradoxical games. Contemp. Phys. **45**(2), 147–157 (2004)
11. Parrondo, J.M.R., Harmer, G.P., Abbott, D.: New paradoxical games based on Brownian ratchets. Phys. Rev. Lett. **85**(24), 5226–5229 (2000)
12. Peacock-lopez, E.: Seasonabiloity as a parrondian game. Phys. Lett. A **35**, 3124–3129 (2011)
13. Rani, M.: Iterative procedure in fractals and chaos. Ph.D. Thesis, Gurukala Kangri Vishwavidyalaya, Hardwar, India (2002)
14. Rani, M., Agarwal, R.: A new experimental approach to study the stability of logistic map. Chaos. Soliton. Fract. **41**(4), 2062–2066 (2009)
15. Rani, M., Kumar, V.: A new experiment with the logistic map. J. Indian Acad. Math. **27**(1), 143–156. MR2224669 (2005)
16. Rani, M., Yadav, A.: Parrondo's paradox in the superior logistic map. Int. J Tech. Res. **1**(2), 1–8 (2016)
17. Romera, M., Small, M., Danca M.F.: Deterministic and random synthesis of discrete chaos. Appl. Math. Comput. **192**(1), 283–297. Zbl 1193.37047 MR2385594 (2007)
18. Singh, S.L., Mishra, S.N., Sinkala, W.: A new iterative approach to fractal models. Commun. Nonlinear Sci. Numer. Simulat. **17**(2), 521–529 (2012)
19. Yadav, A., Rani, M.: Modified and extended logistic map in superior orbit. Proc. Comp. Sci. **57**, 581–586 (2015)
20. Zhang, Y., Whang, Y., Shen, X.: A chaos based image encryption algorithm using alternate structure. Inform. Sci. **50**(3), 334–341. MR2348429 (2007)

Comparative Study of Combination of Swarm Intelligence and Fuzzy C Means Clustering for Medical Image Segmentation

Thiyam Ibungomacha Singh, Romesh Laishram and Sudipta Roy

Abstract The image segmentation issues have been exploited by researchers over the years for diverse application. A hybrid algorithm for image segmentation is proposed in this paper which is the integration of fuzzy c means (FCM) clustering and swarm intelligence. The algorithm is applied to segmentation problems of two medical image modalities, i.e., magnetic resonance imaging (MRI) image, and computed tomography (CT) image. A detailed comparison of the different swarm intelligence based algorithms is presented. The optimization technique is used to generate optimized cluster centers in the image segmentation process. The effectiveness of the algorithms is validated by cluster validity indices.

Keywords Image segmentation · FCM · Swarm intelligence · MRI · CT

1 Introduction

Image segmentation is an active area of research related to digital image processing. It is an important stage of image processing for various applications which includes robotics, medical image segmentation, remote sensing, computer vision problems, etc. Recently, image processing has become an important element in medical diagnosis and development of an automated diagnosis system which is generally referred to as computer-assisted diagnosis (CAD). Image segmentation [1] is a method of identification and categorization of similar areas in an image or it can be considered as the separation of region of interest (ROI) in an image. Segmentation

T. Ibungomacha Singh · R. Laishram (✉)
Manipur Institute of Technology, Takyelpat, Manipur, India
e-mail: romeshlaishram@gmail.com

T. Ibungomacha Singh
e-mail: ibomcha.2007@rediffmail.com

S. Roy
Assam University, Silchar, Assam, India
e-mail: sudipta.it@gmail.com

A. K. Luhach et al. (eds.), *Smart Computational Strategies: Theoretical and Practical Aspects*, https://doi.org/10.1007/978-981-13-6295-8_7

helps in the detailed understanding of the image and it is an inseparable stage of image processing application which cannot be neglected. Hence more dedicated research is required in image segmentation.

In this proposed work, we modeled image segmentation as a clustering problem and one of the most popular clustering algorithms, i.e., fuzzy c means (FCM) [2, 3] clustering algorithm is employed for image segmentation. Further swarm intelligence based optimization algorithms is integrated with the FCM algorithm to improve the segmentation. The combined algorithms are tested on two different medical images which are a brain MRI image and CT liver image. The algorithms are validated using clustering validity indices for its effectiveness. The population-based optimization techniques have been utilized in many engineering and nonengineering problems and prove to be quite successful. This idea generated lots of interest and guided to the development of many nature-inspired algorithms. In this paper, we have considered few of the popular algorithms such as genetic algorithms (GA), particle swarm optimization (PSO), artificial bee colony (ABC), Bat algorithm, gravitational search algorithm (GSA), and gray wolf optimizer (GWO). The details of these algorithms are presented in the subsequent sections.

Although the FCM algorithm has been employed for image segmentation in many earlier work, it is still an important area of research mainly for application in medical imaging systems. An exhaustive review about different methods for segmentation of brain MRI image is presented in [4] and application of FCM algorithm for MRI image segmentation is given in [5]. The authors give a detail description of different variants of FCM for image segmentation. A hybrid clustering technique that combines K-means clustering with FCM is proposed in [6]. The idea brings about a computationally efficient and segmentation result with high precision. The fusion algorithm of FCM and swarm-based algorithms started with the work proposed in [7]. They have incorporated the genetic algorithm (GA) with FCM and applied to Brain MRI images segmentation. The other works of hybrid FCM with GA are also carried out by the authors in [8–10] and many researchers have tried different swarm algorithms with FCM. The fusion of the FCM algorithm with PSO and ABC for image segmentation problem can be found in the literatures [11–14].

2 Fuzzy C Means (FCM) Clustering

The FCM clustering in an image is a technique that allocates pixels to different clusters by looking at the fuzzy membership values of the pixels. It is basically an optimization algorithm that tries to generate optimized cluster centers iteratively by minimizing an objective function or cost function.

For applying FCM algorithm in image segmentation, the 2-D image is first converted into a one-dimensional matrix as $X = [x_1, x_2, \ldots, x_N]$, where x_i is the pixel intensity used as feature and N is the total feature points. The purpose of FCM is to partition the image into 'c' clusters by minimizing a cost function which is defined as follows:

$$J = \sum_{j=1}^{N} \sum_{i=1}^{c} u_{ij}^{m} . \left\| x_j - v_i \right\|^2 \quad (1)$$

where

u_{ij} Fuzzy membership of pixel x_j in ith cluster.
v_i cluster center
m a parameter that controls the partition result.

In iterative manner, the cluster centers and the fuzzy membership values are updated using the following equations:

$$u_{ij} = \frac{1}{\sum_{k=1}^{c} \left(\frac{\left\| x_j - v_i \right\|}{\left\| x_j - v_k \right\|} \right)^{\frac{2}{(m-1)}}} \quad (2)$$

$$v_i = \frac{\sum_{j=1}^{N} u_{ij}^{m} x_j}{\sum_{j=1}^{N} u_{ij}^{m}} \quad (3)$$

3 Optimization Algorithms

In this section, a brief description of the widely used optimization algorithms is presented. The population-based optimization algorithm works to find the solution to the problem by employing many agents in an iterative fashion. Every optimization process requires a fitness function or a cost function for minimization or maximization problem. In our work, the optimization algorithms are incorporated to find the best cluster center by minimizing the objective function defined by FCM in Eq. (1).

3.1 Particle Swarm Optimization (PSO)

The particle swarm optimization (PSO) is an optimization technique that is evolved by understanding swarm behaviors like bird flocking and fish schooling. It was introduced by Kennedy and Eberhart [15]. A population of solution searching agents called particles is used and the positions of the particles indicate the solution of the problem. A best solution of the problem is obtained when the fitness function converges through generations. The updating of velocity and position of the particles are done iteratively using the following equations:

$$v_i(k+1) = w \cdot v_i(k) + c_1 \cdot (p_i(k) - x_i(k)) + c_2 \cdot (p_g(k) - x_i(k)) \tag{4}$$

$$x_i(k+1) = x_i(k) + v_i(k) \tag{5}$$

where $v_i(k)$ and $x_i(k)$ represent the velocity and position of ith particle at kth iteration. $p_i(k)$ and $p_g(k)$ are local best and global best, w, c_1 and c_2 are called inertia and acceleration constants which have to be chosen carefully.

3.2 Artificial Bee Colony (ABC)

The food source searching mechanism of real bees led to the development of a very popular Artificial Bee Colony Algorithm (ABC) optimization algorithm [16]. The bees search for food and the food source quality is ascertained from the nectar amount. The colony of bees is divided into groups based on the assigned works: employed bee, onlooker bee, and scout bees. The details of the working mechanism of each bee can be found in [16].

In the ABC algorithm, a food source position is a candidate for a solution. The quality of the solution is determined from the fitness value. The necessary equations required for the working of artificial bees are given below:

The new food source location or the solution in the is updated using Eq. (6).

$$x_i(t+1) = x_i(t) + \varphi_i(x_i(t) - x_m(t)) \tag{6}$$

$x_i(t)$ food source position or solution at time ‘t’
m random index for selection of neighbor
φ_i a random number in [−1, +1].

For selection of ith solution by the employed bees and the onlooker bees, a probability value p_i is calculated as defined below:

$$p_i = \frac{f(x_i)}{\sum_{m=1}^{S} f(x_m)} \tag{7}$$

where $f(x_i)$ is the fitness function value of solution x_i.

3.3 Bat Algorithm

The object detection principles of bats by using sonar echoes inspired a meta-heuristic algorithm called bat algorithm [17]. The bats are guided by emitting a signal pulse. The information of elapsed time between emission and reflection is used to evaluate the distance of the obstacles or prey. This behavior of the bats can be used to formulate a new technique for tracking solutions in a search space.

The new solutions are obtained by updating the motion of virtual bats using the equations given below:

$$F_i = F_{\min} + (F_{\max} - F_{\min}) \cdot \beta \tag{8}$$

$$V_i(k+1) = V_i(k) + (X_i(t) - X_{\text{best}}(k)) \cdot F_i \tag{9}$$

$$X_i(k+1) = X_i(k) + V_i(k+1) \tag{10}$$

where

$V_i(k)$	velocity of a bat in position $x_i(k)$ at kth iteration
$F_{\min}$, $F_{\max}$	minimum and maximum frequency of the pulse
$X_{\text{best}}(k)$	best position obtained so far
β	a uniformly distributed random number.

A new solution is generated by employing a random walk with direct exploitation according to the equation defined below:

$$x_{\text{new}} = x_{\text{old}} + \delta A(t+1), \tag{11}$$

where $\delta \in [-1, 1]$ is a random number, while A^t is the average loudness of all the best at the time step. The loudness and pulse rate are updated according to equation given below:

$$A_i(k+1) = \alpha \cdot A_i(k), \tag{12}$$

$$r_i(k+1) = r_i^0 \left(1 - e^{-\gamma k}\right), \tag{13}$$

where α and γ are constants.

3.4 *Grey Wolf Optimizer (GWO)*

Grey Wolf Optimizer (GWO) is a meta-heuristic algorithm proposed in [18] that has been developed by observing the conduct of Grey wolves (Canis lupus) which keep a hierarchy in the group and proceeds with a unique methodology while hunting. Mathematically, this hierarchical structure is model by defining four categories of grey wolves which are called hierarchically as alpha (α), beta (β), delta (δ), and omega (ω). The alpha (α) is the leader and dominant wolf that gives the best solution. The hunting mechanism (or solution searching) proceeds sequentially as searching, encircling and attacking of prey. The detailed analysis of GWO model can be found in [18], however basic equations that governed the search mechanism of GWO including the hierarchical structure and hunting mechanism of grey wolves are presented here.

The encircling behaviors of the grey wolves are modeled as follows:

$$\vec{D} = \left|\vec{C} \cdot \overrightarrow{X_p}(i) - \vec{X}(i)\right| \tag{14}$$

$$\overrightarrow{X_P}(i+1) = \overrightarrow{X_P}(i) - \vec{A} \cdot \vec{D} \tag{15}$$

where $\vec{A}$ and $\vec{C}$ are coefficient vectors, $\overrightarrow{X_P}(i)$ represents position (solution) of the prey at ith iteration, and $\vec{X}(i)$ is the position of a grey wolf. The vectors $\vec{A}$ and $\vec{C}$ are determined as follows:

$$\vec{A} = 2\vec{a} \cdot \vec{r_1} - \vec{a} \tag{16}$$

$$\vec{C} = 2\vec{r_2} \tag{17}$$

During the iteration, the elements of $\vec{a}$ are decreased linearly from 2 to 0, $\vec{r_1}$ and $\vec{r_2}$ are random vectors in [0,1]. The position vectors of different category wolves are updated during the hunting process, which are done by the following equations:

$$\begin{aligned} \overrightarrow{D_\alpha} &= \left|\overrightarrow{C_1} \cdot \overrightarrow{X_\alpha} - \vec{X}\right|, \\ \overrightarrow{D_\beta} &= \left|\overrightarrow{C_2} \cdot \overrightarrow{X_\beta} - \vec{X}\right|, \\ \overrightarrow{D_\delta} &= \left|\overrightarrow{C_3} \cdot \overrightarrow{X_\delta} - \vec{X}\right| \end{aligned} \tag{18}$$

$$\overrightarrow{X_1} = \overrightarrow{X_\alpha} - \overrightarrow{A_1} \cdot \overrightarrow{D_\alpha},\ \overrightarrow{X_2} = \overrightarrow{X_\beta} - \overrightarrow{A_2} \cdot \overrightarrow{D_\beta},\ \overrightarrow{X_3} = \overrightarrow{X_\delta} - \overrightarrow{A_3} \cdot \overrightarrow{D_\delta} \tag{19}$$

$$\vec{X}(t+1) = \frac{\overrightarrow{X_1} + \overrightarrow{X_2} + \overrightarrow{X_3}}{3} \tag{20}$$

From (16) and (17), it is observed that vector $\vec{A}$ decreases as $\vec{a}$ is decreased and this influence the grey wolves approaching towards the prey. Finally when $\left|\vec{A}\right| < 1$, the pray can be attacked (or the solution can be found).

3.5 *Gravitational Search Algorithm (GSA)*

An optimization algorithm based on the law of gravity, called gravitational search algorithm (GSA) was proposed in [19]. The gravitational force of attraction that exists between the masses are exploited in this algorithm. The movement among the masses due to gravitational force corresponds to a search mechanism and all the objects will move towards the heavier masses. The position of the masses represents the solution of the problem, and they are navigated by adjusting its gravitational and inertial masses guided by a fitness function. With the progression of the algorithm, the optimum solution is represented by the position of the heaviest mass in the search space. The equations that governed the search process are described below.

Let the position of the *i*th agent in a system with N agents (masses) be defined as follows:

$$X_i = \left(x_i^1,\ \ldots,\ x_i^d,\ \ldots,\ x_i^n\right),\quad \forall i = 1, 2,\ \ldots,\ N$$

where x_i^d is the position of the *i*th agent in the *d*th dimension.

The force due to *j*th mass on *i*th from mass at time '*t*' is defined by

$$F_{ij}^d(t) = G(t)\frac{M_{pi}(t) \times M_{aj}(t)}{R_{ij}(t) + \varepsilon}\left(x_j^d(t) - x_i^d(t)\right), \tag{21}$$

where

$M_{aj}(t)$ *j*th agent's active gravitational mass
$M_{pi}(t)$ *i*th agent's passive gravitational mass
$G(t)$ gravitational constant at time '*t*'
ε a small constant
$R_{ij}(t)$ Euclidean distance between *i*th and *j*th agents.

The cumulative force on *i*th agent is defined as a weighted sum of the forces exerted from other agents:

$$F_i^d(t) = \sum_{j=1, j \neq i}^{N} \text{rand}_j F_{ij}^d(t), \tag{22}$$

where rand_j is a random number in [0, 1].

Using the law of motion, the acceleration of the *i*th agent at time '*t*' is given as follows:

$$a_i^d(t) = \frac{F_i^d(t)}{M_{ii}(t)}, \tag{23}$$

where $M_{ii}(t)$ represents inertial mass of the *i*th agent.

Next, the update equations of the position and its velocity for the next iteration are determined as follows:

$$v_i^d(t+1) = \text{rand}_i \times v_i^d(t) + a_i^d(t) \tag{24}$$

$$x_i^d(t+1) = x_i^d(t) + v_i^d(t+1) \tag{25}$$

where rand_i is a uniform random variable in [0, 1].

The following equations update the gravitational and inertial masses of the agents.

$$M_{ai} = M_{pi} = M_{ii} = M_i, \quad I = 1, 2, \ldots, N$$

$$m_i(t) = \frac{\text{fit}_i(t) - \text{worst}(t)}{\text{best}(t) - \text{worst}(t)}, \tag{26}$$

$$M_i(t) = \frac{m_i(t)}{\sum_{j=1}^{N} m_j(t)}, \tag{27}$$

where $\text{fit}_i(t)$ is the fitness function value of ith agent at time 't', and $\text{best}(t)$ and $\text{worst}(t)$ are the best value and worst value of the fitness function.

4 Result and Discussion

The algorithms discussed in the above sections are evaluated on segmentation of two medical image modalities. The image that we have used in this work are taken from the work done in [20, 21] for brain MRI image and CT scan of a liver image. The segmentation of the brain MRI image is carried out in the image of a normal brain MRI image to separate the three regions of the brain, i.e., white matter (WM), gray matter (GM), and cerebrospinal fluid (CSF). In the second experiment, the segmentation of CT liver with a tumor is carried out. The results of the segmentation using different algorithms are shown in Figs. 1 and 3, respectively, for brain MRI image and CT liver image. The comparison of the convergence of the fitness function of the segmentation techniques is depicted in Figs. 2 and 4 for the two test images, respectively. From the result, it can be clearly observed that the performance of ABC algorithm has better convergence and the segmentation result of ABCFCM in Fig. 1e has comparatively performed better than other algorithms.

The segmentation results of all the methods are validated numerically by three validity indices which are well accepted: fuzzy partition coefficient V_{pc} [22], fuzzy partition entropy V_{pe} [23], and Xie-Beni validity function V_{xb} [24] which are defined in (28) to (30).

$$V_{\text{pc}} = \frac{\sum_j^N \sum_i^c u_{ij}^2}{N} \tag{28}$$

$$V_{\text{pe}} = \frac{-\sum_j^N \sum_i^c \left(u_{ij} \cdot \log u_{ij}\right)}{N} \tag{29}$$

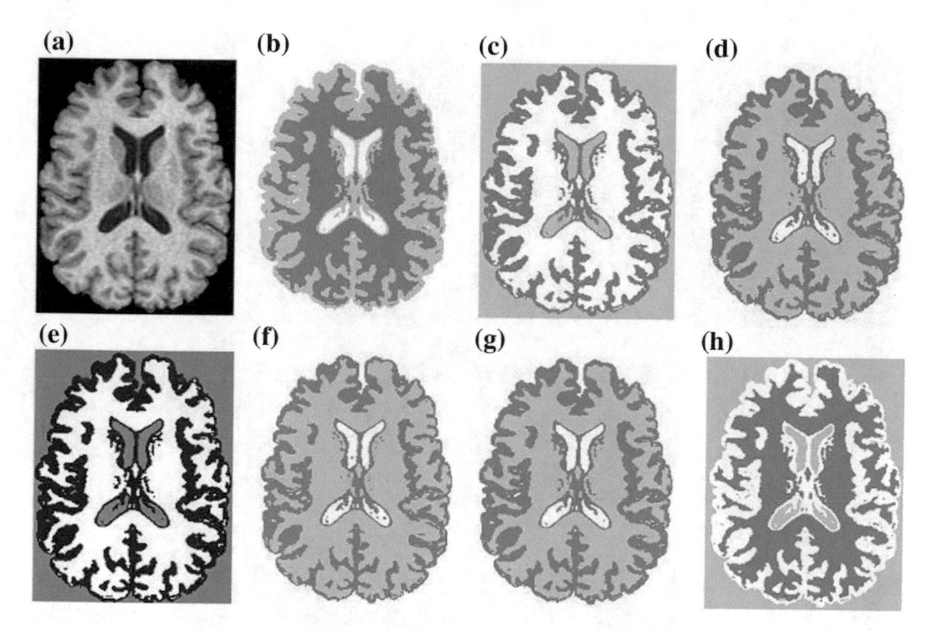

Fig. 1 Segmentation results **a** MRI brain image, **b** FCM, **c** GAFCM, **d** PSOFCM, **e** ABCFCM, **f** GSAFCM, **g** BATFCM, **h** GWOFCM

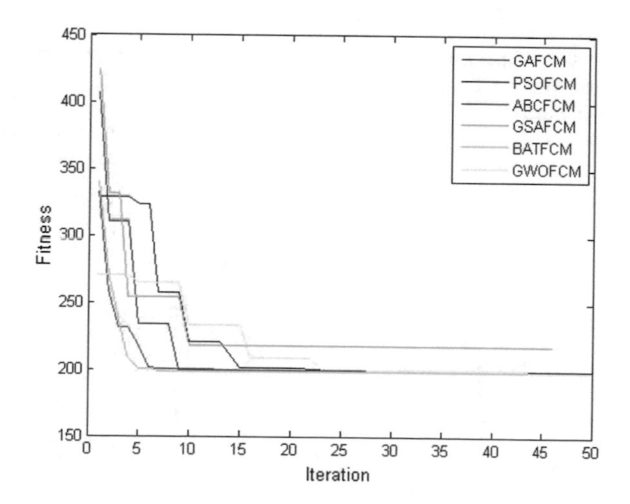

Fig. 2 Fitness function convergence for the segmentation brain MRI image

$$V_{\mathrm{xb}} = \frac{-\sum_j^N \sum_i^c u_{ij} \left\| x_j - v_i \right\|^2}{N * \left(\min_{i \neq k} \{ \| v_k - v_i^2 \| \} \right)} \tag{30}$$

These indices are an indicator of the degree of success of clustering algorithms. The fuzzy partition validity indices V_{pc} and V_{pe} give the fuzziness level for good performance. A maximum V_{pc} or a minimum V_{pe} is desirable for good clustering

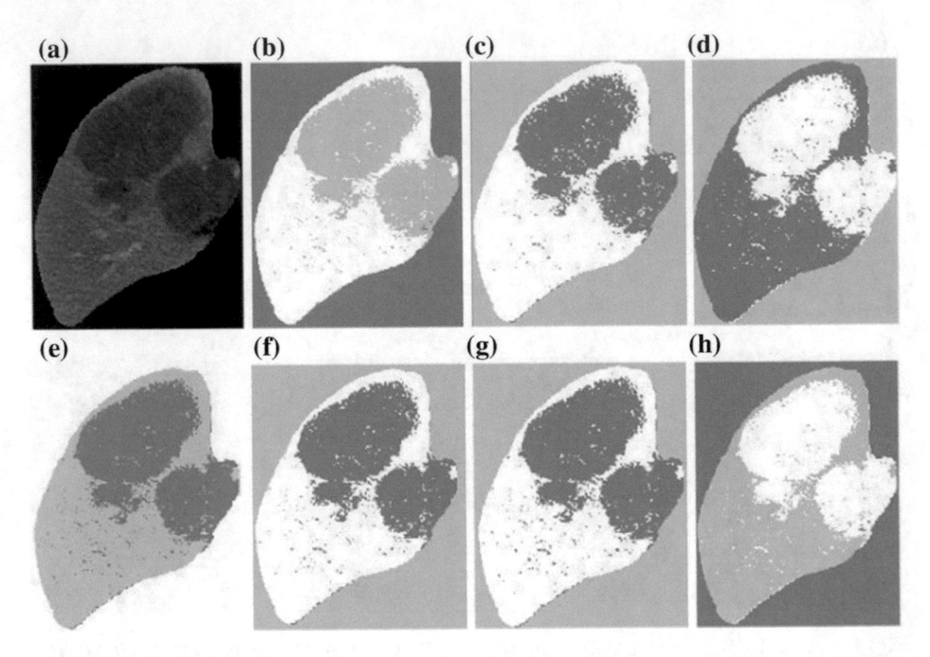

Fig. 3 Segmentation results **a** CT Liver image, **b** FCM, **c** GAFCM, **d** PSOFCM, **e** ABCFCM, **f** GSAFCM, **g** BATFCM, **h** GWOFCM

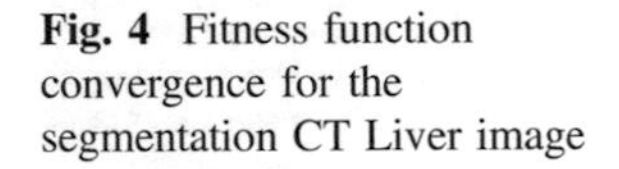

Fig. 4 Fitness function convergence for the segmentation CT Liver image

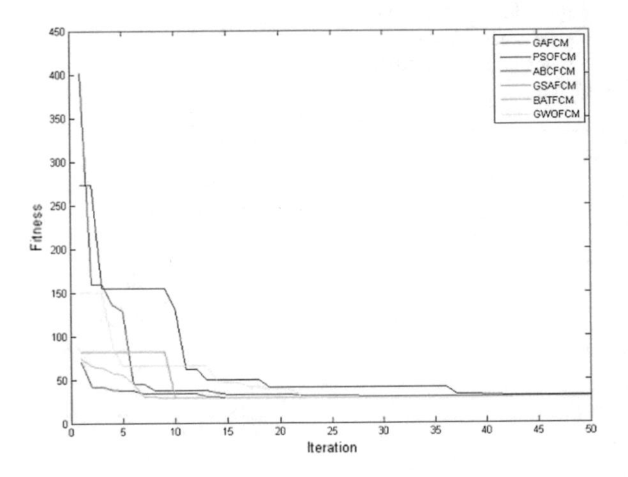

result. The contribution of the features in obtaining fuzzy partition is exploited in the validity function V_{xb}. The consideration of feature value in the definition of V_{xb} makes it more significant and when V_{xb} is minimum the best clustering result is obtained.

The calculated values of the three validity indices for different algorithms are given in Table 1. The numerical values for the three validity indices in Table 1

Table 1 Cluster validity indices with elapsed time

Images	Methods	V_{pc}	V_{pe}	V_{xb}	Time (in s)
Brain MRI	FCM	0.866	0.105	0.0698	2.8
	GAFCM	0.866	0.104	0.0684	21.8
	PSOFCM	0.865	0.104	0.0601	21.7
	ABCFCM	0.865	0.104	0.0595	41.2
	GSAFCM	0.859	0.110	0.0537	21.7
	BATFCM	0.865	0.104	0.0671	21.8
	GWOFCM	0.865	0.104	0.0685	21.7
CT Liver	FCM	0.926	0.059	0.0351	1.77
	GAFCM	0.924	0.061	0.0335	17.1
	PSOFCM	0.924	0.060	0.0329	16.8
	ABCFCM	0.926	0.059	0.0287	21.1
	GSAFCM	0.926	0.059	0.0355	17.5
	BATFCM	0.926	0.059	0.0347	16.9
	GWOFCM	0.926	0.060	0.0348	17.4

further support the superiority of ABC based algorithm. However, the time complexity of ABCFCM is the highest. The accuracy of the segmentation is more important than time complexity. So, the segmentation results showed that ABC algorithm is the best algorithm for the current problem.

5 Conclusion

In this paper, a detailed study of the combination of FCM clustering with different swarm intelligence optimization techniques for medical image segmentation is successfully carried out and presented. The algorithms are evaluated by performing segmentation on brain MRI image and a CT liver image. The performances are further validated by calculating three most popular cluster validity indices. From the results, it was observed the combination of FCM with artificial bee colony (ABC) algorithm has better overall performance compared to other algorithms in both the two different medical image modalities. However, ABC algorithm consumes more CPU time but this should not be a concern since the accuracy of the algorithm is of paramount importance.

References

1. Rafael, C., Gonzalez, G., Richard, E., Woods, W.: Digital Image Processing. Pearson Education (2008)
2. Bezdek, J.C.: Pattern Recognition with Fuzzy Objective Function Algorithms. Plenum, NY (1981)
3. Bezdek, C., Hall, L.O., Clarke, L.P.: Review of MR image segmentation techniques using pattern recognition. Med. Phys. **20**(4), 1033–1048 (1993)

4. Despotović, I., Goossens, B., Philips, W.: MRI segmentation of the human brain: challenges, methods, and applications. Comput. Math. Methods Med. **23** (2015)
5. Balafar, M.A.: Fuzzy c -mean based brain MRI segmentation algorithms. Artif. Intell. Rev. **41**(3), 441–449 (2014)
6. Abdel-Maksoud, E., Elmogy, M., Al-Awadi, R.: Brain tumor segmentation based on a hybrid clustering technique. Egypt. Inf. J. **16**(1), 71–81 (2015)
7. Wang, Y.: fuzzy clustering analysis using genetic algorithm. ICIC Express Lett. **2**(4), 331–337 (2008)
8. Gao, Y., Wang, S., Liu, S.: Automatic clustering based on GA-FCM for pattern recognition. In: Second International Symposium on Computational Intelligence and Design, Changsha, pp. 146–149 (2009)
9. Alder, A., Pramanik, S., Kar, A.: Dynamic image segmentation using fuzzy C-means based genetic algorithm. Int. J. Comput. Appl. **28**(6), 15–20 (2011)
10. Laishram, R., Singh, W.K.K., Kumar, N.A., Robindro, K., Jimriff, S.: MRI brain edge detection using GAFCM segmentation and canny algorithm. Int. J. Adv. Electron. Eng. **2**(3), 1–4 (2012)
11. Laishram, R., Kumar, W.K., Gupta, A., Prakash, K.V.: A novel MRI brain edge detection using PSOFCM segmentation and canny algorithm. In: IEEE International Conference on Electronic Systems, Signal Processing and Computing Technologies (ICESC), pp. 398–401 (2014)
12. Mahalakshmi, S., Velmuruga, T.: Detection of brain tumor by particle swarm optimization using image segmentation. Indian J. Sci. Technol. **8**(22), 1–7 (2015)
13. Balasubramani, K., Marcus, K.: Artificial bee colony algorithm to improve brain MR image segmentation. Int. J. Comput. Sci. Eng. **5**(1), 31–36 (2013)
14. Bose, A.: Kalyani Mali: fuzzy-based artificial bee colony optimization for gray image segmentation. SIViP **10**, 1089–1096 (2016)
15. Eberhart, R.C., Kennedy, J.: A new optimizer using particle swarm theory. In: Proceedings of the Sixth International Symposium on Micromachine and Human Science, Nagoya, Japan, pp. 39–43 (1995)
16. Karaboga, D.: An idea based on honey bee swarm for numerical optimization. Technical Report 06, Erciyes University, Engineering Faculty, Computer Engineering Department (2005)
17. Yang, X.S., Gandomi, A.H.: Bat algorithm: a novel approach for global engineering optimization. Eng. Comput. **29**(5), 1–18 (2012)
18. Mirjalili, S., Mirjalili, S.M., Lewis, A.: Grey wolf optimizer. Adv. Eng. Softw. 69, 46–61 (2014)
19. Rashedi, E., Nezamabadi-Pour, H., Saryazdi, S.: GSA: a gravitational search algorithm, information sciences. **179**(13), 2232–2248 (2009)
20. Pham, D.L., Prince, J.L.: An adaptive fuzzy -means algorithm for image segmentation in the presence of intensity in homogeneities. Pattern Recogn. Lett. **20**(1), 57–68 (1999)
21. Li, B.N., Chui, C.K., Chang, S., Ong, S.H.: Integrating spatial fuzzy clustering with level set methods for automated medical image segmentation. Comput. Biol. Med. **41**(1), 1–10 (2011)
22. Bezdek, J.C.: Cluster validity with fuzzy sets **3**(3), 58–73 (1974)
23. Bezdek, J.C.: Mathematical models for systematic and taxonomy. In: Proceedings of 8th International Conference on Numerical Taxonomy, San Francisco, pp. 143–166 (1975)
24. Xie, X., Beni, G.A.: Validity measure for fuzzy clustering. IEEE Trans Pattern Anal. Mach. Intell. **31**(8), 841–861 (1991)

Block Coordinate Descent Based Algorithm for Image Reconstruction in Photoacoustic Tomography

Anjali Gupta, Ashok Kumar Kajla and Ramesh Chandra Bansal

Abstract Photoacoustic (PA) imaging is a rising, congenital, in vivo biomedical imaging configuration combining equally optics and ultrasonics used for tumor angiogenesis monitoring. A nanosecond laser pulse is accustomed to illuminate biological tissue at a light wavelength typically in the near-infrared (NIR) window when deep light entrance into tissue is desired. The emission pressure rise at the origin is reciprocal to the immersed power and the force wave drives within smooth organic tissues as an audible wave also recognized as Photoacoustic wave. Several photoacoustic image reconstruction algorithms have been developed which includes analytic methods in terms of filtered back-projection (FBP) rather algorithms founded on Fourier transform but these methods generally give suboptimal images. In this paper, we developed an algorithm based on block coordinate descent method for image reconstruction in photoacoustic tomography (PAT). Block coordinate descent (BCD) algorithms optimize the target function over individual segment, at every sub-repetition, whereas keeping all the other segments fixed. This scheme is used to compute pseudoinverse of system matrix and reconstruct image for pressure distribution in photoacoustic tomography. The proposed BCD method is compared with back-projection (BP) method, direct regularized pseudoinverse computation and conjugate gradient based method using simulated phantom data sets.

Keywords Photoacoustic tomography · Reconstruction algorithms · Linear image reconstruction · Block coordinate descent algorithm

---

A. Gupta (✉) · A. K. Kajla · R. C. Bansal
Department of Electronics and Communication Engineering, Arya Institute of Engineering and Technology, Jaipur, India
e-mail: guptaanjali1803@gmail.com

A. K. Kajla
e-mail: kajla_ashok@yahoo.com

R. C. Bansal
e-mail: dr_rcbansal@yahoo.co.in

A. Gupta · A. K. Kajla · R. C. Bansal
RTU, Kota, India

A. K. Luhach et al. (eds.), *Smart Computational Strategies: Theoretical and Practical Aspects*, https://doi.org/10.1007/978-981-13-6295-8_8

1 Introduction

The area of photoacoustic tomography has proficient significant expansion in recent years. Even though various immaculate optical imaging configurations, containing two-photon microscopy, confocal microscopy, and optical coherence tomography have been extremely prosperous, none of these techniques can supply insight over $\sim$1 mm into dispersion organic tissues, consequently they are premised on quasi-ballistic and ballistic photons. Hence there has been a invalid in high-resolution optical imaging over this penetration limit [1, 2]. Photoacoustic tomography, that merges superior ultrasonic resolution and persistent optical oppose in a single modality, has damaged throughout this restriction and filled this faulty.

The region of photoacoustic tomography (PAT) has developed an excellent deal in recent years. PAT is an imaging technology premised on the photoacoustic result. PAT merges strong optical contrast and high ultrasonic resolution in a single configuration, competent of supplying high-resolution functional, structural, and molecular imaging in vivo in optically spreading organic tissue at new nadir. PAT includes ultrasonic detection, optical irradiation, and image structure. The tissue is normally illuminated by a short-pulsed laser beam to generate thermal and audio impulse responses. Regionally immersed light is transformed into heat, that is in addition transformed to a pressure increase via thermoelastic enlargement of the tissue. The beginning pressure rise resolute by the local optical energy displacement also known by specific optical absorption in the unit of J/m^3 and another mechanical and thermal characteristics spread in the tissue as an ultrasonic wave, that is mentioned to as a photoacoustic wave. The photoacoustic wave is discovered by ultrasonic transducers established external the tissue, generating electric signals. The electric signals are then digitized, magnified, and shifted to a computer, where an image is created.

PAT relies on some absorbed photons, moreover scattered or unscattered, to generate photoacoustic signals as long as the photon excitation is serene thermally [3].

Figure 1 shows a schematic scheme to illustrate the photoacoustic effect, where a short duration pulse is an incident on the photoacoustic source and subsequently the incidental light is immersed and transformed into acoustic signals via thermal growth. PAT is extremely sensitive to optical preoccupation. PAT deals on murky context when no absorption endures, the background signal is nothing. Similar dark background allows touchy discovery. In the existence of absorption, the foreground signal is proportionate to the absorption factor. Some little partial alter in optical absorption factor interprets into an equal quantity of fractional alter in PAT signal, that means a relative sensitivity of 100% [4]. The spatial resolution is acquired from ultrasonic identification in the photoacoustic emission phase. Ultrasonic dispersion is much slighter than optical dispersion. The wavelength of the discovered photoacoustic wave is adequately little. As consequence, photoacoustic waves supply superior resolution than optical waves over the smooth depth restrict. Among other factors, the bandwidth and the center frequency of the ultrasonic discovery system

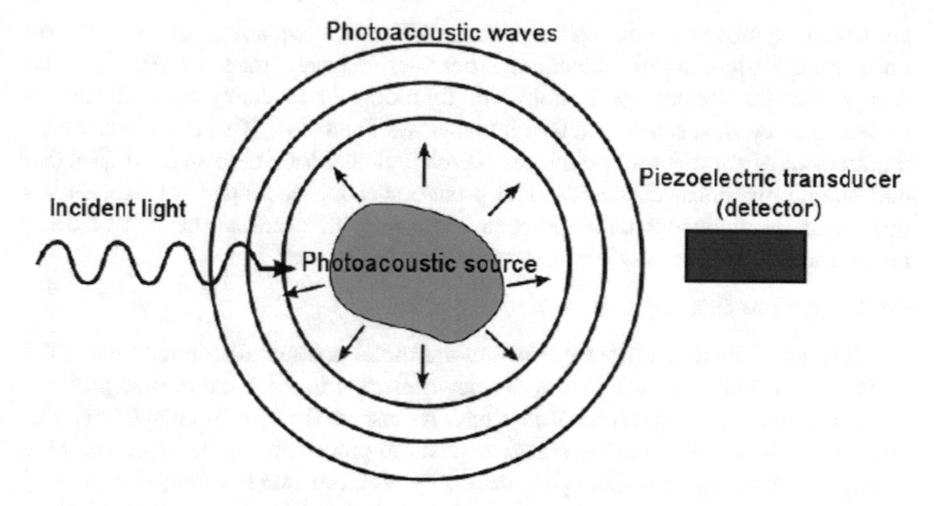

Fig. 1 Fundamental rule of the photoacoustic consequence. Incident light is immersed and transformed into acoustic waves by thermal extension. The acoustic waves are discovered by a piezoelectric transducer [1]

mainly establish the spatial resolution of PAT [5]. The higher the center frequency and the wider the bandwidth, the superior the spatial resolution is.

2 Problem in Photoacoustic Tomography

Photoacoustic imaging primarily deals with two problems, forward and inverse problems. The forward problem involves solving the photoacoustic wave differential equation for a given medium. The inverse problem involves solving for the initial pressure of the medium from the available surface level measurements

(a) Forward Problem

$$\left(\nabla^2 - \frac{1}{v_s^2}\frac{\partial^2}{\partial t^2}\right)p(r,t) = -\frac{\beta}{kv_s^2}\frac{\partial^2 T(r,t)}{\partial t^2}, \tag{1}$$

where, $K = \frac{c_p}{\rho v_s^2 c_v}$ is the isothermal compressibility, ρ is the density of mass, c_v and c_p denotes the particular heat capacities at persistent volume and pressure, v_s implies the speed of sound ($\sim$1480 m/s in water), β denotes the thermal factor of volume increase, p and T denotes the change in pressure (in Pa) and temperature (in K), individually. $p(r,t)$ is the acoustic force at location r and time t.

In the above equation, the left-hand side explains the pressure wave expansion. The source name is described on the right side. The most normally utilized numerical approaches for solving part differential equations in audios are finite-dissimilarity, finite-element, and boundary-element methods [6, 7]. Even though fine for various applications, for time-domain modeling of wideband or high-frequency waves, they can turn into awkward and slow. This is because of the requirement of various grid points per wavelength and little time-steps to diminish undesirable numerical dispersion. The pseudo-spectral technique that expresses a growth of the finite difference procedure can support decrease the first of these issues and the *k*-space way can assist to defeat the second.

(b) Inverse Problem

The inverse problem describes estimating the initial pressure distribution using the surface level measurements. One well-known method to solve the inverse problem is to make use of the system matrix which represents the system completely. The system which describes the PA signal acquisition procedure can be expressed by a Toeplitz matrix of a time-changing causal system. An image (dimension $n \times n$) gets reshaped into a large vector by arranging all the columns, represented by x, note that the dimensions of the image is $n^2 \times 1$. This implies the system matrix (A) contains dimensions of $m \times n^2$. Every column of A mentions the system's output to a corresponding entry in x. In here, the system's response is collected utilizing the *k*-wave pseudo-spectral technique. In other words, the pth column entry of A corresponds to the system's measured response to an impulse at the pth entry of x, i.e., $x(p) = 1$, while rest of the entries are zero. Note that the system response is time-varying, the columns of data are also arranged to result in a large vector having dimensions $m \times 1$. This system matrix (A) will be considered for the reconstruction methods which will be discussed in the following section. An illustration of how the system matrix (A) is being built is given in Fig. 2.

3 Image Reconstruction Methods in Photoacoustic Imaging

The forward model equation of photoacoustic imaging can be represented in form of matrix equation as

$$Ax = b \tag{2}$$

Here A represents a system matrix. The size of A in our study is 30,000 × 40,401. The image to be reconstructed is represented by the vector x. Vector x has the length of 40,401. The measurement vector is represented by b. It has a length of 30,000.

There are several ways to solve Eq. 2 but in essence, these methods can be classified as non-iterative and iterative methods. The following subsections introduce the methods used in this work to solve Eq. 2.

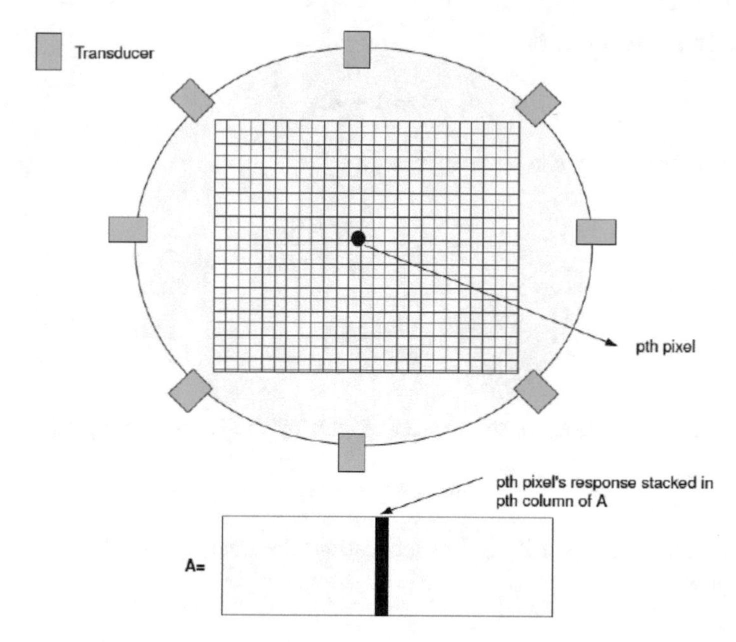

Fig. 2 An illustration of how the system matrix (A) is being built using the forward model which uses the pseudo-spectral method

(a) *Back-Projection (BP) Method*

The back-projection is the simplest image reconstruction scheme where the expression to compute the solution is given by [8],

$$x = A^{\mathrm{T}} b \tag{3}$$

Here, A^{T} is the transpose of system matrix A. Multiplication of A^{T} with measurement vector b is equivalent to back-projection. The back-projection method is computationally very efficient since it is a non-iterative method. Back-projection methods are t be the right choice for quantitative imaging. Moreover, the measurement vector, b is generally corrupted by noise, so this type of back-projection over-amplifies the noise.

(b) *Conjugate Gradient Method*

Conjugate gradient technique is a process to find the numerical solution of a system of linear equations. This conjugate gradient technique is often employed as an iterative procedure. For the system of equations given in 2, this iterative algorithm can be expressed in the following steps [9]:

The residual is given as

$$r_k = b - Ax_k \quad (4)$$

The basis vectors p can be written as,

$$p_k = r_k - \sum_{i<k} \frac{p_i^{\mathrm{T}} A r_k}{p_i^{\mathrm{T}} A p_i} p_i \quad (5)$$

$$\alpha_k = \frac{p_k^{\mathrm{T}} b}{p_k^{\mathrm{T}} A p_k}$$

The iterative solution for the Eq. 2 is given by the following equation:

$$x_{k+1} = x_k + \alpha_k p_k \quad (6)$$

This iterative equation is used to reconstruct the image which is represented by the vector x.

(c) *Regularized Pseudoinverse Computation Based Method*

For the least square problem of Eq. 2, the best fit solution is given as

$$x = A^{+} b, \quad (7)$$

where A^{+} is known as pseudoinverse of A. A^{+} is given as

$$A^{+} = \left(A^{\mathrm{T}} A\right)^{-1} A^{\mathrm{T}} \quad (8)$$

Hence the solution becomes,

$$x = \left(A^{\mathrm{T}} A\right)^{-1} A^{\mathrm{T}} b \quad (9)$$

This method is computationally quite expensive. Computation of A^{+} requires a lot of time. Also, the solution x in Eq. 6 can be noisy. To improve the solution we use a regularization parameter λ such that the regularized pseudoinverse becomes

$$A_{\mathrm{reg}}^{+} = \left(A^{\mathrm{T}} A + \lambda 1\right)^{-1} A^{+} \quad (10)$$

And,

$$x = A_{\mathrm{reg}}^{+} b \quad (11)$$

This equation is used to reconstruct the image which is represented by the vector x.

(d) *Block Coordinate Descent Method*

Block coordinate descent (BCD) technique implements the objective function on one segment (bunch of variables) x_j at every sub-iteration, while all the other segments $x_j \neq x_j$ are kept fixed. The universal convergence of the BCD iterates is established for minimizing a convex non-differentiable equation with definite separability and regularity characteristics [10].

In the case of the linear regression problem [11],

$$f(x) = \frac{1}{2}||b{-}Ax||^2 \tag{12}$$

If we minimize above equation over x_j with all x_j fixed such that $i \neq j$,

$$0 = \nabla_i f(x) = A_i^{\mathrm{T}}(Ax - b) = A_i^{\mathrm{T}}(A_i x_i + A_{-i} x_{-i} - b)$$

i.e., we get,

$$x_i = \frac{A_i^{\mathrm{T}}(b - A_{-i} x_{-i})}{A_i^{\mathrm{T}} A_i} \tag{13}$$

The iterative scheme in Eq. 13 is used to estimate the vector x.

Another approach is to first establish the blockwise pseudoinverse matrix using Eq. 13, i.e.,

$$A_i^{+} = \left(A_i^{\mathrm{T}} A_i\right)^{-1} A_i^{\mathrm{T}}$$

Hence,

$$x_i = A_i^{+}(b - A_{-i} x_{-i}) \tag{14}$$

Further, with use of regularization parameter λ, Eq. 14 can be rewritten as follows:

$$x_i = \left(A_i^{\mathrm{T}} A_i + \lambda I\right)^{-1}(b - A_{-i} x_{-i}) \tag{15}$$

In this study, the Eq. 15 is used for the reconstruction of the image represented by a vector x.

4 Results and Discussions

This section presents the details of the numerical experimental data used in the proposed method, LSQR-based method and other traditional methods used for comparison. In this study, three numerical phantoms were considered, a PAT phantom, a Derenzo phantom, and a blood vessel phantom. For all the results presented here the regularization parameter was set to =0.018 as this was the best-suited value observed for all imaging structures.

(a) *Numerical Blood Vessel Phantom*

Because of the large intrinsic contrast that blood cells give, photoacoustic imaging is broadly utilized for visualizing internal blood vessel formation, both in the brain and also under the skin. Figure 3a represents a blood vessel network utilized as numerical phantoms. The *k*-wave was utilized to produce the simulated PA result for 60 detector circular locations. 40 dB of noise was included in the simulated data. Figure 3 shows the reconstructions obtained through back-projection method, reconstruction obtained through blockwise computation of pseudoinverse and through direct pseudo inverse computation method.

(b) *Derenzo Numerical Phantom*

Derenzo phantom consists of many circular shapes of different size. This can be considered similar to tumors present under the skin. Figure 4 a shows a Derenzo phantom which has been used in the present study. The *k*-wave was utilized to produce the simulated PA data for 60 detector circular locations. 40 dB of noise was included in the simulated data. Figure 4 shows the reconstructions obtained through back-projection method, reconstruction obtained through blockwise computation of pseudoinverse and through direct pseudo inverse computation method.

(c) *PAT Numerical Phantom*

Figure 5 a shows a numerical PAT phantom image. The *k*-wave was utilized to produce the simulated PA data for 60 detector circular locations. Again, 40 dB of noise was included in the simulated data. Figure 5 represents the reconstructions obtained employing back-projection method, reconstruction obtained through blockwise computation of pseudoinverse and through direct pseudo inverse computation method.

Details of Computation Time

One of the important advantages of using BCD method to compute blockwise pseudoinverse for image reconstruction problem is because of its computational efficiency. BCD scheme for pseudoinverse calculation is much faster than direct pseudoinverse calculation using Eq. 10. We have used four blocks BCD methods for pseudoinverse calculation.

The experiments were performed on 128 GB RAM, Z820 hp machine. The following table compares the pseudoinverse computation time (Table 1).

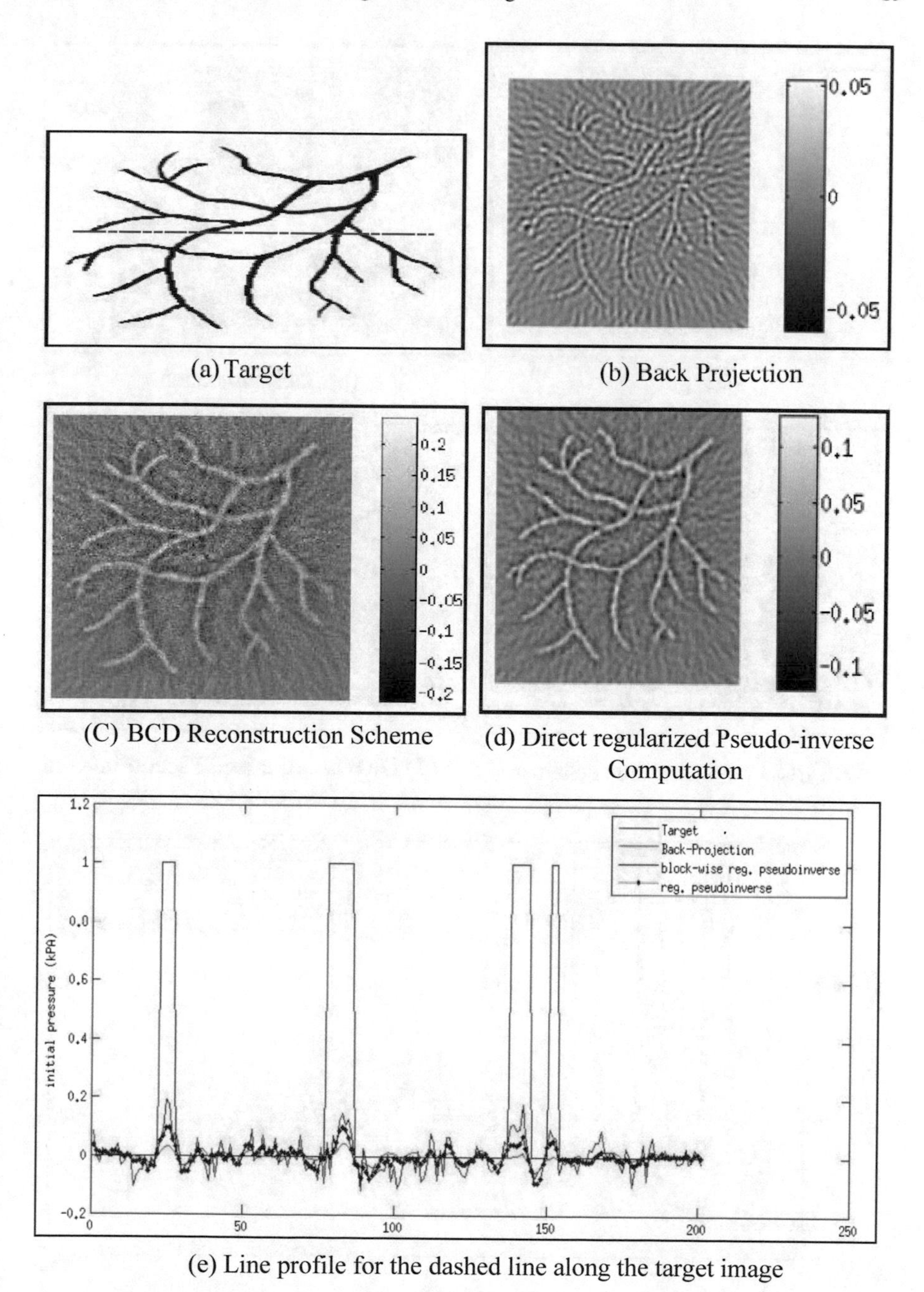

Fig. 3 Reconstructed photoacoustic image of a numerical blood vessel structure phantom as an imaging domain with an initial pressure rise of 1 kPa. **a** Target image, **b** back-projection (BP) reconstruction, **c** BCD scheme reconstruction, **d** reconstruction using regularized direct pseudo inverse computation and finally **e** the graph represents the line profile for the dashed line along the target image

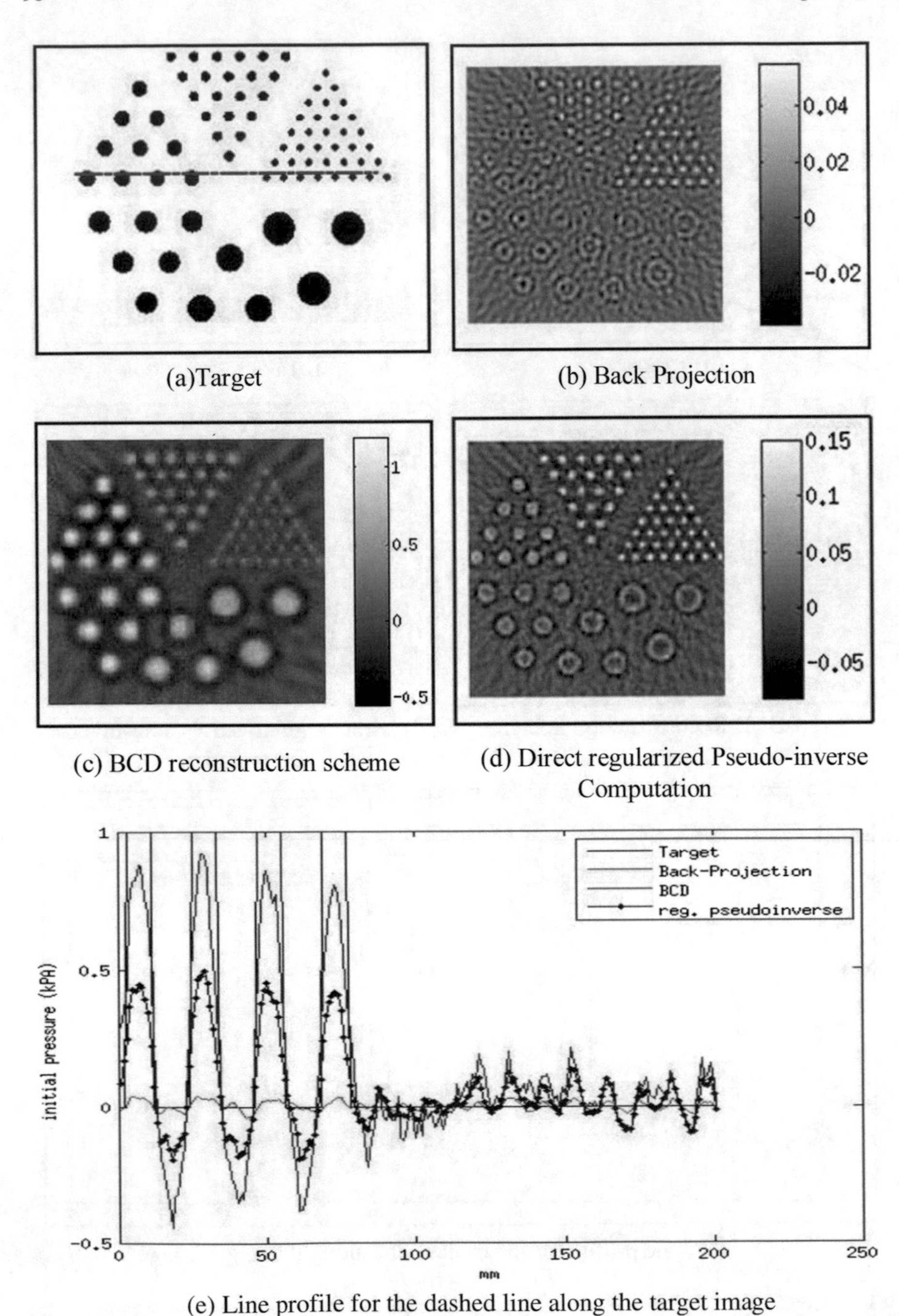

(a)Target

(b) Back Projection

(c) BCD reconstruction scheme

(d) Direct regularized Pseudo-inverse Computation

(e) Line profile for the dashed line along the target image

Fig. 4 Reconstructed photoacoustic image of a numerical Derenzo phantom as the imaging domain with an initial pressure rise of 1 kPa. **a** Target image, **b** back-projection (BP) reconstruction, **c** BCD scheme reconstruction, **d** reconstruction using regularized direct pseudo inverse computation and finally **e** the graph represents the line profile for the dashed line along the target image

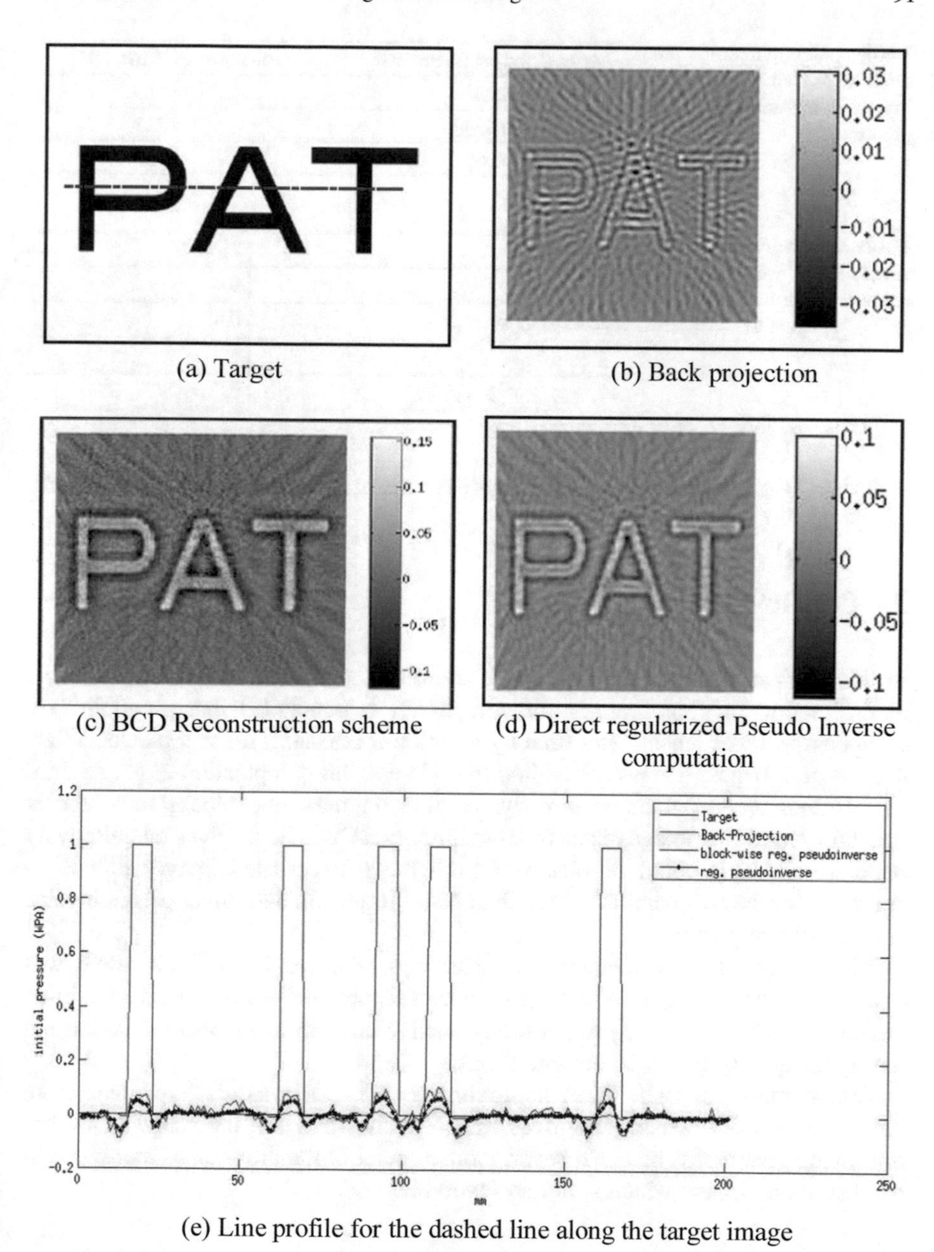

(a) Target

(b) Back projection

(c) BCD Reconstruction scheme

(d) Direct regularized Pseudo Inverse computation

(e) Line profile for the dashed line along the target image

Fig. 5 Reconstructed photoacoustic image of a numerical PAT phantom as the imaging domain with an initial pressure rise of 1 kPa. **a** Target image, **b** back-projection (BP) reconstruction, **c** BCD scheme reconstruction, **d** reconstruction using regularized direct pseudo inverse computation and finally **e** the graph represents the line profile for the dashed line along the target image

Table 1 Computational time recorded for all pseudo inverse computation methods presented

Method for pseudoinverse computation	Time (s)
Direct methods	639
BCD (4 blocks)	213
BCD (9 blocks)	125

Table 2 Computational time simulated for various reconstruction techniques

Method	Time (s)
1. $x = A^{\mathrm{T}}b$	0.4
2. $x = A_{\mathrm{reg}}^{+}b$	0.4
3. $x_i = A_i^{+}b$	4
4. $x_i = A_i^{+}(b - A_{-i}x_i)$	21

Table 2 the computation time for image reconstruction through various methods.

5 Conclusion

In this work, a method based on block coordinate descent is proposed for image reconstruction in photoacoustic tomography. It is concluded that pseudoinverse computation takes much lesser time by dividing it into uniform size blocks. When the 201×201 grid size is divided into four blocks, this computation took one-third of the time when computed directly. A division into nine blocks took almost one-fifth times. The image reconstruction through BCD scheme gives quantitatively better result. This could be observed from the reconstructed images that BCD approach has better quantitativeness than back-projection and direct pseudoinverse computation approach.

The choice of regularization parameter was taken as 0.018. This value was chosen after observing the effects of on image reconstruction. It was observed that however, a lesser value of gives a lesser relative error in image reconstruction but the image quality has to be compromised.

One iteration of BCD takes approximately 21 s. It was observed that more iterations of BCD scheme improves image quality and has the scope of further improving quantitativeness. However, more number of iterations takes more time to reconstruct an image which is not so favorable.

Reference

1. Wang, L.V. (ed.): Photoacoustic Imaging and Spectroscopy. CRC, London (2009)
2. Wang, L.V.: Tutorial on photoacoustic microscopy and computed tomography. IEEE J. Sel. Top. Quantum Electron. **141**, 171179 (2008)
3. Wang, L.V.: Prospects of photoacoustic tomography. Med. Phys. **35**, 12 (2008)

4. Oraevsky, A.A., Wang, L.V.: Photons Plus Ultrasound: Imaging and Sensing, vol. 643, SPIE, Bellingham (2007)
5. Zhang, H.F., Maslov, K., Stoica, G., Wang, L.H.V.: Functional photoacoustic microscopy for high- resolution and noninvasive in vivo imaging. Nat. Biotechnol. **247**, 848–851 (2006)
6. Treeb, B.E., Cox, B.T.: K-wave: MATLAB toolbox for the simulation and reconstruction of photoacoustic wave fields. J. Biomed. Opt. **15**(2), 021314 (2010)
7. Bagchi, S., Roy, D., Vasu, R.M.: Forward problem solution in photoacoustic tomography by discontinuous Galerkin method. J. Opt. Soc. Am. (2011)
8. Xu, M.H., Wang, L.V.: Universal back-projection algorithm for photoacoustic computed tomography. Phys. Rev. E755 (2005)
9. Zhdanov, M.S.: Geophysical Inverse Theory and Regularization Problems, 1st edn. Elsevier Science BV (2002)
10. Zhiwei (Tony) Q., Scheinberg, K., Goldfarb, D.: Efficient Block Coordinate Descent Algorithms for the Group Lasso
11. Shaw C.B.: LSQR—based automated optimal regularization parameter selection fo photoacoustic tomography, SE-360 project, Apr 2013

Applying Machine Learning Algorithms for News Articles Categorization: Using SVM and kNN with TF-IDF Approach

Kanika and Sangeeta

Abstract News articles categorization is a supervised learning approach in which news articles are assigned category labels based on likelihood demonstrated by a training set of labeled articles. A system for automatic categorization of news articles into a standard set of categories has been implemented. The proposed work will use Term Frequency–Inverse Document Frequency (TF-IDF) term weighting scheme for optimization of classification techniques to get more optimized results and use two supervised learning approaches, i.e., Support Vector Machine (SVM) and *K*-Nearest neighbor (kNN) and compare the performances of both classifiers. Each news document is preprocessed and transformed into a term-document matrix (Tsoumakas et al. in Data mining and knowledge discovery handbook. Springer, Berlin, pp 667–685 (2010) [1]). After preprocessing and transforming each news article into a vector of weights, TF-IDF term weighting scheme was used for weighting the word. TF-IDF weighted the words calculating the number of words that appear in a document. An unknown news item is also transformed into a vector of keyword weights, and then categorized into suitable categories such as Sports, Business, and Science and Technology. The system purposed in research work was trained on the collection of approximately 300 categorized news articles extracted from the various Indian newspaper websites and tested on a different set of 60 randomly extracted news items from the same sources (Trstenjak et al. in Proc Eng 69:1356–1364 (2014) [2], Buana et al. Int J Comput Appl 50:37–42 [3]). It has been observed that the performances of both algorithms improve when TF-IDF approach is used.

Keywords News articles · *K*-Nearest neighbor (kNN) · Support vector machine (SVM) · Term frequency–inverse document frequency (TF-IDF)

Kanika (✉) · Sangeeta
I. K. Gujral Punjab Technical University, Kapurthala, India
e-mail: kanikadhanjal@gmail.com

Sangeeta
e-mail: geet.bhagat22@gmail.com

A. K. Luhach et al. (eds.), *Smart Computational Strategies: Theoretical and Practical Aspects*, https://doi.org/10.1007/978-981-13-6295-8_9

1 Introduction

Automated news articles categorization or classification is the automatic classification of news articles or documents under predefined categories or classes. It is one of the applications of text classification. It is a supervised learning approach.

Text classification is a kind of procedure related to Natural Language Processing (NLP). It finds relational mode (classifier) between text's attributes (feature) and text's category according to a labeled training text corpus, and then utilizes the classifier to classify new text corpus. Text classification can be divided into two parts: training and classifying. The purpose of training is to structure classifier, which can be used to classify new texts by the connection between training text and category. Classifying means to make the unknown new text assigned with the known category label [4].

2 Related Work

Trstenjak Bruno et al. purposed a framework based on Term Frequency–Inverse Document Frequency (TF-IDF) approach for text categorization. The authors have given the probability of using *K*-Nearest Neighbor (kNN) classifier with TF-IDF method and purposed a framework for text categorization, which enables the classification of various different categories of source documents according to various parameters, measurement, and analysis of results. The framework was evaluated on the basis of the quality and speed of classification. The results of the experiment demonstrated the positive and negative features of the algorithm. Hence, it has provided the direction for the development of similar kind of systems [2]. Buana Putu Wira et al. combined kNN with *K*-Means on term reweighting to classify Indonesian News. After preprocessing, the weighting of words was done by TF-IDF term weighting scheme. Then, grouping of all the training samples of *K*-Means algorithm and taking cluster samples as the new training samples was done. Then, modified samples were used for classification using kNN algorithm [3]. Bijalwan Vishwanath et al. have used kNN-based learning approach for text categorization, in the first step, and the documents have been categorized using kNN classifier, and then compared with Naïve Bayes classifier and Term graph by returning the most relevant documents. In this paper, the authors concluded that kNN showed maximum accuracy as compared to the Naive Bayes and Term Graph [5]. Rahmawati Dyah et al. have proposed an approach for multi-label classification for Indonesian news articles. In the research work, two approaches, i.e., problem transformation and algorithm adaptation were investigated. The paper focused on four factors, i.e., feature weighting, feature selection method, multi-label classification, approach and single-label classification approach. The experimental results showed that the Support Vector Machine (SVM) algorithm performed the best when it was used with TF-IDF [6].

3 Problem Formulation

As the quantity of electronic documents is increasing day by day, it is becoming more difficult for the classifier to accurately categorizing them into categories according to their contents, i.e., quality is decreasing. Therefore, there is a need for an approach which can increase the efficiency of classifiers. Thus, the proposed work will use TF-IDF term weighting scheme for optimization of two supervised learning approaches, i.e., Support Vector Machine and kNN to categorize news articles into suitable categories such as Sports, Business, and Science and Technology, and compare the performances of both classifiers.

4 Methodology

The steps of the research methodology are explained as follows (Fig. 1):

- Input training and testing new documents into the system.
- Processing of the input news documents is done, i.e., removing stop words, whitespaces, punctuations, numbers, and performing stemming of training and testing documents etc.
- Document Term Matrix of preprocessed news documents is created using the TF-IDF approach as purposed in this research [7].
- Classification of testing news documents to different categories, i.e., Sports, Business, and Science and Technology is performed using the classifiers SVM and kNN. The performance of kNN can be improved by changing the value of *k*.

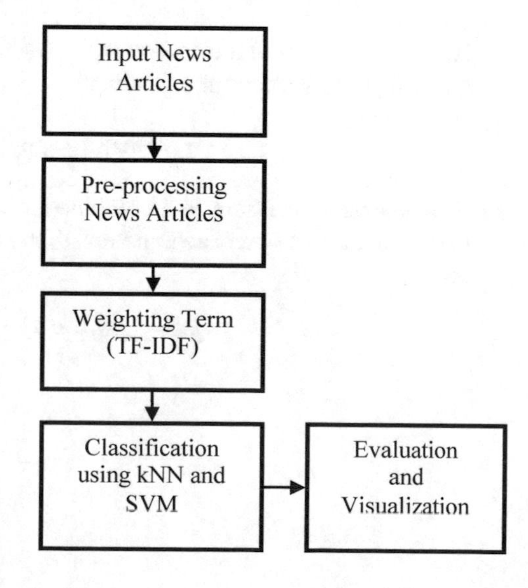

Fig. 1 Research overview diagram

Table 1 Confusion matrix [8]

		Actual/truth	
		Positive	Negative
Predicted	Positive	tp	fn
	Negative	fp	tn

- There are many different ways of measuring the performance of a classifier and evaluate its performance. For supervised classification with two classes, each and every performance measure is based on four numbers obtained after applying the classifier to the test dataset. These numbers are called true positives tp, false positives fp, true negatives tn, and false negatives fn. The entries are the counts in a table with two rows and two columns which is as Table 1.

Table 1 is called as a confusion matrix. The term used in matrix table, i.e., true positive, etc., is a standard, but whether columns correspond to truth/actual and rows to predicted or vice versa is not a standard. The entries in a confusion matrix table are the counts, i.e., integers. The total of the four entries tp + tn +fp + fn = n, is the number of test cases. Many different summary statistics can be computed from these entries by considering the type of application [8]:

- **Accuracy**: The accuracy of a classifier is the degree of closeness of measured value of a quantity to the actual value of a quantity.

$$\text{Accuracy } a = (\text{tp} + \text{tn})/n \quad (1)$$

- **Precision**: The precision of a classifier is the fraction of retrieved articles that are relevant to find.

$$\text{Precision } p = \text{tp}/(\text{tp} + \text{fp}) \quad (2)$$

- **Recall**: The recall of a classifier is the fraction of articles that are relevant to the query that are successfully retrieved.

$$\text{Recall } r = \text{tp}/(\text{tp} + \text{fn}) \quad (3)$$

- ***F*-measure**: A measure is a combination of precision and recall, i.e., it is the harmonic mean of precision and recall, the traditional *F*-measure or balanced *F*-score:

$$F\text{−measure } F = 2.p.r/(p + r) \quad (4)$$

5 Experiments and Results

5.1 Experimental Data

News articles have been collected from newspaper websites of Tribune, The Hindu, Times of India, Hindustan Times and Indian Express, etc. There are three categories of news articles, i.e., Sports, Science and Technology, and Business. There are 100 training documents for each category and 20 testing documents for each category (Table 2).

5.2 User Interface

A user interface has been developed using Shiny package in R—for automated news categorization [9] (Fig. 2).

Table 2 Experimental data

Category of news articles	Training documents	Testing documents
Business	100	20
Sports	100	20
Science and Technology	100	20

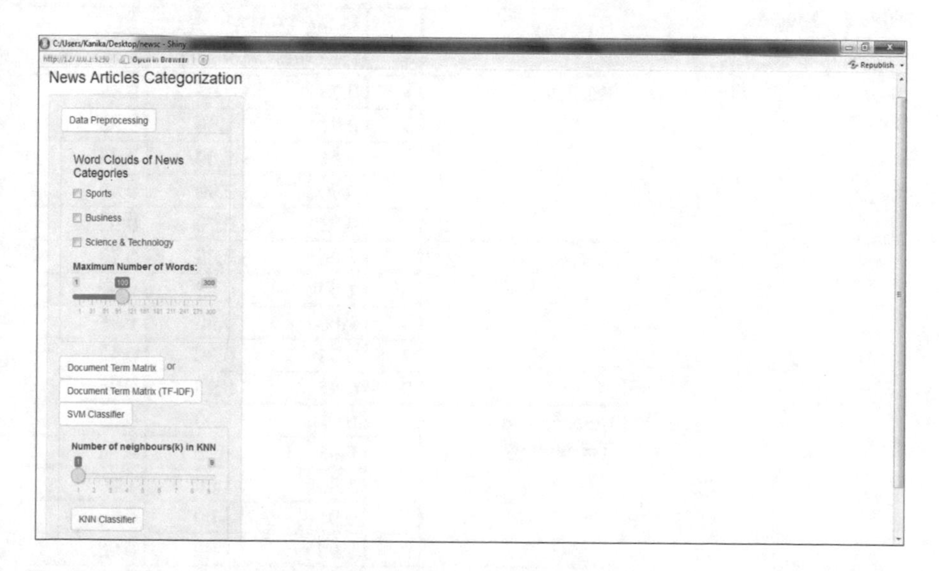

Fig. 2 GUI for news articles classification

The user can directly interact with the application by the following widgets:

- Preprocessing of documents.
- Visualizing Word Clouds of news categories by selecting any one category at a time.
- Creating a document term matrix using term weighting schemes TF-IDF and TF for training and testing data respectively.
- Applying classification either SVM or kNN. Change the values of k for kNN and visualizing the results of classifier, i.e., confusion matrix and accuracy etc.

5.3 Experiments and Analysis

- In this research, it has been observed that kNN is easier to implement. It performs well where there is a large training database, or many combinations of predictor variables. It needs a lot of storage to store all training data. It has long computational times. On the other hand with SVM, training is relatively easy. It scales relatively well to high dimensional data. The more samples in the training set the more complex and slow the classification process. A good kernel function is needed for it, i.e., similarity defined by the kernel function should have a good correlation with the type of the problem.
- The accuracy which is being observed in research has been summarized in Tables 3 and 4.

Table 3 Category classification using kNN

Category		Using TF-IDF		
	K	Precision	Recall	F-measure
Business	1	0.76	0.95	0.84
	3	0.91	1.00	0.95
	5	**0.91**	**1.00**	**0.95**
	7	0.90	0.90	0.90
	9	0.95	0.90	0.92
Sports	1	1.00	0.65	0.79
	3	1.00	0.85	0.92
	5	**1.00**	**0.90**	**0.95**
	7	0.95	0.90	0.92
	9	0.95	0.95	0.95
Science and Technology	1	0.91	1.00	0.95
	3	0.95	1.00	0.98
	5	**1.00**	**1.00**	**1.00**
	7	0.95	1.00	0.98
	9	0.95	1.00	0.98

Note: The values in bold are signifying the highest accuracy

Table 4 Category classification using SVM

Category	Using TF-IDF		
	Precision	Recall	F-measure
Business	0.95	0.90	0.92
Sports	0.90	0.95	0.93
Science and Technology	1.00	1.00	1.00

It has been observed that the precision, recall, and *F*-measure of kNN algorithm increase when we apply TF-IDF term weighting scheme. The performance of kNN improves when we improve the number of neighbors (*k*) from 1 to 9. For $k = 5$ (best case), kNN shows impressive performance for all of the three categories. For Science and Technology category, precision, recall, and *F*-measure reaches to 1.

It has been observed that the precision, recall, and *F*-measure of SVM algorithm improve when we apply TF-IDF term weighting scheme. For Science and Technology category, precision, recall, and *F*-measure reaches to 1.

Figure 3 graph shows the comparison of precision, recall, and *F*-measure of kNN (best case for $k = 5$) and SVM. For Science and Technology category, precision, recall, and F-measure of both classifiers reaches to 1 (Table 5).

Thus, both the classifiers show an impressive improvement in their accuracy when we use the TF-IDF term weighting scheme. The accuracy of kNN improves

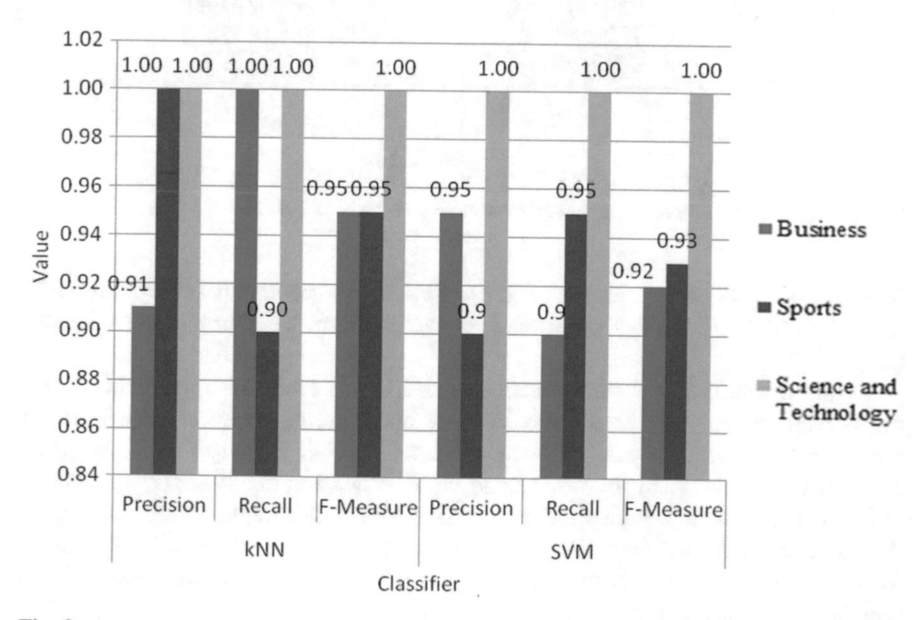

Fig. 3 A graph showing the comparison of precision, recall, *F*-measure of kNN and SVM using TF-IDF

Table 5 Accuracy comparison table of kNN and SVM

Method		(Without TF-IDF)	(Using TF-IDF)
kNN	*K*	Accuracy (%)	Accuracy (%)
	1	65	86.66
	3	60	95
	5	53.33	**96.66**
	7	45	93.33
	9	40	95
SVM		90	**95**

Note: The values in bold are signifying the highest accuracy

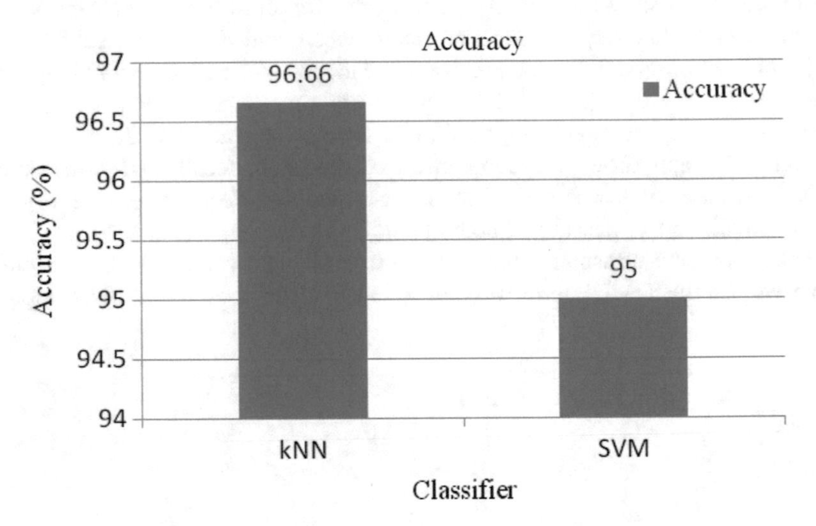

Fig. 4 A graph showing the comparison of accuracy of kNN and SVM

from 53.33 to 96.66% (best case for $k = 5$) and the accuracy of SVM improves from 90 to 95% when TF-IDF approach is used for term weighting of news articles (Fig. 4).

In this research, it has been observed that kNN is easier to implement. It performs well where there is a large training database, or many combinations of predictor variables. It needs a lot of storage to store all training data. It has long computational times. On the other hand with SVM, training is relatively easy. It scales relatively well to high dimensional data. The more samples in the training set the more complex and slow the classification process. The results of experiments for different cases have been visualized on the user interface and some of the screenshots of results are as Figs. 5, 6, 7, and 8.

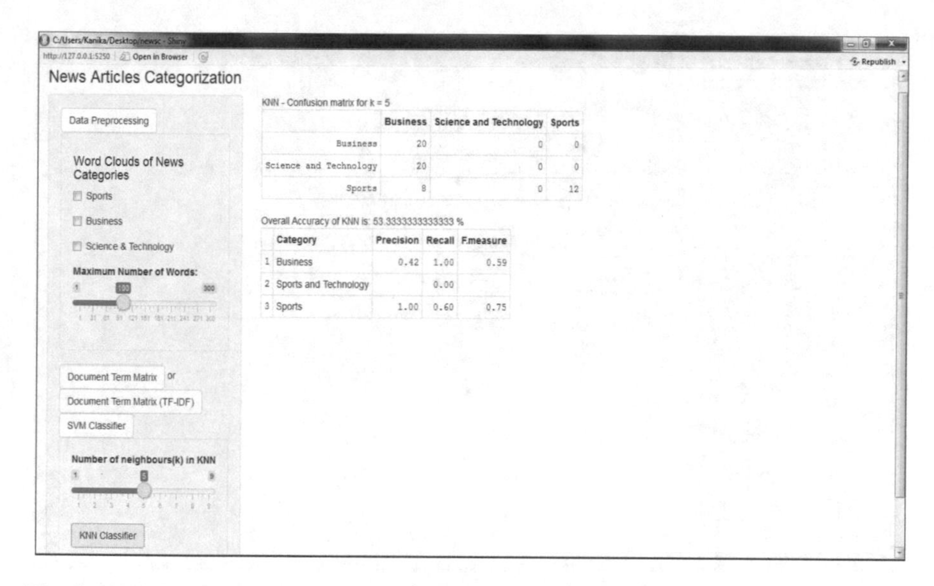

Fig. 5 kNN classification (k = 5) performed on testing news articles (without TF-IDF)

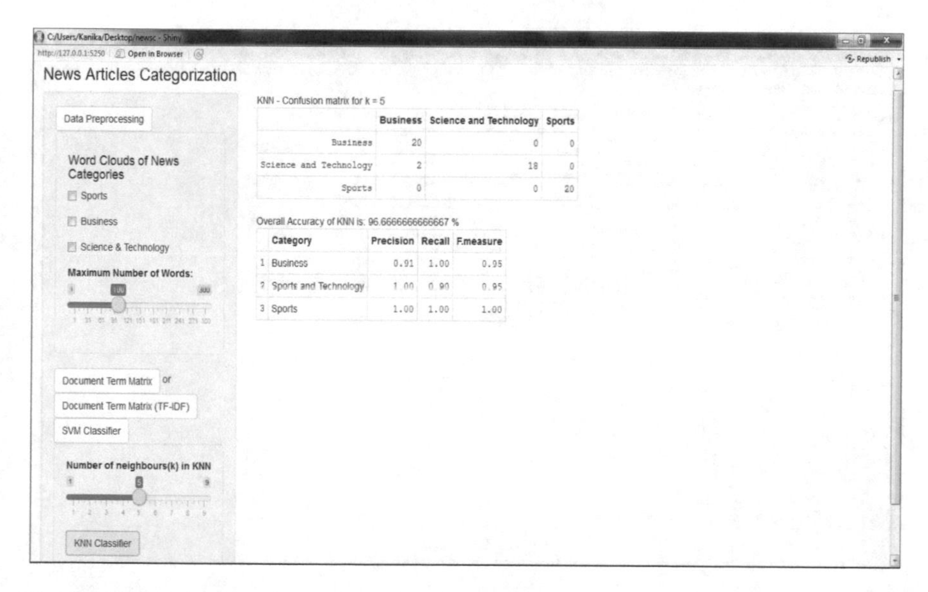

Fig. 6 kNN classification for (best case k = 5) performed on testing news articles using TF-IDF

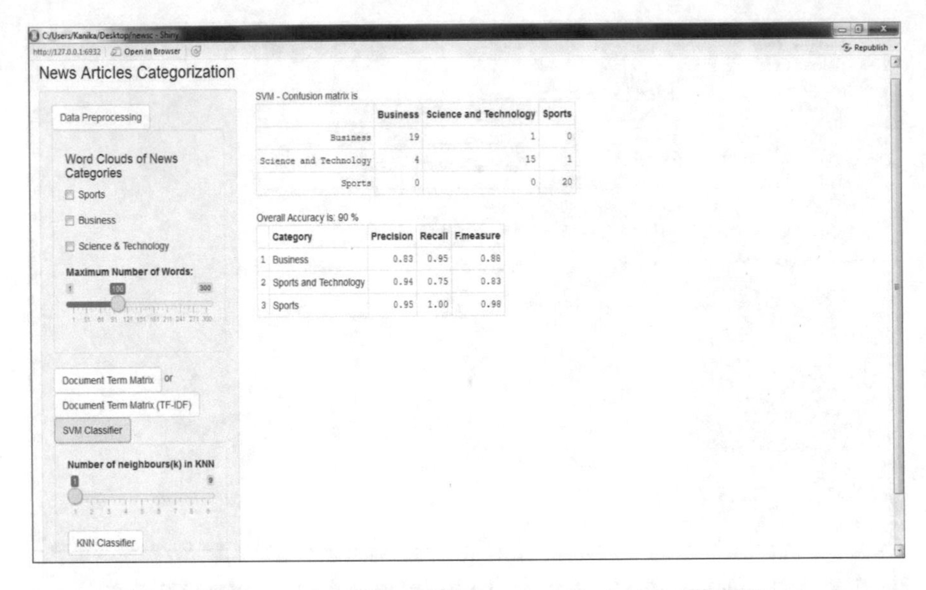

Fig. 7 SVM classification performed on testing news articles (without TF-IDF)

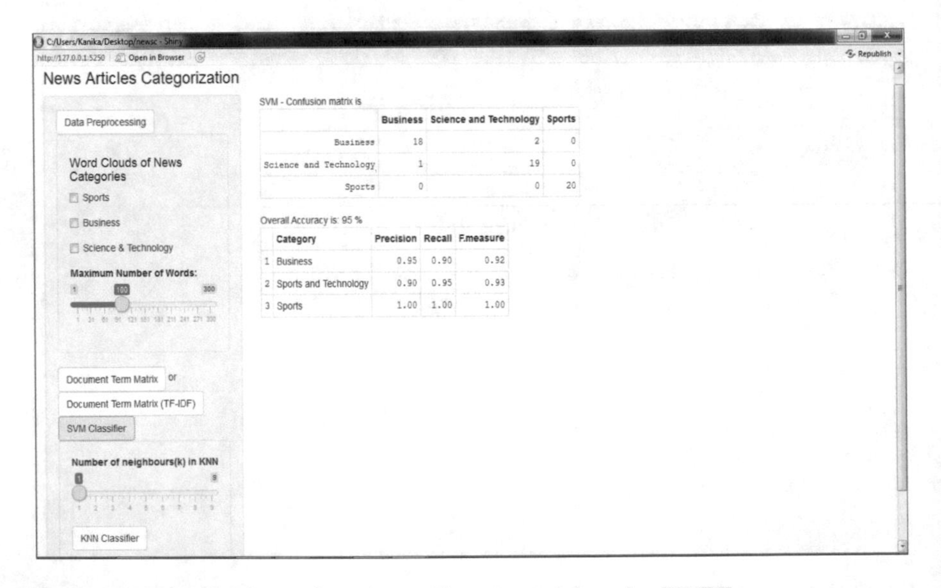

Fig. 8 SVM classification performed on testing news articles using TF-IDF

6 Conclusion

This research analyses news articles categorization by kNN algorithm and SVM algorithm. An interactive machine learning application is developed using the R programming language with Shiny web framework. The research work shows that how the performance of both classifiers, i.e., kNN and SVM algorithms can be improved for classification of news articles by using TF-IDF term weighting scheme on the training corpus having three categories Business, Sports, and Science and Technology and also compared their performances on the same set of testing documents. The performance of kNN improves by increasing the number of neighbors, i.e., *k* and SVM shows impressive performance when the linear kernel is used. The accuracy of SVM improves from 90 to 95% using the TF-IDF approach and the accuracy of kNN also improves for different values of k and it reaches to 96.66.

The TF-IDF scheme can also be applied with some other supervised algorithms to enhance their performance. For future work, more categories of news articles can be used. Some other term weighting schemes can also be used to check their performance against classifiers.

References

1. Tsoumakas, G., Katakis, I., Vlahavas, I.: Mining multi-label data. In: Data Mining and Knowledge Discovery Handbook, pp. 667–685. Springer, Berlin (2010)
2. Trstenjak, B., Mikac, S., Donko, D.: kNN with TF-IDF based framework for text categorization. Proc. Eng. **69,** 1356–1364 (2014). In: 24th DAAAM International Symposium on Intelligent Manufacturing and Automation, 2013
3. Buana, P.W., Jannet Sesaltina, D.R.M., Putra Darma Gede Ketut, I.: Combination of K-nearest neighbour and K-means based on term re-weighting for classify Indonesian News. Int. J. Comput. Appl. (0975-8887) **50**(11), 37–42 (July 2012)
4. Ikonomakis, M., Kotsiantis, S., Tampakas, V.: Text classification using machine learning techniques. WSEAS Trans. Comput. **4**(8), 966–974 (Aug 2005)
5. Bijalwan, V., Kumar, V., Kumari, P., Pascual, J.: kNN based machine learning approach for text and document mining. Int. J. Database Theory Appl. **7**(1), 61–70 (2014)
6. Rahmawati, D., Khodra, L.M.: Automatic multilabel classification for Indonesian news articles (2015). IEEE 978-1-4673-8143-7/15
7. Wen, Zhang, Taketoshi, Yoshida, Xijin, Tang: A comparative study of TF*IDF, LSI and multi words for text classification. Expert Syst. Appl. J. **38**(3), 2758–2765 (2011)
8. Elkan, C.: Evaluating Classifiers, elkan@cs.ucsd.edu (Jan 20, 2012)
9. Package shiny. https://cran.r-project.org/web/packages/shiny/shiny.pdf

Syntactic Analysis of Participial-Type Complex Sentence in Punjabi Language

Sanjeev Kumar Sharma and Misha Mittal

Abstract Complex sentences are always a challenge for processing in natural language. These sentences play a major role in all the languages. Complex sentences are mainly composed of dependent and independent clauses. Therefore, for syntactic analysis of these sentences, one has to check the external structure, as well as internal composition of the clause. In this research article, the author has developed an algorithm for detecting and correcting the syntactic errors present in participial-type complex sentences of Punjabi language. The algorithm developed involves the identification and separation of dependent and independent clauses, and then detection of grammatical mistakes from their structure and in the last step, errors are corrected using grammatical rules of Punjabi language. The author tested the system on a set of participial sentences having grammatical mistakes and observed that the system was able to rectify all the errors.

Keywords Punjabi grammar checker · Complex sentences · Participial type

1 Introduction to Punjabi Language

Punjabi language is the member of Indo-Aryan language and it is one of the widely spoken languages from the family of Indo-Aryan languages. There are more than 100 million speakers of Punjabi in India and Pakistan. Out of these 100 million, 30 million are from India and 70 million are from Pakistan. Other than India and Pakistan, Punjabi is spoken in Canada, United Kingdom, and in some parts of

S. K. Sharma (✉)
Department of Computer Science and Applications, DAV University, Jalandhar, India
e-mail: Sanju3916@rediffmail.com

M. Mittal
SBS State Technical Campus, Ferozepur, India
e-mail: Mittalmisha14@gmail.com

A. K. Luhach et al. (eds.), *Smart Computational Strategies: Theoretical and Practical Aspects*, https://doi.org/10.1007/978-981-13-6295-8_10

United States of America by the immigrants from India and Pakistan. It is the official language of Punjab state of India and a lot of efforts are being done for the technical development of Punjabi language [1].

2 Introduction to Complex Sentences

Complex sentences are composed of different types of clauses which are categorized as dependent clause and independent clauses [2]. A complex sentence contains at least one dependent clause with an independent clause. Both these clauses are joined by using subordinate conjunctions. Consider the following example:

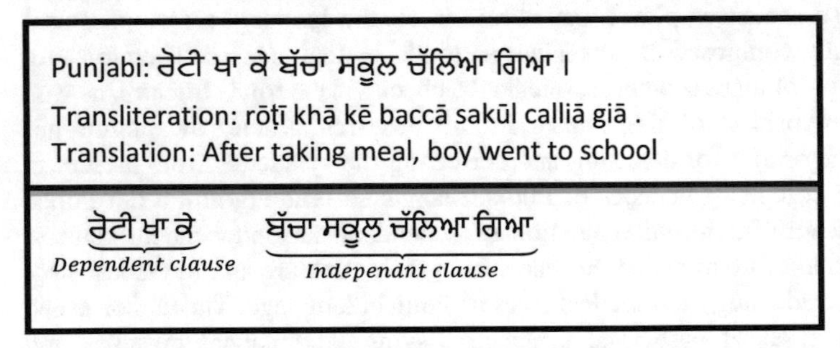

As marked in the above sentence, there are two clauses. One is dependent clause, i.e., ਰੋਟੀ ਖਾ ਕੇ (rōṭī khā kē) and other is independent clause, i.e., ਬੱਚਾ ਸਕੂਲ ਚੱਲਿਆ ਗਿਆ (baccā sakūl calliā giā). Further subdivision of complex sentences has been provided in Fig. 1.

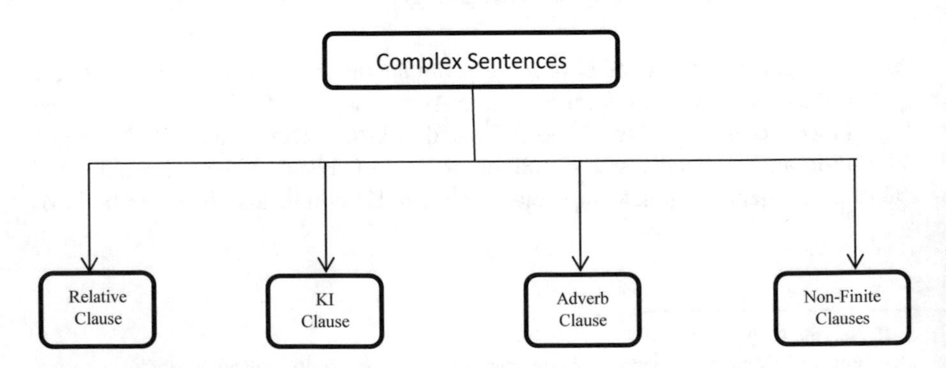

Fig. 1 Classification of complex sentences

2.1 Introduction to Nonfinite Clause

Nonfinite clause is a dependent clause and is called nonfinite due to the presence of nonfinite verb in the Punjabi sentences [2]. These nonfinite verbs pressurize the clause to act as dependent clause. Phrases constructed by such verbs are also called subordinate verbal phrase. Nonfinite verbs are identified by the presence of one of the suffixes, i.e., DIA, IAA, KE, NE, NON, and N. On the basis of these suffixes, the predicate bound clauses can be further subdivided in to three types: participial (having nonfinite verb with DIA and IAA as suffix with or without interjection), infinitival (having NE, NON, or N as suffix with or without proceeded by negative particle) and conjunctival (having KE proceeded by root form of the verb).

3 Participle Sentences

These sentences contain dependent and independent clauses. The dependent clause has a subject and an incomplete action or condition performed on this subject in the form of nonfinite verb and the independent clause explains the consequence of the action performed on subject in the dependent clause [2, 3]. Since the dependent clause has only subject and nonfinite verb, and nonfinite verb does not inflect for any grammatical category, there is no need to check the grammar of the dependent clause. The independent clause, on the other hand, contains a subject and a predicate and hence, the following agreement must be checked:

- Modifier and noun agreement
- Noun and adjective agreement
- Subject and verb agreement
- Order of modifier of noun in noun phrase
- Order of words in a verb phrase.

Participial type of sentence is further categorized into the following two categories.

3.1 Sentence Having Separate Subject of Dependent and Independent Clauses

In such sentences, both dependent and independent clauses have their own subject. Consider the following example:

ਰੱਸੀ ਟੱਪਦਿਆਂ ਮੁੰਡਾ ਡਿਗ ਪਿਆ ।

(rassī ṭappdiāṃ muṇḍā ḍig piā)

ਰੱਸੀ ਟੱਪਦਿਆਂ — *Dependent clause*

ਮੁੰਡਾ ਡਿਗ ਪਿਆ — *Independent clause*

In the above sentence, ਰੱਸੀ (rassī) is the subject of dependent clause (ਰੱਸੀ ਟੱਪਦਿਆਂ (rassī ṭappdiāṃ)) and ਮੁੰਡਾ (muṇḍā) is the subject of independent clause (ਮੁੰਡਾ ਡਿਗ ਪਿਆ (muṇḍā ḍig piā)). In this type of sentence, all the components of independent clause must grammatically agree to one another. Therefore, all the grammatical agreements in the independent clause must be checked with respect to its own subject and not with the subject of the dependent clause.

3.2 Sentence Having Common Subject Between Dependent and Independent Clauses

The sentences falling in this category share a common subject between dependent and independent clauses. Consider the following example:

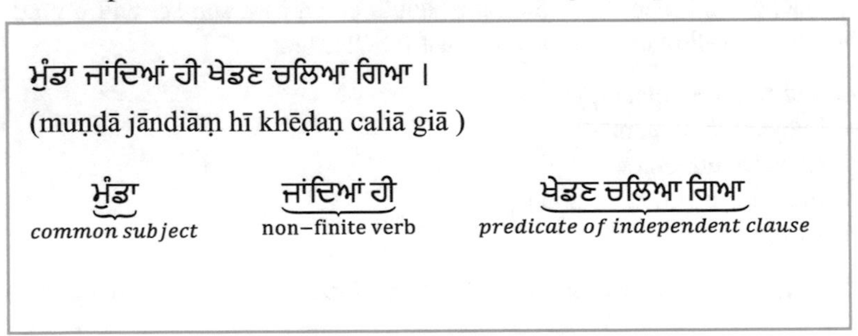

As shown in the above example, ਮੁੰਡਾ (muṇḍā) is the common subject shared by both clauses one is dependent clause, i.e., ਜਾਂਦਿਆਂ ਹੀ (jāndiāṃ hī) and second is independent clause, i.e., ਖੇਡਣ ਚਲਿਆ ਗਿਆ (khēḍaṇ caliā giā). Therefore, dependent and independent clauses generated from the above complex sentence are:

Punjabi: ਮੁੰਡਾ ਜਾਂਦਿਆਂ ਹੀ

Transliterated: (muṇḍā jāndiāṃ hī)

Independent clause:

Punjabi: ਮੁੰਡਾ ਖੇਡਣ ਚਲਿਆ ਗਿਆ

Transliterated: (muṇḍā jāndiāṃ hī khēḍaṇ caliā giā)

Therefore, the predicate of independent clause, i.e., ਖੇਡਣ ਚਲਿਆ ਗਿਆ (khēḍaṇ caliā giā) must be grammatically in agreement with this common subject, i.e., ਮੁੰਡਾ (muṇḍā).

4 Automatic Grammar Checking Techniques

There are many researchers who worked on the development of a syntactic analyzing system. There are mainly three approaches used by different researchers. These three approaches include syntax-based approach, rule-based approach, and statistical-based approach.

4.1 Syntax-Based Approach

This approach was used by Bernth [4] for development of EasyEnglish analyzer to check discourse and document level errors, Hein [5] used this approach for development of ScarCheck (a Swedish Syntactic analyzer), Ravin [6] developed a Text-Critiquing System to check grammar error and style weakness, Young-soog [7] used this approach for improvement in Korean proofreading system, Martin et al. [8] for development of Brazilian Portuguese Syntactic analyzer, Bondi et.al. [9] for deviant language, Arppe for Swedish [10], Bustamante [11] for GramCheck and Kabir et al. [12] for development of Urdu Syntactic analyzer.

4.2 Statistics Based Syntactic Analysis

This approach has been used by Alam et al. [13] in the development of Syntactic analyzer for Bangla and English language, Carlberger et al. [14] for Granska (a

Swedish Syntactic analyzer), Ehsan and Faili [15] for Persian language, Temesgen and Assabie [16] for Amharic language, Paggio [17] in SCARRIE project, Izumi [18] for spoken English, Helfrich [19] for multiple languages, Hermet [20] used web as linguistic resource, Kinoshita et al. [21] for CoGroo (a Brazilian Portuguese Syntactic analyzer) and Henrich and Reuter [22] proposed a Language Independent Statistical Grammar (LISG) checking system.

4.3 Rule-Based Syntactic Analysis

This technique was used by Schmidt-wigger [23] for developing a Syntactic analyzer for German language. Kann [14] used handwritten error rules in CrossCheck (a Swedish Syntactic analyzer), Naber [24] used purely rule-based approach for development of English Syntactic analyzer, Rider [25] used hand constructed error rules in English Syntactic analyzer, Faili [15] used this technique for development of Persian Syntactic analyzer, Tesfaye [26] used this rule-based technique for Afan Oromo Syntactic analyzer, Jiang et al. [27, 28] developed a rule-based Chinese Syntactic analyzer, Kasbon et al. [29] used rules for development of Syntactic analyzer of Malay language, Singh and Lehal [30] used rules for development of Punjabi Syntactic analyzer; Bal and Shrestha [31] used this technique to develop Nepali Syntactic analyzer.

5 Rules to Share the Common Subject

5.1 Predicate Having Its Own Subject

If the predicate of independent clause has a subject, i.e., it starts with a noun phrase, then this sentence does not have a common subject, otherwise, the sentence will share the subject of dependent clause with independent clause. Consider the following example:

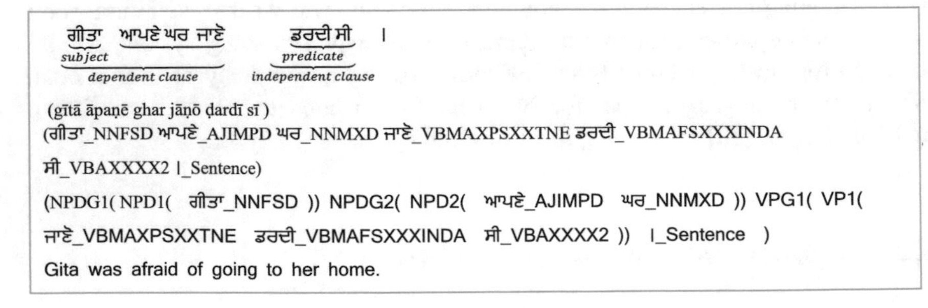

As shown in the above example, the predicate of independent clause does not contain a noun phrase and hence, the subject ਗੀਤਾ (gītā) of dependent clause will be used as common subject.

5.2 *Dependent Clause Having Consecutive Noun Phrases*

If dependent clause contains two consecutive noun phrases then the first noun phrase will act as the subject of independent clause. Consider the following example:

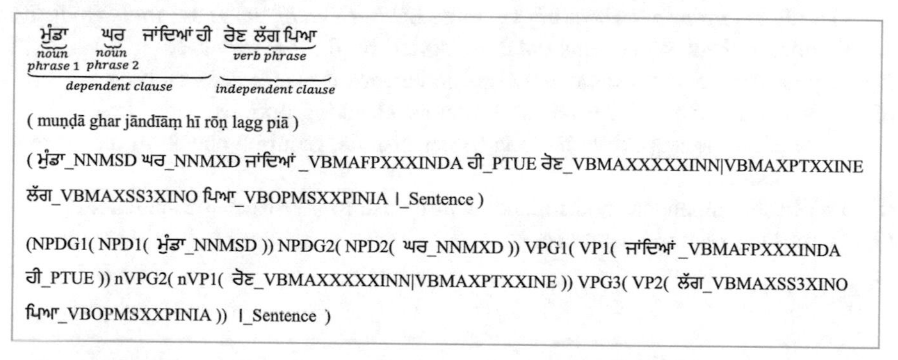

In the above sentence, dependent clause contains two noun phrases; ਮੁੰਡਾ (muṇḍā) and ਘਰ (ghar). Therefore, first noun phrase, i.e., ਮੁੰਡਾ (muṇḍā) will be the common subject.

6 Algorithm Used

In this research, the author used divide-and-conquer type approach in which long participial-type complex sentence is split in to its constituents, i.e., dependent and independent clauses. After split, these components are checked for grammatical mistake. The author used rule-based approach for syntactic analysis of dependent and independent clauses. Various approaches have been used by different authors for converting the complex sentence into simple sentence. These approaches includes approach used by Daelemans for Dutch and English [32], A multilingual method by Puscasu [33], CRF by Lakshami for Malayalam [34], and Van for English [35], machine translation for sentence simplification by Wubben [36] and Lucia [37], etc. In this research, the author used a combination of rule-based approach (used by Poornima et. al. [38] and Leffa [39]) and morphological features

[40, 41] of the language are used to split the sentence in dependent and independent clauses. The algorithm used for syntactic analysis of participial-type complex sentence is as follows:

Databases used: Grammar checking rules.
Input: Tokenized and annotated Punjabi sentence
Output: Updated Tokenized and annotated Punjabi sentence

(1) Using the information of clause boundary, separate the dependent and independent clauses and store them in separate tables.
(2) If independent clause contains a subject then go to step 4 else go to step 3.
(3) Insert the subject of dependent clause in independent clause.
(4) Get all the grammar checking options, having *OnOff* value set to 1, from the Grammar checking options database sorted by the *Priority* field.
(5) Repeat steps 6–8 for dependent and independent clause separately.
(6) Repeat steps 7 and 8 for all the grammar checking options.
    7) Call the respective method to perform the required check on the current clause.
(8) Rejoin the dependent and independent clause to construct the sentence again.
(9) Output the rectified sentence.

Flow Chart

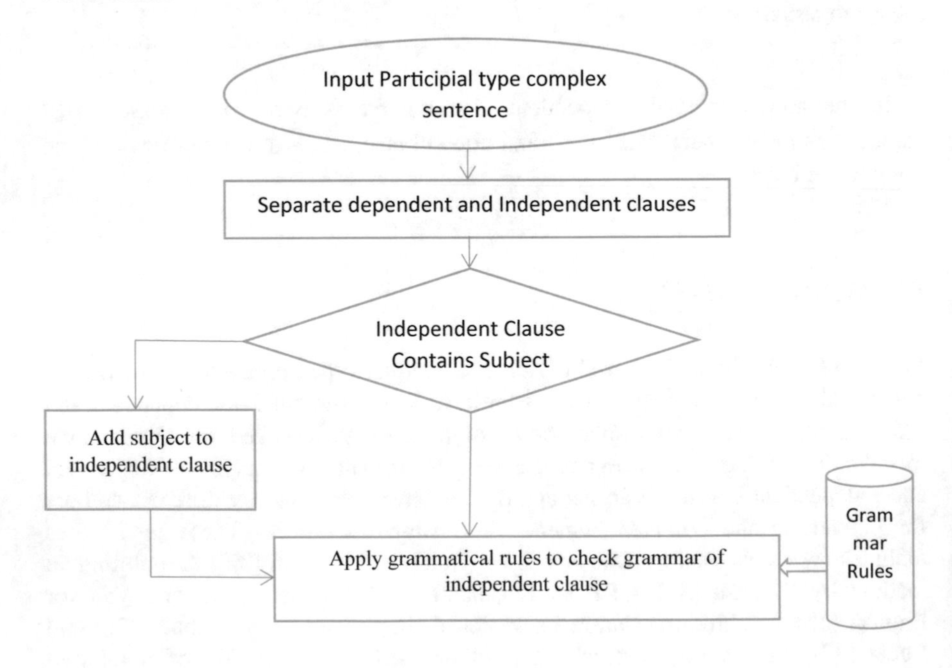

ਮੁੰਡਾ ਘਰ ਜਾਂਦਿਆਂ ਹੀ ਖੇਡਣ ਚਲੇ ਗਏ ।
dependent clause *independent clause*

(muṇḍā ghar jāndiāṃ hī khēḍaṇ calē gaē .)

Complete architecture of syntactic analysis of the above sentence is shown in Fig. 2.

As shown in Fig. 2, first, the sentence is split down into two parts at the nonfinite verb; one part will contain the common subject of independent clause along with

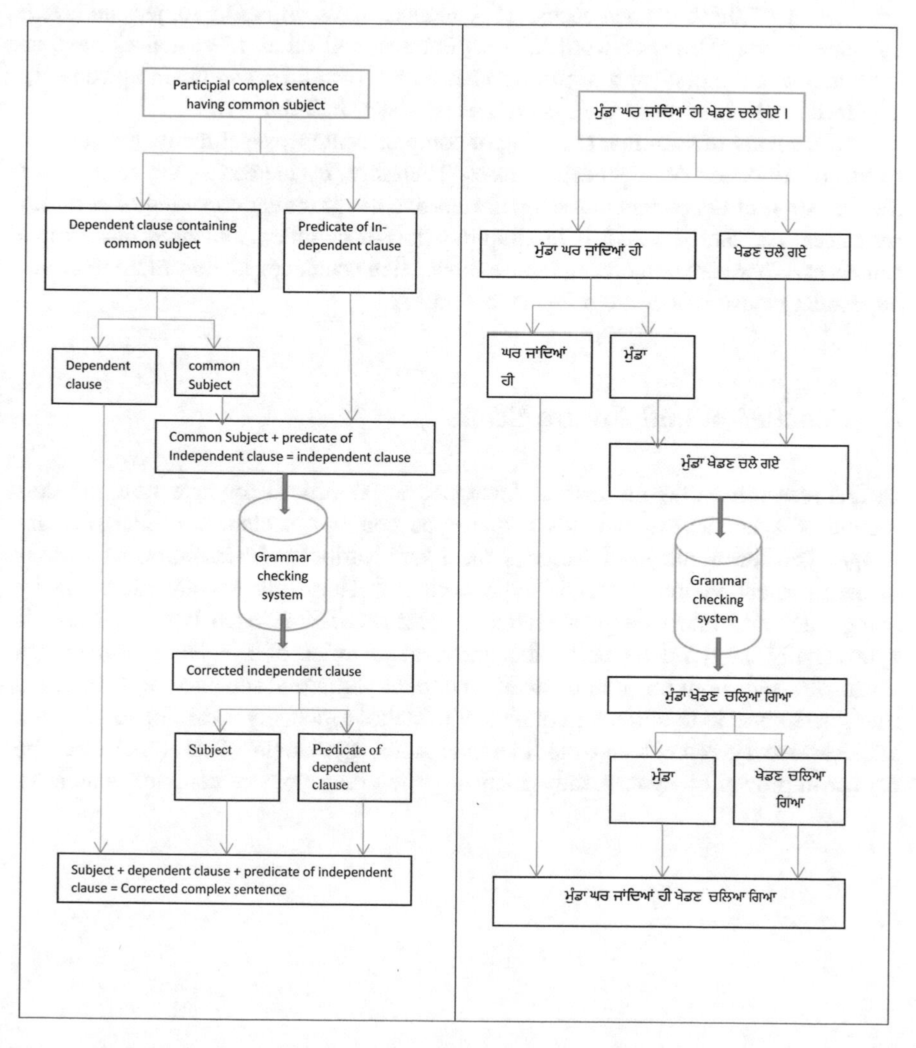

Fig. 2 Architecture of grammar checker for participial-type system

the dependent clause and the second part contains the predicate of independent clause. In the next step, the common subject is copied from the first part and joined to the predicate of the independent clause in the second part. This results in generation of independent clause. This independent clause is then checked for different types of grammatical errors.

7 Result and Discussion

To test the system author selected 30 Participial-type Complex Sentence having errors. Out of these 30 sentences, 15 sentences have common subject and verb agreement error (Order of word in Verb phrase), and other 15 sentences have an error in use of the first form of the verb before KE (Modifier and Noun agreement). The results obtained from the system are tabulated in Table 1.

The accuracy of grammar checking of complex sentences is directly proportional to the identification of dependent clauses. Therefore, by increasing the accuracy of identification of dependent clauses will improve the grammar checking of complex sentences. Another problem is in simplification of complex sentences into simple sentences. There are some complex sentences like sentences having more than one dependent clause which are difficult to simplify.

8 Conclusion and Future Scope

In this research work, the author developed an algorithm for detection and correction of syntactic errors in Participial-type complex sentences of Punjabi language. The technique used includes the identification and separation of various clauses present in the Participial-type sentence. These clauses are identified by using rules and morphological features. After separation, each type of clause is syntactically analyzed by using the grammatical rules of Punjabi language. On testing the algorithm for grammatically incorrect participial sentence, it is observed that system is able to correct grammatical mistakes especially error due to common subject–verb agreement and modifier and noun agreement. This system can be further improved by syntactically analyzing the other types of complex sentences.

Table 1 Results obtained from syntactic analysis of participial-type complex sentences

Sentence type	Error type	Number of errors in input sentences (*A*)	No. of errors corrected by the system (*B*)	No. of errors not corrected by the system (*C*)	Precision $\frac{B+C}{A} \times 100$	Recall $\frac{B}{A} \times 100$	*F* score $\frac{\text{Precision} \times \text{Recall}}{\text{precision} + \text{recall}} \times 2$
Complex sentences containing nonfinite clause	Common subject and verb agreement Error (Order of word in Verb phrase)	15	15	0	100	100	100
	Use of first form of verb before KE. (Modifier and Noun agreement)	15	15	0	100	100	100

References

1. Lehal, G.S.: A survey of the state of the art in Punjabi language processing. Language in India, www.languageinindia.com, “Strength for Today and Bright Hope for Tomorrow”, vol. 9 (10 Oct 2009). ISSN 1930-2940. Print
2. ਡਾ.ਬਲਦੇਵ ਸਿਘ ਚੀਮਾ: ਪੰਜਾਬੀ ਵਾਕ ਪ੍ਰਬੰਧ (ਥਣਤਰ ਅਤੇਕਾਰਜ), ਪਬਲੀਕੇਸ਼ਨ ਬਊਿਰੋਪੰਜਾਬੀ ਯੂਨੀਵਰਸਟਿੀ ਪਟਆਿਲ਼ਾ | (2005)
3. ਬੂਟਾ ਸਿਘ ਬਰਾਰ: ਪੰਜਾਬੀ ਵਆਿਕਰਨ (ਸਧਿਾਂਤ ਅ ਤੇ ਵਹਿਾਰ), ਪਬਲੀਕੇਸ਼ਨ ਬਊਿਰੋ ਪੰਜਾਬੀ ਯੂਨੀਵਰਸਟਿੀ ਪਟਆਿਲ਼ਾ | (2008)
4. Bernth, A.: EasyEnglish: a tool for improving document quality. In: Proceedings of the Fifth Conference on Applied Natural Language Processing. Association for Computational Linguistics, pp. 159–165 (1997)
5. Hein, A.S.: A chart-based framework for grammar checking initial studies. In: Proceedings of 11th Nordic Conference in Computational Linguistic, pp. 68–80 (1998)
6. Ravin, Y.: Grammar errors and style weaknesses in a text-critiquing system. Natural Language Processing: The PLNLP Approach, pp. 65–76. Springer US, Boston (1993)
7. Young-Soog, C.: Improvement of Korean proofreading system using corpus and collocation rules. Language, pp. 328–333 (1998)
8. Martins, R.T., Hasegawa, R., Montilha, G., De Oliveira, O.N.: Linguistic issues in the development of ReGra: a grammar checker for Brazilian Portuguese. Nat. Lang. Eng. **4**(04), 287–307 (1998)
9. Bondi, J., Johannessen, K.Hagen, Lane, P.: The Performance of a Grammar Checker with Deviant Language Input. Mounton de Gruyter, Berlin and New York (2000)
10. Arppe, A.: Developing a grammar checker for Swedish. In: The 12th Nordic Conference of Computational Linguistics, pp. 13–27 (2000)
11. Bustamante, F.R., León, F.S.: GramCheck: a grammar and style checker. In: Proceedings of the 16th Conference on Computational Linguistics-Volume 1. Association for Computational Linguistics, pp. 175–181 (1996)
12. Kabir, H., et al.: Two pass parsing implementation for an Urdu grammar checker. Proceedings of IEEE international multi topic conference. (2002)
13. Alam, M., Naushad UzZaman, J., Khan, M.: 2006. N-gram based statistical grammar checker for Bangla and English. In: Proceedings of Ninth International Conference on Computer and Information Technology (ICCIT 2006)
14. Carlberger, J., Domeij, R., Kann, V., Knutsson, O.: A Swedish grammar checker. Submitted to Comp. *Linguistics, oktober* (2002)
15. Ehsan, N., Faili, H.: Towards grammar checker development for Persian language. IEEE International Conference on Natural Language Processing and Knowledge Engineering (NLP-KE), pp. 1–8 (2010)
16. Temesgen, A., Assabie, Y.: Development of Amharic grammar checker using morphological features of words and N-gram based probabilistic methods. In: IWPT-2013, p. 106 (2013)
17. Paggio, P.: Syntactic analysis and error correction in the SCARRIE project. In: Proceedings der Nordiska datalingvistdagarna (NoDaLiDa’99), pp. 152–161 (1999)
18. Izumi, E., Uchimoto, K., Saiga, T., Supnithi, T., Isahara, H.: Automatic error detection in the Japanese learners’ English spoken data. In: Proceedings of the 41st Annual Meeting on Association for Computational Linguistics-Volume 2. Association for Computational Linguistics, pp. 145–148 (2003)
19. Helfrich, A., Music, B.: Design and evaluation of grammar checkers in multiple languages. In: Proceedings of the 18th Conference on Computational Linguistics-Volume 2. Association for Computational Linguistics, pp. 1036–1040 (2000)
20. Hermet, M., Désilets, A., Szpakowicz, S.: Using the web as a linguistic resource to automatically correct lexico-syntactic errors. In: Proceedings of the LREC’08. Marrakech, Morroco (2008)

21. Kinoshita, J., Salvador, L.N., Menezes, C.E.D.: CoGrOO: a Brazilian-Portuguese grammar checker based on the CETENFOLHA corpus. In: The Fifth International Conference on Language Resources and Evaluation, LREC (2006)
22. Henrich, V.: LISGrammarChecker: Language Independent Statistical Grammar Checking (Doctoral dissertation, Reykjavík University) (2009)
23. Schmidt-Wigger, A.: Grammar and style checking for German. Proceedings of CLAW. Vol **98**, (1998)
24. Naber, D.: A rule-based style and grammar checker. Thesis, Technical Faculty, University of Bielefeld, Germany (2003)
25. Rider, Z.: Grammar checking using POS tagging and rules matching. In: Class of 2005 Senior Conference on Natural Language Processing (2005)
26. Tesfaye, D.: A rule-based Afan Oromo Grammar Checker. IJACSA Editorial (2011)
27. Yin, D., Jiang, P., Ren, F., Kuroiwa, S.: Chinese complex long sentences processing method for Chinese-Japanese machine translation. IEEE, New York (2007). Print
28. Jiang, Y., Wang, T., Lin, T., Wang, F., Cheng, W., Liu, X., Zhang, W.: A rule based Chinese spelling and grammar detection system utility. In: IEEE International Conference on System Science and Engineering (ICSSE), pp. 437–440 (2012)
29. Kasbon, R., Amran, N., Mazlan, E., Mahamad, S.: Malay language sentence checker. World Appl. Sci. J. (Special Issue on Computer Applications and Knowledge Management) **12**, 19–25 (2011)
30. Gill, M.S., Lehal, G.S.: A grammar checking system for Punjabi. In: 22nd International Conference on Computational Linguistics: Demonstration Papers. Association for Computational Linguistics, pp. 149–152 (2008)
31. Bal, B.K., Shrestha, P.: Architectural and system design of the Nepali grammar checker. PAN Localization Working Paper (2007)
32. Daelemans, W., Hothker, A., Sang, E.T.K.: Automatic sentence simplification for subtitling in Dutch and English. In: Proceedings of the 4th Conference on Language Resources and Evaluation, Lisbon, Portugal, 1045–1048 (2004). Print
33. Puscasu, G.: A multilingual method for clause splitting. In: Proceedings of the 7th Annual Colloquium for the UK Special Interest Group for Computational Linguistics (CLUK 2004), Birmingham, UK. Print
34. Lakshmi, S., Ram Sundar Vijay, R., Sobha, D.L.: Clause boundary identification for malayalam using CRF. In: Proceedings of the Workshop on Machine Translation and Parsing in Indian Languages (MTPIL-2012), pp. 83–92, COLING 2012, Mumbai, Dec 2012. Print
35. Van Nguyen, V., Le Nguyen, M., Shimazu, A.: Using conditional random fields for clause splitting. In: Proceedings of the Pacific Association for Computational Linguistics, University of Melbourne Australia (2007)
36. Wubben, S., Van Den Bosch, A., Krahmer, E.: Sentence simplification by monolingual machine translation. In: Proceedings of the 50th Annual Meeting of the Association for Computational Linguistics: Long Papers-Volume 1. Association for Computational Linguistics, pp. 1015–1024 (2012)
37. Lucia, S.: Translating from Complex to Simplified Sentences. Lecture Notes in Computer Science, vol. 6001, pp. 30–39. Springer, Berlin (2010). Print
38. Poornima, C., Dhanalakshmi, V., Kumar Anand, M., Sonam, P.K.: Rule based sentence simplification for English to Tamil machine translation system. Int. J. Comput. Appl. (09758887), **25**(8) (July 2011). Print
39. Leffa, V.J.: Clause processing in complex sentences. In: Proceedings of the First International Conference on Language Resources and Evaluation. Granada, Espanha, vol. 2, pp. 937–943 (1998). Print
40. Chandni, Narula R., Sharma S.K.: Identification and separation of simple, compound and complex sentences in Punjabi language. Int. J. Comput. Appl. Inf. Technol. **6**(II) (Aug-Sept 2014). Print
41. Kaur, N., Garg, K., Sharma, S.K.: Identification and separation of complex sentences from Punjabi language. Int. J. Comput. Appl. **69**(13) (May 2013). Print

Perception-Based Classification of Mobile Apps: A Critical Review

Mamta Pandey, Ratnesh Litoriya and Prateek Pandey

Abstract Nowadays, small computing development, especially mobile application development phenomenon is increasing very rapidly. After downloading/installing/using application, the users can share their experience related to rating, gratification, dissatisfaction, etc. with the app on various distribution platforms. Since 2008, these platforms are very popular for the mobile app users. However, their real potential for the development process is not yet well understood. This paper presents a critical analysis of perception-based user reviews of mobile apps. We analysed over 16,000 reviews of different mobile apps available on the Google Play store. This paper also introduces several challenging issues to classify mobile app reviews into different categories such as reviews related to the requirement, user interface, design, testing, trust, maintenance and battery. We investigated the frequency of occurrence and impact of the issues on the success of mobile applications. This analysis necessitates a tailor-made version of the software development model which suits mobile app development.

Keywords Classification · Mobile application · Software development · User reviews · User's requirement

1 Introduction

Since 2008, traditional software development or conventional software development shifted to mobile software development. The growth of mobile devices is increasing rapidly, so mobile apps become extremely popular in different fields

---

M. Pandey · R. Litoriya (✉) · P. Pandey
Jaypee University of Engineering & Technology, Guna 473226, Madhya Pradesh, India
e-mail: ratneshlitoriya@yahoo.com

M. Pandey
e-mail: mamta.pandey07@gmail.com

P. Pandey
e-mail: prateek.pandey@juet.ac.in

A. K. Luhach et al. (eds.), *Smart Computational Strategies: Theoretical and Practical Aspects*, https://doi.org/10.1007/978-981-13-6295-8_11

such as business, education, communication, etc. There are different categories of apps available in various digital distribution channels. The revenue of mobile apps is also increasing in the market. But the development of mobile software arena is not so much mature, and the development engineering process is not well understood. So that different types issues occurs [1]. There are different types of app store such as Google Play, Apple app store, Blackberry store and Windows Phone store where users can search, find and install software applications with a few efforts. The popularity of this app store is increasing day by day. As of June 2016, the user spends their 90% time in apps [2]. Over the last 5 years, millions of mobile app downloads occurred [3]. Handheld devices such as Android phones users can search the app in play store and install them with a few efforts. Mobile apps user provides the reviews as a feedback by giving star ratings and text message also called reviews. Various studies highlight the importance of app reviews in making successful apps [4]. The users are able to express their sentiment or they can tell to the developer if there is some feature or missing feature. Current researches have emphasized the role of user reviews for various stakeholders. Most of the studies related to the app reviews indicate the requirement extraction posted by the users [5]. There are also a bunch of non-informative reviews which contain no conclusive information, some symbolic notations and spam. It is a complex task for the developer and user to filter and identify conclusive reviews. The main contribution of this work is threefold. First, it collects categories of apps, select 10 apps randomly from the app categories, second, find reviews corresponding to apps and cauterization of user reviews according to the classification. Third, calculate the correlation of different types of issues and rating and impact of user reviews on app success. Here, we choose Google Play store for the data collection, because the market of Google play is increasing day by day and it is very popular among the users. This paper has been organized as follows: Sect. 2 provides the research background, Sect. 3 provides empirical study, Sect. 5 provides analysis method, Sect. 6 provides result and discussion, Sect. 7 concludes the work and provides future work.

2 Related Work

Smartphone apps are very popular among peoples. They are stored in different distributed channels such as Google, Apple, Blackberry and Windows app stores. Apple apps are hosted on iTunes, while the main source for others is Android apps. In the entire digital distribution channels, star rating is a valuable way to the overall measuring of mobile apps. Continuous growing market of apps shows that app developers have to confirm that the user applauds of their app affirmation. From a research attitude, analysing the quality of mobile apps is also important. However, previous research in app quality has targeted on the developers' aspect of characteristics. There are thousands of mobile app developers and millions of apps [6]. App stores are a good place where various apps available such as Google Play store,

Apple app store, Blackberry app store, Amazon, etc. In this paper, we used Google Play store because of huge amount of apps available in the Google Play store [7]. Mudambi et al., pointed out that user reviews and ratings play an important role for taking the decision regarding installing an app or purchasing online goods [8]. Herman et al. mined information from 30,000 Blackberry apps and found that there is a strong correlation between an app's star rating and its download [9]. Vasa et al., analysed and found priority of user reviews [10]. Mobile app market took a new form of software repository and dissimilar from traditional software. Kim et al., observed testing procedures to amend the quality of mobile apps [11, 12]. Agrawal et al., studied how to better recognize the unexpected app performance. Our research focus on: (1) Types of issue mostly occurs in mobile apps, (2) reason of issues, (3) impact of the overall rating of apps. Herman et al. analysed that quantity of downloading and rating of apps are strongly correlated. Kim et al., found that rating is also a determinant point in the user's point of view for installing or downloading of apps [13]. Pandey et al., identified fifteen critical issues in mobile apps and applied Fuzzy-DEMATEL (Decision Making Trial and Evaluation Laboratory) method to analyze issues in apps and divide these issues into cause and effect groups [14]. Although the nature of data studied in the above works is similar to ours, the techniques used and research perspective is different.

3 Empirical Study

The section of the experimental study presents study design, analysis of mobile app reviews research approach. The objective of this study is to inspect the impact of user reviews-related factors on mobile app success. These factors are related to different type of issues occur in the apps. Here, we focused on the Android platform because it is very popular among users.

We crawled 310 apps of free version from Google Play store and selected these apps because we could find these apps from Google Play store, and the corresponding data such as number of ratings, reviews, category of apps and number of apps in each category, etc. We consider these factors because they are external factor. These factors reflect the quality of apps. Table 1 presents the categories and number of apps we considered, respectively. Due to the rapid development of mobile software, the market of mobile software is growing very fast and it is very popular among young researchers. The app developer deploy their apps in the app market such as Google play store, Windows play store, Apple app store, etc [7]. The app user search for the apps installed and uses those apps. After using apps, they share their own experience that is called user reviews. The analysis of user satisfaction is helpful in improving the apps. User reviews have become an important source of information about the experience of users.

We performed broadscale experimental analysis for the explanation of the following research questions.

Table 1 Informative and non-informative reviews

Sr. No.	Category of apps	No. of apps used	Considered review	No. of informative	No. of non inform
1	Android wear	10	5000	3822	1298
2	Books	10	5000	3262	1738
3	Beauty	10	5000	3011	1989
4	Business	10	5000	2842	2158
5	Communication	10	5000	3917	1083
6	Dating	10	5000	2605	2395
7	Education	10	5000	3388	1612
8	Entertainment	10	5000	3926	1074
9	Event	10	5000	2835	2166
10	Finance	10	5000	3847	1153
11	Food & drink	10	5000	3676	1324
12	Health	10	5000	3683	1317
13	House	10	5000	3289	1711
14	Libraries	10	5000	3903	1097
15	Lifestyle	10	5000	2863	2137
16	Maps	10	5000	3751	1249
17	Medical	10	5000	3168	1832
I8	Music	10	5000	3201	1799
19	News	10	5000	3103	1897
20	Parenting	10	5000	3153	1847
21	Personalization	10	5000	3789	1211
22	Photography	10	5000	2317	2683
23	Productivity	10	5000	2618	2382
24	Shopping	10	5000	3707	1293
25	Social	10	5000	1868	3132
26	Sport	10	5000	3294	1706
27	Tools	10	5000	3796	1204
28	Travel	10	5000	1886	3114
19	Video players	10	5000	2294	2706
30	Weathers	10	5000	3015	1985
31	Comics	10	5000	2846	2154

4 Explanatory Analysis

Mobile apps reviews have explored first, the quantity of app reviews having conclusive information. The amount of informative reviews again classified in terms of different kinds of issues such as requirement-related, user interface-related, testing-related, trust-related, maintenance-related and energy consumption-related issues. These types of issues are again classified on the bases of keywords. All classified issues related to the apps have been included in Table 1 (number of informative reviews).

RQ 1: How can we classify reviews of mobile applications?

Our process to answer RQ1 is divided into several sections: Data collection, our approach and results.

4.1 Data Collection

In this review paper, 310 apps have been taken for study and the selection of apps has been done on random bases. All kinds of ratings (1–5 stars) are included in this study. We have collected around 500 reviews for each selected app, thus in total 155,000 reviews are considered.

4.2 Approach and Result

Find all details of apps such as a number of downloads, user rating and number of reviews, etc. thereafter, classified reviews into two categories: informative and non-informative reviews. Some comments provide conclusive information, those comments are called as informative reviews, similarly some comments do not provide meaningful information, those comments are called as non-informative reviews. Out of 155,000 reviews, there are 98,674 informative reviews and 56,446 non-informative reviews. The informative and non-informative reviews for mobile apps used in our research are listed in Table 1.

RQ 2: How many types of issues generally occurred in mobile apps?

4.3 Classification of User Reviews

We classify app reviews on the following bases.

The classification of mobile app reviews is illustrated in Fig. 1. Informative reviews is again classified in six catteries such as requirement issues, user interface issues, testing issues, trust issues, maintenance issues and battery issues.

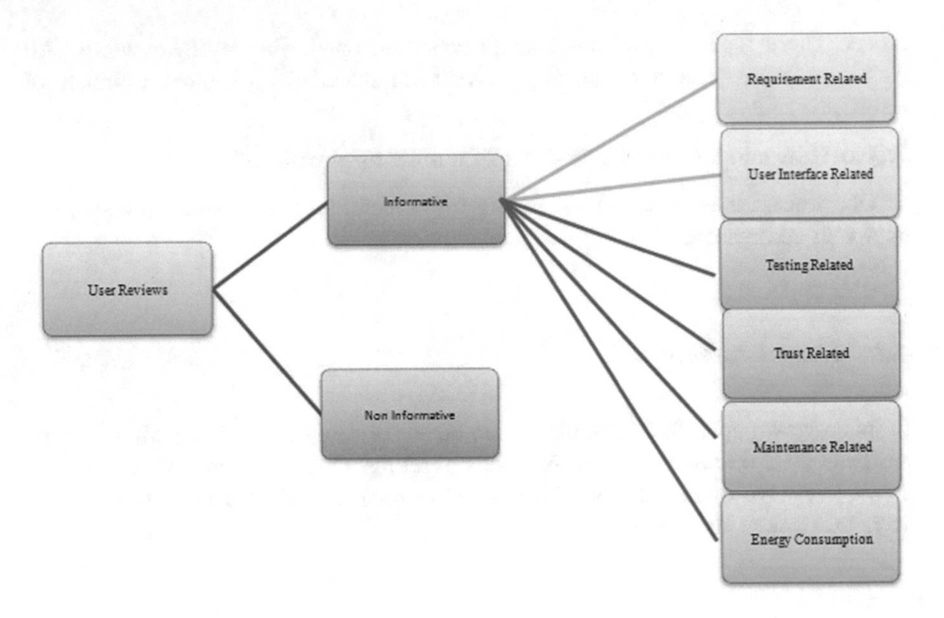

Fig. 1 Classification of mobile app reviews

4.3.1 Requirement

Various studies have focused on the requirement engineering for mobile apps. If we talk about traditional software, there are different phases of requirement engineering such as requirement elicitation, requirement analysis, requirement specification, requirement verification and requirement management [14, 16]. Sentiment analysis and natural language processing (NLP) are used by some researchers to analyse the user reviews [15, 16]. Rein and Munch [17] prioritizes to feature with the help of a case study. Finkelstein et al. [18] observed the set of features from the release notes present in different variety of app stores for various apps. Sarro et al. [19] recognized feature migration life cycles among apps. Different literatures have analysed app reviews and tried to find out, which kind of issues generally the users have to face [20, 21]. In this section, we identified such reviews those indicate to requirement terms such as missing features and add features. The conclusion of the article was the abstract requirement of mobile apps generated by user reviews. There are 58,419 requirement-related issues listed in Table 2.

4.3.2 User Interface

Based on some past work, there has been quite a bit of the previous study in the area of user interface, requirements, testing, trust, maintenance and battery for mobile

Table 2 Descriptive statics of the analysed apps

Issues	Min	Median	Mean	Max
Requirement related	1139	1973	1884.48	2524
testing related	1139	1973	1899.26	2524
User interface related	116	321	353.33	911
Trust related	38	286	349.074	694
Maintenance related	20	34	53.11	175
Energy consumption related	0	0	202	1889

apps. There have been a few works that have been done in the area of user interface for mobile application. One of the most important issues during the development of mobile application is interface design. The app users usually are affected by destitute interface design that actively interrupts the usability of those apps. Due to the unique feature of mobile devices such as low resolution, small screen size and insufficient data entry method such kind of challenge occurs. Lee and Neamtiu [22, 23] present that content, context and customizations are the crucial aspects of designing successful interface for mobile software.

4.3.3 Testing

Various analyses have suggested that different techniques have been proposed to improve the testing of mobile application. A considerable number of app development firms are following traditional methods of testing which are not suitable for app testing. Machiry et al. [24] proposed another tool that is Dynodroid, input generated dynamically and tested by this tool. These both tools are used for user interface testing and system events. EvoDroid tool is used for investigating the Android apps. 70–80% result can be achieved by this tool [25]. An approach that leverages user reviews to formulate which Android gadgets an app is suited to was investigated by Khalid et al. [26]. There are 24 apps studied by a researcher and analysed by various types of bugs [27]. There are several challenges faced by developers every time they start working on a new mobile app [28, 29]. Mobile apps have different characteristics than desktop apps so mechanized testing tools are not adequate to certain execution path due to the inability to generate inputs, and it cannot investigate everything further along that execution path. Here, we identify issues related to testing on the bases of app reviews. We classify issue related to testing based on different keywords such as existing feature is not working, cannot download, bugs or errors, etc.

4.3.4 Trust

User trust of mobile application can be defined as a belief of user on the mobile app that could fulfil the expectations of the user. In our study, it is defined as customer or public belief on a mobile app that could fulfil task according to various expectations of the user. It is very clear that trust plays a very important role in the use of apps because the trust behaviour of user solely depends on the apps could perform as expectation of the users, such as demand of personal information by various mobile apps (e.g. financial apps, some shopping apps demand for the account no, personal details) [30, 31]. The trust of user in mobile apps is not an objective, it is highly subjective. Trust is influenced by many factors; it is hard to measure directly. In this paper, during data collection, we find some terms those indicated to poor trust on mobile apps such as "balance from my account has been debited but not credited yet", such kind of reviews mostly occurred in financial category apps. We find two types of terms: (a) amount deducted but not credited, (b) this app is cheater/bluff into.

4.3.5 Maintenance

The maintenance of mobile apps is one of the most unexplored areas in the field of software engineering rather than desktop application. Issues in mobile apps have not been focussed in previous software maintenance studies. For example, most of the mobile apps display warning, and has been a frame up in the previous studies, these warnings require an expressive amount of maintenance [32], said that number of previous studies discovered the maintenance of mobile apps from different point of views, e.g. reusability, notification and advertise-related maintenance. The code of various types of mobile apps is reused and correlated by Mojica et al. [33]. They concluded that 23% of the classes are inherited by the base of Android API and 27% of the classes inherit by domain-specific base class. The comparison of mobile apps and desktop apps bug fixing is analysed by Hindle [34]. The response time of mobile apps is earlier than other apps. Software analytics platform for mobile apps SAMOA is introduced by Minelli and Lanza, and they analysed 20 apps. They summarized that mobile apps are smaller than desktop apps in size.

4.3.6 Energy Consumption

Different studies have indicated that energy consumption is a big issue for mobile gadgets; enough amount of analysis has suggested practices to measure and reduce the energy consumption of mobile apps. Hindle et al. [35, 36] presented one of the first works related to the measurement of energy consumption of mobile gadgets. Another analysis has presented a system that perform programme analysis to provide per instruction energy modelling and energy consumption for apps [39]. A tool Green Droid presented by which battery consumption bug automatically identified

Table 3 Correlations among different app issues

Sr. No	App issues	Correlation
1	All issues and rating	−0.32
2	Requirement and rating	−0.159
3	User interface and rating	−0.601
4	Trust and rating	−0.340
5	Maintenance and rating	−0.062
6	Energy consumption and rating	−0.12

[37]. Other studies performed experimental research on energy consumption in order to provide app developers with ways of minimizing it. Classification of bugs related to energy consumption based on 39,000 app user reviews have been presented by Cohen [38]. To better understanding of energy consumption of apps, Mockus analysed 405 mobile apps [39]. They have different varieties of findings such as (a) idle state mobile apps spent more energy or consume more energy, (b) Heavy resource consumed by networking component. In our study, during the data collection, we found some terms related to power or energy consumption: (a) eat battery, (b) consume power, (c) battery down.

RQ 3: How diverse issues effects upon user rating of mobile app?

Mobile app issues own a significant role in high and low of app rating. In this paper, we analysed different types of app issues, and description of various types of issues have been mentioned in Table 2. This issue affects the overall rating of app. Here, we calculate the correlation of user rating and different types of app issues have been presented in Table 4.

Correlations of app issues with rating have been shown in Table 3. Correlation coefficients are always values between −1 and 1, where −1 shows a perfect linear negative correlation, 1 shows a perfect linear positive correlation. A negative correlation coefficient means that for any two variables, an increase in one is associated with a decrease in another variable.

5 Analysis Method

5.1 *Dependent and Independent Variables*

We grouped our issues into five types: requirement, UI, testing, trust, maintenance and energy consumption-related issues. The independent variable is represented by the success of the considered apps (if the value of independent variable will increase, success rate of app will also increase). Here, we consider it as a dependent variable to “user rating” of the app.

5.2 Model Construction

Before applying Multi linear Regression Model (MLR), we have justified that our data respects MLR assumptions, i.e. linear relationship, little multi colinearity and no autocorrelation.

Multiple linear regressions are used to find the relationship between two or multiple explanatory variables and a response variable by fitting a linear equation for observation of data [40]. The form of the model is

$$y = b_0 + b_1x_1 + b_2x_2 + b_3x_3 + \cdots b_nx_n \tag{1}$$

Here, y denotes the dependent variable (also called response variable) whose values we predicted, and $x_1, x_2, x_3 \ldots x_n$ denote the independent variable (also called explanatory variable). In our MLR model, the response variable is user rating and explanatory variables are different types of app issues. Our MLR model is built as in [41].

5.3 Identify Colinearity

Colinearity may be complicated avert the apperception of a flawless set of an independent variable for a statistical model. To identify colinearity here, we used the variance inflation factor (VIF) whose score threshold was set to 5 in our study.

6 Results and Discussion

The results of our study have been shown in Table 4. As it can be noticed, there is a statistically significant relationship between the app's rating and different types of issues. The regression coefficient of these issues is negative demonstrating that app having large number of reviews with a high number of issues tends to be no any specific process followed for app development. Our results do not investigate any success factor of mobile apps. With respect to users perception such as reviews, download and rating, we could detect lacks in mobile apps.

Table 4 Model statics for fitting data

Issues	Adjusted R^2	P-value
Requirement related	−0.035	0.985
User interface related	−0.032	0.757
Testing related	−0.008	0.390
Trust related	−0.035	0.990
Maintenance related	−0.033	0.806
Energy consumption related	−0.027	0.636

7 Conclusions and Future Work

User reviews play a very important role in analysing the quality of mobile apps. After using the apps, the user post their experience whatever they felt about apps. There are different types of problem faced by the users during the use of apps. Here, we focus only such issue that occurs frequently. In this paper, we describe the six types of issues, which mostly occur in mobile apps. This approach can be used to provide a general idea to the developer so that the developer could know about the issues. We observed that there is no suitable approach followed by the developers for mobile app development due to which different types of issues occur in apps. Although there are well-known different types of traditional approach such as waterfall, spiral, etc. available for software development directly, they cannot apply to mobile computing. For future work, we would like to propose a tailor-made version of the mobile app development model.

References

1. Wasserman, A.I.: Software engineering issues for mobile application development mobile revolution. In: Proceedings of the FSE/SDP workshop on Future of Software Engineering Research (2010). https://doi.org/10.1145/1882362.1882443
2. https://yahoodevelopers.tumblr.com/post/127636051988/seven-years-into-the-mobile-revolution-content-is
3. Gartner: Number of mobile app downloads worldwide from 2009 to 2017 (in millions). Technical report, Gartner Inc. (2015)
4. Li, H., Zhang, L., Zhang, L., Shen, J.: A user satisfaction analysis approach for software evolution. In: 2010 IEEE International Conference on Progress in Informatics and Computing (PIC), China (2010). https://doi.org/10.1109/pic.2010.5687999
5. Iacob, C., Harrison, R.: Retrieving and analyzing mobile apps feature requests from online reviews. In: Proceedings of the 10th Working Conference on Mining Software Repositories, San Francisco, CA, USA (2013). https://doi.org/10.1109/msr.2013.6624001
6. Butler, M.: Android: changing the mobile landscape. IEEE Pervasive Comput. (2011). https://doi.org/10.1109/mprv.2011.1
7. Number of apps available in leading app stores, https://www.statista.com/statistics/276623/number-of-apps-available-in-leading-app-stores
8. Mudambi, S.M., Schuff, D.: What makes a helpful online review? A study of customer reviews on amazon.com. J. MIS Q. (2010)
9. Harman, M., Jia, Y., Zhang, Y.: App store mining and analysis: MSR for app stores. In: Proceedings of the 9th Working Conference on Mining Software Repositories, Switzerland (2012). https://doi.org/10.1109/msr.2012.6224306
10. Vasa, R., Hoon, L., Mouzakis, K., Noguchi, A.: A preliminary analysis of mobile app user reviews. In: Proceedings of the 24th Australian Computer-Human Interaction Conference (2012). https://doi.org/10.1145/2414536.2414577
11. Kim, H., Choi, B., W.E. Wong: Performance testing of mobile applications at the unit test level. In: 3rd IEEE International Conference on Secure Software Integration and Reliability Improvement (2009). https://doi.org/10.1109/ssiri.2009.28
12. Jha, A.K.: A risk catalog for mobile applications. Ph.D. dissertation, Florida Institute of Technology (2007)

13. Agarwal, S., Mahajan, R., Zheng, A., Bahl, V.: Diagnosing mobile applications in the wild. In: Proceedings of the Ninth ACM SIGCOMM Workshop on Hot Topics in Networks (2010). https://doi.org/10.1145/1868447.1868469
14. Pandey, M., Litoriya, R., Pandey, P.: Analyzing Diverse Issues in Mobile App Development: A Fuzzy DEMATEL Approach. Int. J. Engg. Technol, **7** (4.36), 868–879 (2018)
15. Guzman, E., Maalej, W.: How do users like this feature? A fine grained sentiment analysis of app reviews. In: IEEE 22nd International Requirements Engineering Conference, Sweden (2014). https://doi.org/10.1109/re.2014.6912257
16. Maalej, W., Nabil, H.: Bug report, feature request, or simply praise? On automatically classifying app reviews. In: 23rd IEEE International Requirements Engineering Conference (2015). https://doi.org/10.1109/re.2015.7320414
17. Rein, A.D., Munch, J.: Feature prioritization based on mock purchase: a mobile case study. In: Proceedings of the Lean Enterprise Software and Systems Conference (2013)
18. Finkelstein, A., Harman, M., Jia, Y., Sarro, F, Zhang, Y.: Mining app stores: extracting technical, business and customer rating information for analysis and prediction. UCL Department of Computer Science (2013)
19. Sarro, F., Subaihin, A.A.A., Harman, M., Jia, Y., Martin, W., Zhang, Y.: Feature lifecycles as they spread, migrate, remain, and die in app stores. In: Proceedings of the 23rd International Requirements Engineering Conference (2015). https://doi.org/10.1109/re.2015.7320410
20. Fu, B., Lin, J., Li, L., Faloutsos, C., Hong, J. Sadeh, N.: Why people hate your app: making sense of user feedback in a mobile app store. In: Proceedings of the 19th ACM SIGKDD International Conference on Knowledge Discovery and Data Mining (2013). https://doi.org/10.1145/2487575.2488202
21. Pagano, D., Maalej. W.: User feedback in the app store: an empirical study. In: 21st IEEE International Requirements Engineering Conference (2013). https://doi.org/10.1109/re.2013.6636712
22. Lee, Y.E., Benbasat, I.: Interface Design for Mobile Commerce, Communications of the ACM (2003). https://doi.org/10.1145/953460.953487
23. Hu, C., Neamtiu, I.: Automating GUI testing for Android applications. In: Proceedings of the 6th International Workshop on Automation of Software Test (2011). https://doi.org/10.1145/1982595.1982612
24. Machiry, A., Tahiliani, R, Naik, M.: Dynodroid: an input generation system for Android apps. In Proceedings of the 20139 th Joint Meeting on Foundations of Software Engineering (2013). https://doi.org/10.1145/2491411.2491450
25. Kim, H., Choi, B., Wong, W.E.: Performance testing of mobile applications at the unit test level. In: IEEE International Conference on Secure Software Integration and Reliability Improvement (2009). https://doi.org/10.1109/ssiri.2009.28
26. Khalid, H., Nagappan, M., Shihab, E., Hassan, A.E.: Prioritizing the devices to test your app on: a case study of Android game apps. In: Proceedings of the 22nd ACM SIGSOFT International Symposium on Foundations of Software Engineering (2014). https://doi.org/10.1145/2635868.2635909
27. Bhattacharya, P., Ulanova, L., Neamtiu, I., Koduru, S.C.: An empirical analysis of bug reports and bug fixing in open source Android apps. In: Proceedings of the 2013 17th European Conference on Software Maintenance and Reengineering (2013). https://doi.org/10.1109/csmr.2013.23
28. Joorabchi, M.E., Mesbah, A., Kruchten, P.: Real challenges in mobile app development. In: 2013 IEEE International Symposium on Empirical Software Engineering and Measurement, USA (2013). https://doi.org/10.1109/esem.2013.9
29. Rosen, C, Shihab, E.: What are mobile developers asking about? A large scale study using stack overflow. J. Empir. Softw. Eng. (2016). https://doi.org/10.1007/s10664-015-9379-3
30. McKnight, D.H., Choudhury, V, Kacmar, C.: Developing and validating trust measures for e-commerce: an integrative typology. J. Inf. Syst. Res. (2002). https://doi.org/10.1287/isre.13.3.334.81

31. Ruiz, I., Nagappan, M., Adams, B., Berger, T., Dienst, S., Hassan, A.: On ad library updates in Android apps. IEEE Softw. (2014). https://doi.org/10.1109/ms.2017.265094629
32. Ruiz, I., Nagappan, M., Adams, B., Berger, T., Dienst, S., Hassan, A.: Understanding reuse in the Android market. In IEEE International Conference on Program Comprehension (2012)
33. Mojica Ruiz, I.J., Adams, B., Nagappan, M., Dienst, S., Berger, T., Hassan, A.E.: A large-scale empirical study on software reuse in mobile apps. IEEE Softw. (2014)
34. Hindle, A.: Green mining: a methodology of relating software change to power consumption. In: Proceedings of the 9th IEEE Working Conference on Mining Software Repositories (2012). https://doi.org/10.1109/msr.2012.6224303
35. Hindle, A., Wilson, A., Rasmussen, K., Barlow, E.J., Campbell, J.C., Romansky, S.: GreenMiner: a hardware based mining software repositories software energy consumption framework. In: Proceedings of the 11th Working Conference on Mining Software Repositories (2014). https://doi.org/10.1145/2597073.2597097
36. Hao, S., Li, D., William, Halfond, G.J., Govindan, R.: Estimating mobile application energy consumption using program analysis. In: Proceedings of the 35th International Conference on Software Engineering (2013)
37. Banerjee, A., Chong, L.K., Chattopadhyay, S., Roychoudhury, A.: Detecting energy bugs and hotspots in mobile apps. In: Proceedings of the 22nd ACM SIGSOFT International Symposium on Foundations of Software Engineering (2014). https://doi.org/10.1145/2635868.2635871
38. Cohen, J.: Applied Multiple Regressions—Correlation Analysis for the Behavioural Sciences (2003)
39. Mockus, A.: Organizational volatility and its effects on software defects. In: Proceedings of the eighteenth ACM SIGSOFT International Symposium on Foundations of Software Engineering (2010). https://doi.org/10.1145/1882291.1882311
40. Fox, J.: Applied Regression Analysis and Generalized Models, 2nd edn. SAGE Publications Inc. (2008)
41. Pandey, M., Litoriya, R., Pandey, P.: An ISM Approach for Modeling the Issues and Factors of Mobile App Development. Transf. Int. J. Software Eng. Knowl. Eng. **28** (07), 937–953 (2018)

An Improved Approach in Fingerprint Recognition Algorithm

Meghna B. Patel, Satyen M. Parikh and Ashok R. Patel

Abstract There are several reasons like displacement of finger during scanning, environmental conditions, behavior of user, etc., which causes the reduction in acceptance rate during fingerprint recognition. The result and accuracy of fingerprint recognition depends on the presence of valid minutiae. This paper proposed an algorithm, which identify the valid minutiae and increase the acceptance rate and accuracy level. The work of the proposed algorithm is categorized into two parts: preprocessing and post-processing. The proposed algorithm enhanced most of the phases of preprocessing for removing the noise, and make the clear fingerprint image for feature extraction and enhanced the post-processing phases for eliminating the false extracted minutiae, to extract exact core point detection, and matching valid minutiae. The developed proposed algorithm is tested using FVC2000 and FingerDOS databases for measuring the average FMR = 1% and FNMR = 1.43% and accuracy 98.7% for both databases.

Keywords Orientation estimation · Image enhancement · Thinning · Fingerprint recognition

1 Introduction

To identify an authorized person is the vital task in the world of internet. Password-based and ID-based authentications are traditional approach, which becomes incompetent to suit the high-security requirements applications like ATM

M. B. Patel (✉) · S. M. Parikh
A.M.Patel Institute of Computer Studies, Ganpat University, Gujarat, India
e-mail: meghna.patel@ganpatuniversity.ac.in

S. M. Parikh
e-mail: parikhsatyen@yahoo.com

A. R. Patel
Florida Polytechnic University, Lakeland, FL, USA
e-mail: arp8265@gmail.com

A. K. Luhach et al. (eds.), *Smart Computational Strategies: Theoretical and Practical Aspects*, https://doi.org/10.1007/978-981-13-6295-8_12

machines, cash terminals, access control system, and so on. Biometric-based authentication substitutes the traditional approaches. It verifies and identifies an individual based on his behavioral and physiological biometric data. As per historical data, the fingerprint recognition systems are used by every forensic and low enforcement agency. Nowadays, government programs and services are integrated with biometrics and become a creator and consumer of biometric technology, as well as, it is used in civilian and commercial applications because of cheaper, small capture device, and robust development of fingerprint recognition system [1].

A fingerprint image contains mainly two types of features: ridge flow information and minutiae. In fingerprint image, ridge flow information is defined by the ridges and valleys pattern. The minutiae are mainly referred to as ridge ending and ridge bifurcation, which contains the disconnection in fingerprint impression. There are overall 150 types of minutiae types identified. In the analysis of fingerprints, the ridge flow information and minutiae play a vital role in showing that the two fingers are not the same.

In a full fingerprint, average 70–150 minutiae are there. The numbers of minutiae in a fingerprint image are differing from one finger to another, and problems affected to matching processes and success level of fingerprint recognition are: The same finger images captured during different sessions can be basically different, finger plasticity creates the distortion (cold finger, dry/oily finger, high or low moisture), the overlapping is done among query and template finger and match partial fingerprint because of displacement of finger on scanner, different pressure, and skin conditions like cuts to fingerprint, manual working that would damage or affect fingerprints(construction, gardening), noisy fingerprint image, or poor quality image can create problems for feature extraction because of the missing features. Yet, the displacement of fingerprint and lower quality images make the fingerprint recognition a challenging task [2].

This paper is managed in the following manner. Section 2 contains the literature survey regarding the preprocessing and post-processing phases. The preprocessing phase limitation shown in literature survey is solved by using the best technique or by enhancement in that technique that is done by us is addressed in Sect. 3. The steps involved in the proposed work for post-processing is involved in Sect. 4. Section 5 contains the information of databases and experimental environment. The measurement standards are described in Sect. 6. Section 7 shows the experimental result, and then summary of an algorithm is shown in Sect. 8.

2 Literature Review

The work of the preprocessing is mainly used for image enhancement, and post-processing is used for feature extraction and matching. This is a very useful way to develop reliable fingerprint recognition system. Each paragraph shows the literature review of the phases of preprocessing and post-processing and found the limitation, and address the best suitable technique to work ahead.

There are many image enhancement techniques for fingerprint image, which are proposed in the literature [3]. Some of the methods enhance the image from gray scale and some of based on image binarization. For gray scale image, normalization process is the first step in preprocessing for enhancing the image. Pixel-wise operation is done in normalization [2]. Block-wise normalization is implemented by Kim and Park [4], and adaptive normalization is proposed by Shi and Govindaraju [5]. The most widely used method is used in normalization, which is developed by Hong et al. [6] for removing the noise.

Orientation estimation is the second phase in preprocessing. It shows that at the time of enrollment, the fingerprint is not correctly oriented vertically, and it can be moved away or displaced from the vertical orientation. In literature [7] states that the fingerprints made up of two types of features like minutia or singular points is considered as local features, and the frequency and orientation of ridge pattern is considered as global features. To verify or identify an individual, local features are used while the global features describe local orientation of ridges and valleys using orientation fields. The orientation estimation process plays an important role in fingerprint recognition. The orientation field indicates the direction of ridges. The impact of orientation estimation process is affected by the other succeeding processes like image enhancement, singular point detection, and classification of an image and matching. The fingerprint recognition generates the false result, if an incorrect orientation estimation is done. The gradient-based orientation method is reliable to find out ridge orientation using gradient of gray intensity was proposed in [8, 9]. The spectral estimation methods, waveform projection, and filter-bank-based approach are used for the same purpose. Yet, it is stated that compared to gradient-based methods, these three methods did not give the accurate results, as well as limited number of filters, and several methods include comprehensive comparison process that increase the computational expensive.

Image enhancement is the third phase in preprocessing. Image enhancement is possible on binary image or gray scale images [10]. If it is done on binary image, it will add some spurious information and loses important information, which creates a problem during minutia extraction. The image enhancement is done directly on gray scale image for getting reliable local orientation estimation and ridge frequency. Image enhancement can be implemented on gray scale image in three different ways using Gabor Filter, FFT based, and Histogram Equalization [11]. The Histogram Equalization work is based on pixel orientation at the initial phase. It improves the quality of image but cannot change the structure of ridge [12]. Another method is FFT based, which worked on only frequency while Gabor filter used based on frequency, as well as orientation map. Hong et al. [6] proposed an algorithm for orientation estimation followed by Gabor filter, which gives reliable result even for degraded images. But for online recognition system, it is computationally expensive [13]. In the literature [14] reveals that for fingerprint image enhancement, Gabor filter remains effective. As well as in [12] proposed the anisotropic filter to enhance the gray scale image and prove that it improves the image enhancement process. O'Gorman et al. [15] used an anisotropic smoothening kernel and proposed contextual filters-based fingerprint image enhancement algorithm, which is suitable for fingerprint images.

Binarization (Thresholding) is the third phase in preprocessing. There are mainly two ways of Thresholding algorithm [16]. One is Global Thresholding and another is Adaptive Thresholding. The global thresholding can be used when an entire image of the pixel value and the background remains constant. It is not used in the situation where the background enlightenment is not clear. The variations are come in contrast ratio for the same person's fingerprint at the time of device is pressed. In that situation with the single intensity threshold, the image binarization does not happen with constant image contrast. To overcome this situation, it is found that when compared to conventional thresholding technique adaptive is better [17]. The adaptive thresholding is used for to binarize the image to separate it in ridge part and non-ridge part based on image distribution pixel values.

Thinning is the last phase in preprocessing. Based on the literature review, thinning algorithms either directly works on gray scale image without binarization [18–20] or maximum algorithms required binary image as an input images [21–23]. There are three types of algorithm that works for binary images: (1) sequential algorithms, (2) parallel algorithms, and (3) medial axis transform algorithms [21]. Sequential and parallel algorithms are iterative thinning algorithms while medial axis algorithms are non-iterative thinning algorithms. The non-iterative algorithms are efficient in computation time but are responsible for creating noisy skeleton while in iterative thinning algorithm, the parallel thinning algorithms are frequently discussed because they are fast and efficient. The [24, 25] implemented Zhang-Suen's algorithm and proved that it takes the minimum computational time and gives better result.

Minutiae extraction is the first phase in post-processing. To extract minutiae in thinned binarized image, two ways are used: using crossing number and morphology operators. In thinned binarized image, the crossing number is the most accepted technique [26–28]. It is adapted more compared to others because of intrinsic simplicity and computational efficiency. While the other method based on morphology is used if the image is preprocesses using morphology operator for reducing the effort during post-processing. The morphology operators are used to eliminate false minutiae like bridge, spur, and so on at the time of preprocessing. The after minutiae extraction is done with hit or miss transform [29].

Removal of false point is the third phase in post-processing. Sometimes, due to lack of ink and over inking create a spurious minutia like false ridge break and cross-connection of ridges. That is why it is worthless to believe that the preprocessing phase eliminates all the noise and make the image noise free [30]. Because of that reason, still, fingerprint image is not considered as clear image for matching. This spurious minutia is considered as false minutiae. The performance of the fingerprint identification system will increase after removing the false minutiae.

Core point detection is the second phase in post-processing. Core point detection is used for classification of fingerprint based on the global patterns of ridges. It is useful in matching when finger image is rotated. For that, global pattern core and delta points are used to rotate the image. The Orientation Field, Multi-Resolution, Curvature, and Poincare Index are the existing core point detection algorithms mentioned in [31]. The existing methods does not provide accurate and estimated outcome for noisy images. The literature [32] reveals that some of the methods of orientation estimation are good in performance, where as some failed with fingerprint patterns and not enough to identify true singular point and detect forged core points due to blurs, scars, crashes, etc., as well as because of the post-processing method use the local characteristics of core point. Using the fingerprint's computed direction field, the Poincare Index is widely accepted in the existing algorithm. The limitation of Poincare Index method is giving poor result in detection of false core point for lower quality images [32].

Minutiae matching is the last phase in post-processing. A minutia matching is used to determine that minutiae set of input finger and template finger are from the same finger or not. In matching, it not only compares these two set of minutia but also consider some problems like shifting of fingerprint, rotation of fingerprint, and so on. The pattern-based [33] and minutiae-based approach [34] are the famous approach for fingerprint matching.

The proposed algorithm follows the preprocessing and post-processing phases. The limitations of preprocessing shown in the literature survey are solved in preprocessing that is shown as an overview in the related work section.

3 Related Work on Preprocessing

3.1 Normalization Using Mean and Variance

The discrepancy in gray level, besides, the ridges and valleys is inaccurate due to the poor quality images. The normalization using mean and variance reduce the variation in gray level values. Hong et al. [6] proposed the normalization process and is a widely accepted method. The method identify that the desire mean M0 = 100 and VAR0 = 100 is suitable for fingerprint images and remove the noise which is occurred due to given pressure on scanner.

3.2 Enhancement in Gradient-Based Algorithm for Orientation Estimation

As shown in literature survey, the gradient-based approach is the best method for orientation estimation. To find out the ridge orientation and to remove the inconsistency gradient-based approach [35] is followed. When applying it, inconsistency is occurred because the angle between −PI/4 and PI/4 range cannot contain all the directions in a fingerprint image. To remove the inconsistency found the equation that changes the angle range from [−PI/4, PI/4] to [0 to PI]. The paper [36] shows the implementation and performance comparison of gradient-based orientation method.

3.3 Image Enhancement Using O'Gorman Filter

As suggested in literature survey that Gabor filter is computationally expensive so we use O'Gorman filter, which is one type of directional filter suitable for enhancing fingerprint images. This filter does not need frequency but used direction of pixel for generating the filter. In this, we first find 7 × 7 matrix in 0° direction. After that rotate it based on orientation of every pixel. After that, compute coordinate of rotated filter and based on that we find out the weight for it from 0° direction filter. The problem arise with rotate formula is, every times the value of i' and j' cannot be integer. This limitation is solved by interpolation to find out the weight in 0° direction filter. The paper [37] shows the implementation and performance comparison of Gabor filter, existing O'Gorman filter and improved O'Gorman filter.

3.4 Binarization Using Adaptive Thresholding with 9 × 9 Matrix

To solve the limitation of global thresholding, adaptive thresholding is one way of dynamic threshholding technique in which the different threshold values for different local areas are used. It is decided according to pixel wise. Here, T depends on neighborhood of (x, y). In [38] shows the comparison of global thresholding, Otsu thresholding, and adaptive thresholding with different window sizes like 5 × 5, 9 × 9, and 15 × 15 neighborhoods were used to calculate the threshold (T) for each pixel. The comparison show proves that 9 × 9 window size gave the best qualitative result. Because when we decrease the size of window, the gray scale portion is removed, which means that the important ridge lines are removed and when we increase the size of window, it darkens the image means extra noise can be added, which creates false end point and bifurcation minutiae, and lead to failure for extracting the true minutiae during feature extraction process.

3.5 Enhancement in Zhang-Suen's Algorithm for Thinning

In this phase, it enhanced the Zhang-Suen algorithm using the fix ridge. The Zhang-Suen's algorithm [39] used 3 × 3 matrix as a mask for binary image, which moves down an entire image and go through all pixels of the image. After implementing Zhang-Suen's algorithm, some new noisy points are created. To remove this noisy point, it followed the fix ridge algorithm after thinning process. To mark current pixel as foregroud check the condition like, if current pixel is background and from its four neighbors more than three neighbors are foreground then mark current pixel as foreground. The fix ridge operation is applied two times for deleting most of the noisy points. In paper [40] shows the implementation and performance comparison of Hildtch, Zhang-Suen, and enhanced Zhang-Suen's algorithm.

4 Proposed Algorithm

As discussed in the related work in Sect. 3 about preprocessing, our research work is carried out for post-processing. The proposed work introduces the post-processing, as well as overall enhancement done in fingerprint recognition system.

The outcome of preprocessing phase becomes an input for the post-processing. The post-processing involves minutiae extraction, remove false minutiae, core point detection, and matching processes.

4.1 Minutiae Extraction

As mentioned in the literature survey, crossing number is the best technique for minutiae extraction. In this phase, the minutiae-like bifurcation and endpoints are extracted and is used for verification and identification purpose. The thinned image becomes an input for minutiae extraction. It uses the thinned image with eight-connected ridge flow pattern. The below Fig. 1 shows the 3 x 3 window which is used for extracting minutiae. Where, P is the point which is consider to determine as a minutiae. P_1–P_8 is 8 points which is nearby the point. According to Rutovitz [41] for ridge pixel P, the crossing number is given by using the below Eq. (1) [42].

$$C_n(P) = \left(\frac{1}{2}\right) \sum_{i=1}^{8} |P_i - P_{i+1}| \tag{1}$$

In Eq. (1), P_i is considered as a binary pixel either 0 or 1 value in the neighborhood of P and $P_1 = P_9$. The P is measured as the crossing number $C_n(P)$ as half of cumulative successive differences among pairs of adjacent pixels associated with eight neighborhoods of P. Here, we are using two minutiae ridge ending and ridge bifurcation for matching the finger images. In that condition, only check if $C_n(P) = = 1$, then it is treated as ridge ending and if $C_n(P) = = 3$, then as ridge bifurcation. Figure 4 shows the result of minutiae extraction.

Fig. 1 3 × 3 matrix

P4	P3	P2
P5	P	P1
P6	P7	P8

4.2 Core Point Detection

As shown in literature survey, the Poincare Index method is widely accepted method for core point detection. The below Eq. (2) shows the equation of Poincare Index. We have followed the Poincare technique and also enhanced it using a method to detect the core from some essays [43, 44]. The Poincare Index method check the direction of eight neighbours and count the difference means changes of it. In the improved version, we used two matrixes to check the neighbors' direction change, one is 2 × 2 matrix and another is bigger than this one, it is 9 × 9 matrix, in this matrix 24 pixel will be checked, shows in Figs. 2 and 3. As well as Fig. 4 shows the result of core point detection.

$$\text{Poincare}\,(i,j) = \frac{1}{2\pi}\sum_{k=0}^{N=1}\Delta(k) \quad \Delta(k) = \begin{cases} \delta(k) & \text{if}|\delta(k)| \leq -\frac{\pi}{2} \\ \pi + \delta(k) & \text{if}|\delta(k)| \prec \frac{\pi}{2} \\ \pi - \delta(k) & \text{otherwise} \end{cases} \tag{2}$$

$$\delta(k) = \theta(X(k+1) \bmod N, Y(k+1) \bmod N) - \theta(X_k, Y_k)$$

where (*i*, *j*) is the point be checked.

4.3 Remove False Minutiae

The feature extracted after minutiae extraction phase is fed into remove false minutiae algorithm which scan minutiae's joined branches on specified above start by the centralize minutiae on window to found the false minutiae which are listed in Table 1 [45]. The result is shown in Fig. 4.

Fig. 2 2 × 2 matrix

i, j-1	i-1, j-1
i, j	i+1, j

Fig. 3 9 × 9 matrix

		P5	P6	P7	P8	P9		
	P4						P10	
P3								P11
P2								P12
P1				i, j				P13
P24								P14
P23								P15
	P22						P16	
		P21	P20	P19	P18	P17		

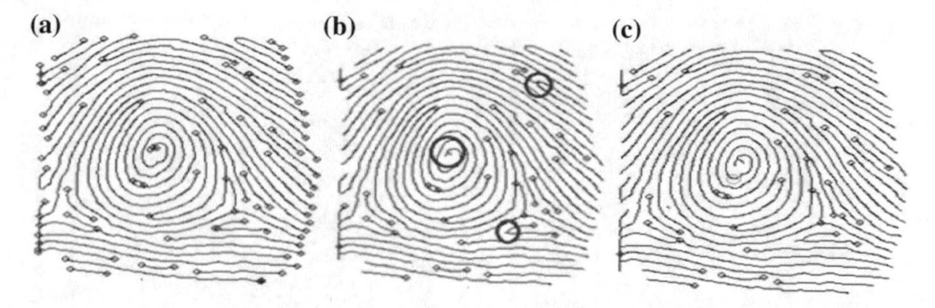

Fig. 4 **a** Result of minutiae extraction, **b** result of removal of false minutiae, **c** result of core point detection

Table 1 Types of false minutiae [45]

Case	Shape	Description	State
M1		Spike slices occurred in valley	The *D* is considered as the average distance between two equivalent ridges because the distance within ridge ending and ridge bifurcation is smaller than *D*
M2		Connect two ridges by false spike	In same ridge if two bifurcation are shown and the distance is smaller than *D* then remove the both ridge bifurcation
M3		In same ridge two close bifurcation are shown	
M4		In a same orientation and by short distance broken points are separated in two ridges	Distance within two ridge ending smaller than *D* and their directions corresponding with small angel discrepancy and no any ridge ending found in between them then they considered as false minutiae and removed the broken ridge part
M5		Same as M4 but one portion is too short in broken ridge	
M6		The third ridge is found between two ridges	
M7		a very short ridge found in the threshold window	The distance within two ridge ending is very short than *D* is measured as false minutiae and removed it

4.4 Minutiae Matching

Minutiae matching is the last step of post-processing. For the proposed work, we followed the minutiae-based approach. To improve the performance of matching process, in proposed work combine two approaches. The two approaches are approach-based on alignment [46] and approach based on local minutiae: radius: 50px from central minutiae [47]. The approach based on alignment each minutia of one image is compared with all the minutiae of other images and try to find the similarity between them and if the matched minutiae are more than threshold, the

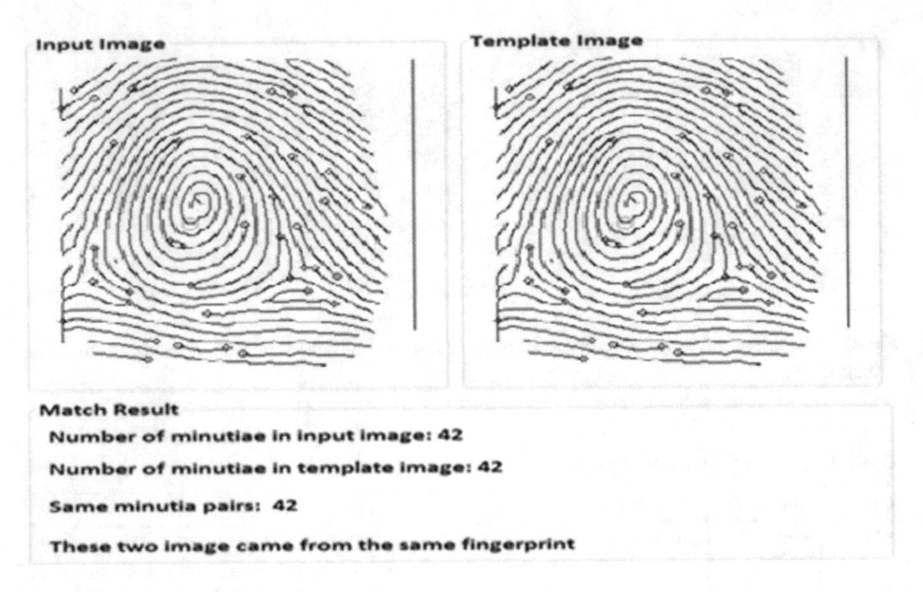

Fig. 5 Result of matching process

result is positive else negative. In approach based on local minutiae: radius: 50px from central minutiae mainly focused on the distance of the local minutiae from the central point. It also focusses on some criteria like type of a minutia, the distance to core point, the slope angle and different angle within minutia and core point, how many ridges within minutia, and core point. Figure 5 shows the result of minutiae matching.

5 Data Set

FVC2000 [48] and FingerDOS [49] databases are used in the research for evaluation of the proposed algorithm. The FVC2000 Set B images are 8-bit gray scale images. It has DB1_B, DB2_B, DB3_B, and DB4_B set, which contains a total of 320 finger images captured by different scanners like Low-cost Optical Sensor, Low-cost Capacitive Sensor, Optical Sensor, and Synthetic Generator, respectively. Each set contains a total of 80 images for 10 users with 8 impressions of finger with different sizes like 300 × 300, 256 × 364, 448 × 478, and 240 × 320 with 500 dpi resolution. The FingerDOS database contains 3,600 fingerprint images for 60 users, which is captured using optical finger sensor, i.e., SecuGen iD-USB SC. The database encloses the 8-bit gray scale images of thumb, index, and middle finger of both right and left hands of the users. Each set contains total 600 images for 60 users with 10 impressions of each finger with 260 × 300 size with resolution of 500 dpi. The database contains 56% male and 44% female fingerprint images of average age of 22 year-old. The implemented proposed algorithm is tested on a machine with an Intel(R) Core (TM)2 Duo CPU T6570 @ 2.10 GHz, 1.99 GB RAM and is developed using the JAVA language.

6 Performance Measurement Standards

To check the performance of fingerprint recognition, FMR and FNMR are estimated based on imposter and genuine attempts. In genuine attempts, the finger impression is matched with the impressions of other similar finger and in imposter attempts, every finger impression is matched with all the remaining finger's first impression. The number of imposter attempt for FVC2000 is ${}_{10}C_2 \times 1 \times 1 = 45$ for each cluster and for FingerDOS ${}_{60}C_2 \times 1 \times 1 = 1770$ for each cluster of that database. Like that for counting the genuine attempt for FVC2000 database is ${}_{8}C_2 \times 10 = 280$ and for FingerDOS database it is ${}_{10}C_2 \times 60 = 2700$.

FMR (False Match Rate): It is also called as False Accept Rate (FAR). It can be calculated as

$$\text{FMR} = \frac{\text{Number of false acceptance}}{\text{Number of impostor attempts}} \times 100\% \tag{3}$$

FNMR (False Non-Match Rate): It is also called as False Reject Rate (FRR). It can be calculated as

$$\text{FNMR} = \frac{\text{Number of false rejection}}{\text{Number of genuine user attempt}} \times 100\% \tag{4}$$

Accuracy:

$$\text{Accuracy} = 100 - \frac{(\text{FMR} + \text{FNMR})}{2} \tag{5}$$

Execution Time:

The time required for fingerprint recognition process is measured by calculating the time of all phases. The time required to calculate each process is the difference between execution time of last line and first line of process.

7 Experimental Result

Tables 2, 3, 4, 5.

Table 2 Summary report of FMR on FVC2000 and FingerDos databases

	FVC2000				FingerDOS					
					Left hand			Right hand		
	DB1_B	DB2_B	DB3_B	DB4_B	Index	Middle	Thumb	Index	Middle	Thumb
Total comparisons	45	45	45	45	1770	1770	1770	1770	1770	1770
Matched impressions	0	0	1	3	0	0	0	0	0	0
Unmatched impressions	45	45	44	42	1770	1770	1770	1770	1770	1770
FMR	0	0	2	6	0	0	0	0	0	0
Average FMR (%)	2				0					

Table 3 Summary report of FNMR on FVC2000 and FingerDOS Databases

	FVC2000				FingerDOS					
					Left hand			Right hand		
	DB1_B	DB2_B	DB3_B	DB4_B	Index	Middle	Thumb	Index	Middle	Thumb
Total comparisons	280	280	280	280	2700	2700	2700	2700	2700	2700
Matched impressions	271	271	280	280	2636	2621	2578	2656	2653	2691
Unmatched impressions	9	9	0	0	64	79	122	44	47	9
FNMR	3	3	0	0	2	2	4	1	1	0
Average FNMR (%)	1.2				1.67					

Table 4 The percentage of FMR, FNMR, and Accuracy

Database	Recognition accuracy		
	FMR	FNMR	Accuracy
FVC2000	2	1.2	98.4
FingerDOS	0	1.67	99.16

Table 5 Summary report of execution times of different processes of fingerprint recognition using FVC2000 and FingerDOS databases (in milliseconds)

Processes	FVC2000				FingerDOS					
					Left hand			Right hand		
	DB1_B	DB2_B	DB3_B	DB4_B	Index	Middle	Thumb	Index	Middle	Thumb
Normalization	68	70	141	64	37	17	37	43	46	43
Smoothing	144	149	289	128	75	36	68	86	84	80
Orientation detection	122	133	333	149	39	59	41	32	31	41
Edge detection	527	521	1298	365	205	111	224	246	212	210
Ridge detection	193	189	339	168	93	36	75	89	108	108
Fixed ridge	68	66	105	58	30	14	23	36	38	39
Thinning	288	304	520	194	125	65	141	134	147	157
Minutia detection	3	3	7	2	2	0	2	2	2	2
Remove false minutia	714	669	4023	943	127	116	185	118	175	305
Core point detection	211	219	514	178	101	55	101	96	100	105
Total execution time	2341	2327	7573	2254	838	516	903	887	947	1094
Average execution time	3624				864					

8 Conclusion

The proposed work is an attempt to enhance the preprocessing and post-processing algorithm of the fingerprint recognition and gradually improve the overall performance. In preprocessing phase, gradient-based orientation estimation method enhanced after removing inconsistency, as well as enhanced O'Gorman filter to improve the ridge and valley contrast and connect broken ridges. Further, it implemented an adaptive threshold using 9×9 matrix and enhanced the Zhang-Suen's thinning algorithm in the facet of removal of pixel criteria for preserving the connectivity of pattern, remove noisy points, and for sensitivity of the binary image. In post-processing phase, an attempt is extended to extract ridge ending and ridge bifurcation minutiae from finger image using crossing number followed by enhanced Poincare Index method used for core point detection, remove the false minutiae, and matching the minutiae based on combination of alignment based and distance from local minutiae matcher methods. To test the experimental result and performance of the proposed algorithm carried out using FVC2000 and

FingerDOS databases and prove that the FMR and FNMR for FVC2000 is 2 and 1.2 and for FingerDOS is 0 and 1.67, respectively. The accuracy of FVC2000 is 98.4% and FingerDOS is 99.16%. The average execution time of FVC2000 is less than 3.62 and for FingerDOS is 0.86 s.

References

1. Maltoni, D., Maio, D., Jain, A.K., Prabhakar S.: Fingerprint analysis and representations. In: Handbook of Fingerprint Recognition, 2nd edn, pp. 97–166. Springer, London (2009)
2. Maltoni, D., Maio, D., Jain, A.K., Prabhakar, S.: Fingerprint matching. In: Handbook of Fingerprint Recognition, 2nd edn, pp. 167–233. Springer, London (2009)
3. Shalash, W.M., Abou-Chadi Fatma, E.Z.: Fingerprint image enhancement with dynamic block size. In: 23rd IEEE National Radio Science Conference, vol. 0, pp. 1–8. IEEE (2006)
4. Kim, B., Kim, H., Park, D.: New enhancement algorithm for fingerprint images. In: IEEE Proceedings 16th International Conference on Pattern Recognition, vol. 4651, no. 2, pp. 1051–1055 (2002)
5. Shi, Z., Govindaraju, V.: A chaincode based scheme for fingerprint feature extraction. Pattern Recogn. Lett. **27**, 462–468 (2006)
6. Hong, L., Wang, Y.F., Jain, A.K.: Fingerprint image enhancement: algorithm and performance evaluation. In: IEEE Transactions on Pattern Analysis and Machine Intelligence, vol. 20, no. 8, pp. 777–789. IEEE (1998)
7. Dass, S.C.: Markov random field models for directional field and singularity extraction in fingerprint images. IEEE Transact. Image Process **13**(10), 1358–1367 (2004)
8. Wang, Y., Hu, J., Han F.: Enhanced gradient-based algorithm for the estimation of fingerprint orientation fields. Appl. Math. Comput. **185**, 823–833 (2007)
9. Biradar, V.G., Sarojadevi, H.: Fingerprint ridge orientation extraction: a review of state of the art techniques. Int. J. Comput. Appl. **91**(3), 8–13 (2014)
10. Mario, D., Maltoni, D.: Direct gray-scale minutiae detection in fingerprints. In: IEEE Transactions on Pattern Analysis and Machine Intelligence, vol. 19, no. 1, pp. 27–40. IEEE (1997)
11. Omran, S.S., Salih, M.A.: Comparative study of fingerprint image enhancement methods. J. Babylon Univ./Eng. Sci. **22**(4), 708–723 (2014)
12. Greenberg, S., Aladjem, M., Kogan, D., Dimitrov.: Fingerprint image enhancement using filtering techniques. In: 15th IEEE International Conference on Pattern Recognition, vol. 3, pp. 326–329. IEEE (2000)
13. Rajkumar, R., Hemachandran, K.: A review on image enhancement of fingerprint using directional filters. Assam Univ. J. Sci. Technol: Phys. Sci. Technol. **7**(II), 52–57 (2011)
14. Wang, J.: Gabor filter based fingerprint image enhancement. In: Fifth International Conference on Machine Vision (ICMV 2012): Computer Vision, Image Analysis and Processing, Proceeding in SPIE, vol. 8783, Id. 878318 (2013)
15. O'Gormann, L., Nickerson, J.V.: An approach to fingerprint filter design. In: Pattern Recognition, vol. 22, no. 1, pp. 29–38. Elsevier (1989)
16. Otsu, N.: A threshold selection method from gray-level histogram. In: IEEE Transactions on Systems, Man, and Cybernetics, vol. 9, no. 1, pp. 62–66. IEEE (1979)
17. Roy, P., Dutta, S., Dey, N., Dey, G., Chakraborty, S., Ray, R.: Adaptive thresholding: a comparative study. In: IEEE International Conference on Control, Instrumentation, Communication and Computational Technologies, pp. 1182–1186. IEEE (2014)
18. Cai, J.: Robust filtering-based thinning algorithm for pattern recognition. Comput. J. **55**(7), 887–896 (2012)

19. Witkin, A.: Scale-space filtering: a new approach to multiscale description. In: IEEE International Conference on Acoustics, Speech, and Signal Processing, vol. 9, pp. 150–153. IEEE (1984)
20. Sezgin, T., Davis,R.: Scale-space based feature point detection for digital ink. In: ACM SIGGRAPH 2007 Courses, Article no. 36, pp. 29–35. ACM (2007)
21. Lam, L., Lee, S.W., Suen, C.Y.: Thinning methodologies—a comprehensive survey. In: IEEE Transaction on Pattern Analysis and Machine Intelligence, vol. 14, no. 9, pp. 869–885. IEEE (1992)
22. Chatbri, H., Kameyama, K.: Sketch-based image retrieval by shape points description in support regions. In: IEEE 20th International Conference on Systems, Signals and Image Processing, pp. 19–22 (2013)
23. Chen, Y., Yu, Y.: Thinning approach for noisy digital patterns. In: Pattern Recognition, vol. 29, no. 11, pp. 1847–1862. Elsevier (1996)
24. Kocharyan, D.: A modified fingerprint image thinning algorithm. Am. J. Softw Eng. Appl. **2** (1), 1–6 (2013)
25. Luthra, R., Goyal, G.: Performance comparison of ZS and GH Skeletonization algorithms. Int. J. Comput. Appl. **121**(24), 32–38 (2015)
26. Farina, A., Vajna, Z.M.K., Leone, A.: Fingerprint minutiae extraction from skeletonized binary images. In: Pattern Recognition, vol. 32, no. 5, pp. 877–889. Elsevier (1999)
27. Mehtre, B.M.: Fingerprint image analysis for automatic identification. In: Machine Vision and Applications, vol. 6 no. 2–3, pp. 124–139. Springer (1993)
28. Ratha, N.K., Chen, S., Jain., A.K.: Adaptive flow orientation-based feature extraction in fingerprint images. In: Pattern Recognition, vol. 28, no. 11, pp. 1657–1672. Springer (1995)
29. Thakkar, D.: Minutiae Based Extraction in Fingerprint Recognition. https://www.bayometric.com/minutiae-based-extraction-fingerprint-recognition/. Accessed on Oct 2017
30. Jain, L.C., Halici, U., Hayashi, I., Lee, S.B., Tsutsui, S.: Intelligent biometric techniques in fingerprint and face recognition. In: CRC Press International Series on Computational Intelligence. ACM (1999)
31. Iwasoun, G.B., Ojo, S.O.: Review and evaluation of fingerprint singular point detection algorithms. Br. J. Appl. Sci. Technol. **4**(35), 4918–4938 (2014)
32. Gnanasivam, P., Muttan, S.: An efficient algorithm for fingerprint preprocessing and feature extraction. In: Procedia Computer Science, vol. 2, pp. 133–142. Elsevier (2010)
33. Tudosa, A., Costin, M., Barbu, T.: Fingerprint recognition using Gabor filters and wavelet features. In: Scientific Bulletin of the Politehnic University of Timisoara, Romania, Transactions on Electronics and Communications, vol. 49, no. 1, pp. 328–332 (2004)
34. Ravi, J., Raja, K.B., Venugopal, K.R.: Fingerprint recognition using minutia score matching. Int. J. Eng. Sci. Technol. **1**(2), 35–42 (2009)
35. Jain, A.K., Prabhakar, S., Hong, L., Pankanti, S.: Filterbank-based fingerprint matching. In: IEEE Transactions on Image Processing, vol. 9, no. 5, pp. 846–859. Elsevier (2000)
36. Patel, M.B., Parikh, S.M., Patel, A.R.: Performance improvement in gradient based algorithm for the estimation of fingerprint orientation fields. Int. J. Comput. Appl. **167**(2), 12–18 (2017)
37. Patel, M.B., Patel, R.B., Parikh, S.M., Patel, A.R.: An improved O'Gorman filter for fingerprint image enhancement. In: IEEE International Conference on Energy, Communication, Data Analytics and Soft Computing of Recent Technology and Engineering, Proceeding IEEE (2017)
38. Patel, M.B., Parikh, S.M., Patel, A.R.: Performance improvement in binarization for fingerprint recognition. IOSR J. Comput. Eng. **19**(3), 68–74, version II (2017)
39. Zhang, T.Y., Suen, C.Y.: A fast parallel algorithm for thinning digital patterns. In: Communication of the ACM, vol. 27, no. 3. pp. 236–239. ACM (1984)
40. Patel, M.B., Parikh, S.M., Patel, A.R.: An improved thinning algorithm for fingerprint recognition. Int. J. Adv. Res. Comput. Sci. **8**(7), 1238–1244 (2017)
41. Rutovitz, D.: Pattern recognition. J. Roy. Stat. Soc. **129**(4), 504–530 (1966)
42. Zhao, F., Tang, X.: Preprocessing and post processing for skeleton-based fingerprint minutiae extraction. In: Pattern Recognition, vol. 40, no. 4, pp. 1270–1281. Elsevier (2007)

43. Dey, T.K., Hudson, J.: PMR: point to mesh rendering, a feature based approach. In: IEEE Visualization 2002, pp. 155–162. IEEE (2002)
44. Nie, G., Wang, J., Wu, Z., Li, Y., Xu, R.: Fingerprint singularity detection based on continuously distributed directional image. Comput. Eng. Appl. **42**(35) (2006)
45. Garg, M., Bansal, H.: Fingerprint recognition system using minutiae estimation. Int. J. Appl. Innov. Eng. Manage. **2**(5), 31–36 (2013)
46. Mehtre B.M., Murthy N.N., Kapoor S.: Segmentation of fingerprint images using the directional image. In: Pattern Recognition, vol. 20, no. 4, pp. 429–435. Elsevier (1987)
47. Jim Z.C., Lai, Kuo, S.C.: An improved fingerprint recognition system based on partial thinning. In: 16th IPPR Conference on Computer Vision, Graphics and Image Processing (2003)
48. Maio, D., Maltoni, D., Capelli, R., Wayman, J.L., Jain, A.K.: FVC2000: fingerprint verification competition. In: IEEE Transactions on Pattern Analysis and Machine Intelligence, vol. 24, no. 3, pp. 402–412. IEEE (2002)
49. Francis-Lothai, F., Bong, D.B.L.: Fingerdos: a fingerprint database based on optical sensor. Wseas Transact. Inf. Sci. Appl. **12**(29), 297–304 (2015)

Analysis of Neck Muscle Fatigue During Cervical Traction Treatment Using Wireless EMG Sensor

Hemlata Shakya and Shiru Sharma

Abstract The aim of this paper is to analyze the muscle fatigue during cervical traction treatment using wireless EMG sensor. Neck pain is a common problem in our daily life. It is a common cause of the various factors due to bad posture, office workers, computer professional, etc. Ten male–female patients who provided informed consent participate in this study. Cervical traction treatment with a tension of 7 kg was given to each subject for 15 min of cervical traction. The collected EMG data was used to extract various features of neck muscles in the time domain. Analysis of extracted parameters measures showed a significant difference in the muscle activity. The results show the effectiveness of continuous traction treatment in the decrease of neck pain.

Keywords Neck pain · Cervical traction · Electromyography · Feature extraction

1 Introduction

Neck pain is a general problem of our daily life. Neck pain affects around 70% of individuals in middle age. It is a typical reason for various factors due to bad posture, computer professional, and sporting activities [1].

It is a subgroup of various neck disorders, and cervical traction is a part of the treatment program. It is a treatment technique which is mainly used for neck pain, muscular dysfunction, radiculopathy, and spondylosis. Spinal traction means drawing or pulling on the vertebral column, and it is called cervical traction. The controls of most traction machines include manual adjustment of traction weight and timing of traction pull phases [2]. Figure 1 shows the seven vertebrae C1, C2,

H. Shakya (✉) · S. Sharma
School of Biomedical Engineering, Indian Institute of Technology (BHU), Varanasi 221005, India
e-mail: hemlata.shakya19@gmail.com

S. Sharma
e-mail: ssharma.bme@itbhu.ac.in

A. K. Luhach et al. (eds.), *Smart Computational Strategies: Theoretical and Practical Aspects*, https://doi.org/10.1007/978-981-13-6295-8_13

C3, C4, C5, C6, and C7, and C5–C6 nerves is most commonly affected joints. Between these, bones is an intervertebral circle which acts as shock absorbers and when healthy, they are strong but at the same time bendable so you can node, twist, sideband, and extend your neck.

Traction is generally used in the cervical spine to reduce strain on the cervical nerve roots in the patient with disc herniation, degenerative circle disease.

Cervical traction can be applied manually physically by a trained by health care professional, and there are various types of traction unit, and that can be obtained to provide relief symptoms.

There are various ways to apply cervical traction to the neck.

- Manual Cervical Traction: This traction is performed by the physical therapist.
- Mechanical Traction: This instrument gives distraction force.
- Gravitational Traction: Gravity provides the distraction force.

The different mechanical factors significant to neck position, length of traction, traction force, the angle of pull, and the location of the patient [3].

In these papers discussed above all proven studies the role of traction therapy in relieving of neck pain, muscle spasm, and radiculopathy.

A review of the literature on cervical traction by Angela Tao NG shows that cervical traction is more efficient in the reduction of cervical pain and arm pain [4]. Safoura Hosienpour, concluded that physiotherapy is more useful in the short time, after several weeks acupuncture and strengthening exercise were more effective in reducing pain [5]. In the survey conducted by Reem S. Dawood, cervical traction posture is similarly effective in improving cervical shape, pain force, and function neck disability in patients with mechanical neck issue compared to exercise

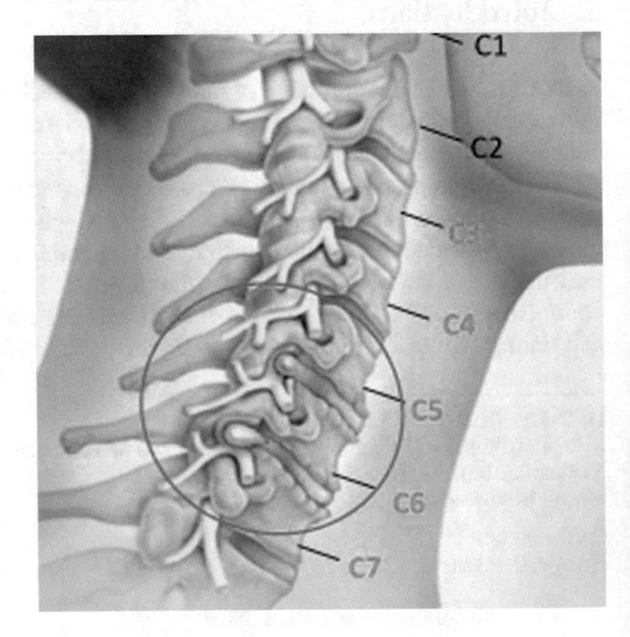

Fig. 1 Positions of cervical pain

program alone which was the less successful [6]. In Sharma and Patel study, this paper the TENS was more efficient in the management of cervical radiculopathy along with isometric neck exercise, in reducing both neck and arm pain [7]. In Subhash Chandra Rai, concluded that TENS and exercise are more effectual in the management of cervical radiculopathy [8]. In Akinbo et al. observed that the ideal weight to exert tension on the neck is 10% of the total body weight [9].

In Atteya, cervical traction showed the significant decrease of EMG activity during the pull phase of traction as well as after traction in biofeedback traction modality in comparison to the conventional traction modality [10]. Nanno also indicated that the traction is successful in relieving pain [11]. In the survey by Murphy, no significant difference between groups was noted in EMG recording at rest and within 10 min of traction, The EMG activity in subjects with neck pain during and few hours after cervical traction provides relief in pain [12]. In FG. Delacerda, this paper showed that the electrical activity of the upper trapezius muscle increases significantly as the angle of application of cervical traction pull increased [13].

Therefore, the study aimed to examine the effect of neck muscle stress in various subjects for the reduction of neck pain. Acquired EMG signal was calculated using time domain features are mean absolute value (MAV), root mean square (RMS), and standard deviation (SD).

1.1 Electromyography

Electromyography (EMG) is an instrument for recording the electrical activity of the muscles. EMG signal represents neuromuscular activity. It can be recorded information from the human tissue through the surface electrodes. It is a noninvasive technique used to amplify show and record changes in the skin surface when muscular surface system. EMG signal is produced by physiological variation in the state of muscle fiber membranes during voluntary and involuntary contractions. The signal is a biomedical signal had become the main research region in the medical field including the extensive variety of expertise from the doctor, specialist to PC researcher [14]. The surface EMG signal is relative amplitude 0–10 mV (peak to peak), and frequency range lies between 0 and 500 Hz [10].

In rehabilitation engineering, EMG signal is one of the main neural control sources for power upper limb prosthesis. The uses of surface EMG signal analysis in muscular function evaluation mainly focuses on time domain. It reflects the specific indices of EMG signal amplitude changes and is commonly used as EMG time domain [15].

The fatigue EMG signal shows up this type of aspect the amplitude of EMG signal increases and power range moves toward the lower frequency. It is essential for the detection of muscle exhaustion and the estimation of the level of weakness [16].

2 Methodologies

Ten (10) patients of both males and females, participate in the investigation with age ranged between 23 and 75 years using wireless EMG sensor. These patients having neck pain complaint were visiting therapy unit for traction treatment.

Figure 2 shows the wireless EMG system with the sensor. The wireless EMG sensor is used in the recording of EMG data. The wireless EMG sensor was placed on (C5–C6) muscle recording the EMG signal during traction. The subjects treated with 15 min of cervical traction with a tension of 7 kg. Cervical traction was managed in a sitting position as generally done in a clinic (Fig. 2, 3). This study was approved by the ethics committee of IMS BHU.

The collected EMG data was analyzed to acquire information about multiple features of muscle, in the time domain, i.e., mean absolute value, root mean square, and standard deviation using MATLAB software platform. Time domain parameter measures the significant changes in the muscle activity.

2.1 *Time Domain Features*

The EMG signal is a time series signal, and the EMG signal analyzed the different parameters MAV, RMS, SD obtained in MATLAB. Time domain mainly relates to the amplitude value and time value.

(i) **Mean Absolute Value**: It is a set of data is the average distance between each data value and the mean. It can be given by

Fig. 2 Wireless EMG system with sensor

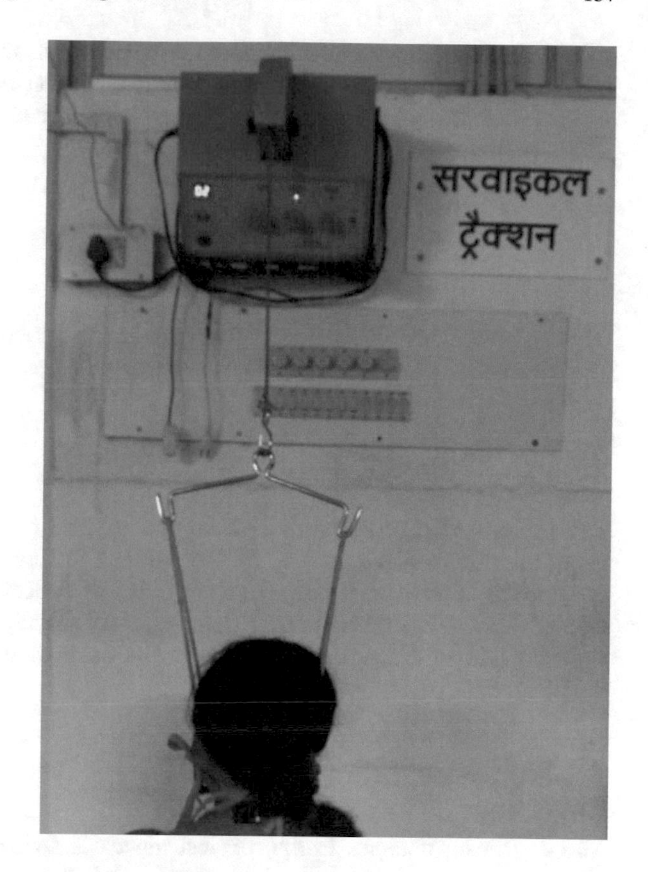

Fig. 3 Clinical photograph during cervical traction treatment

$$\text{MAV} = \frac{1}{N}\sum_{i=1}^{n} |x_n| \quad (1)$$

where x_n is the representation of the EMG signal and the parameter N is the number of the sample.

(ii) **Root mean Square**: RMS is a mathematical value is the square root of the mean value of the squared function of the instantaneous value. This is calculated by

$$X_{\text{RMS}} = \sqrt{\frac{1}{N}}\sum_{n=1}^{N} |x_n|^2 \quad (2)$$

where N represents the number of the sequence or segment and x_n corresponds to the values of the sequence or segment.

(iii) **Standard Deviation**: Standard deviation means the spread of data from the mean. In signal processing, standard deviation represents noise and other interference. It is used to find the threshold level of the muscle contraction velocity [17].

$$\mathrm{SD} = \sqrt{\frac{1}{N}\sum_{i=1}^{N}(x-\mu)^2} \tag{3}$$

where N represents the length of the sequence or segment and μ is the mean value.

3 Results

Ten patients with neck pain had followed up. The collected data were analyzed using time domain parameters. During clinical testing, every subject was sitting position for 15 minutes per day for one week. The collected data was used further to extract various features in the time domain (MAV, RMS, and SD). The result shows some significant differences in the calculated parameters (Fig. 4).

4 Discussion

The significant changes in the cervical muscle activities during applied traction. Time domain parameters were calculated using the recorded EMG data through wireless EMG sensor for 15 min during the first day and last day of the week. Traction method can be used with the intention of stretching the muscle and widening the intervertebral foramina. The importance of therapy is that the angle of pull, head position, and placement of the force can be controlled [11].

Few related studies with EMG during cervical traction showed different outcomes. Yang found that cervical traction may have short-term neck pain-relieving effect [18]. Bukhari et al. concluded that traction therapy would manage pain and disability more effectively than treated with manual traction and exercise therapy [19]. Elnaggar et al. showed that the continuous cervical traction had a significant effect on neck pain and arm pain reduction, and a significant improvement in nerve function [20].

The present study showed a significant variation between the pre-treatment and post-treatment. In this investigation the changes in muscle activity during traction treatment. During the use of treatment, there is a muscle strain, and skin is stretching. The time domain parameters were extracted for muscle fatigue analysis. Table 1 shows that the decrease in time domain value was identified during traction. This study data shows that time domain features are gradually decreased during traction in a sitting position. All patients treated by cervical traction were

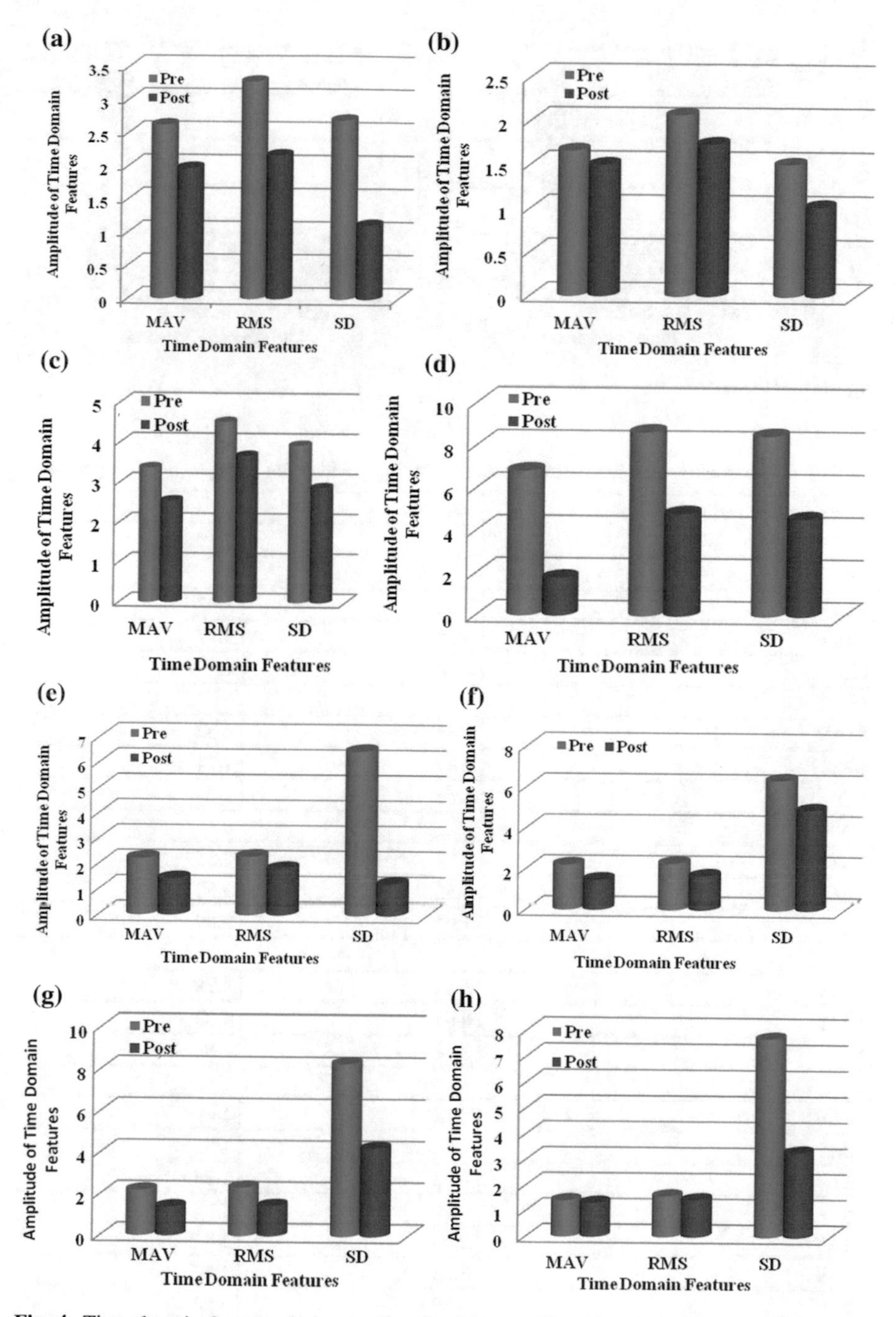

Fig. 4 Time domain features during traction in sitting position for one week (pre and post day)

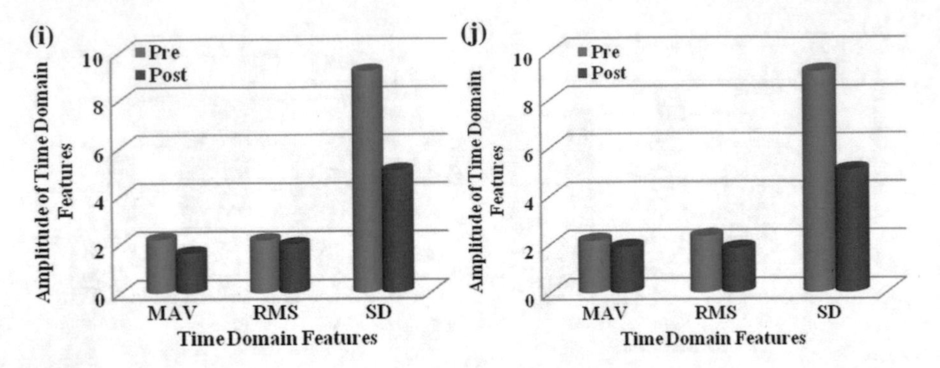

Fig. 4 (continued)

Table 1 Time domain features during traction in sitting position for 1 week (pre and post day)

Time domain features	Session	
	Pre	Post
(a)		
MAV	2.62	1.97
RMS	3.28	2.17
SD	2.70	1.12
(b)		
MAV	1.66	1.50
RMS	2.07	1.74
SD	1.51	1.03
(c)		
MAV	3.37	2.55
RMS	4.52	3.68
SD	3.94	2.89
(d)		
MAV	6.81	1.78
RMS	8.64	4.83
SD	8.48	4.6
(e)		
MAV	2.22	1.42
RMS	2.3	1.83
SD	6.48	1.25
(f)		
MAV	2.20	1.47
RMS	2.28	1.67
SD	6.36	4.92
(g)		

(continued)

Table 1 (continued)

Time domain features	Session	
	Pre	Post
MAV	2.19	1.40
RMS	2.34	1.47
SD	8.30	4.26
(h)		
MAV	1.42	1.33
RMS	1.60	1.46
SD	7.71	3.32
(i)		
MAV	2.21	1.65
RMS	2.18	2.01
SD	9.20	5.06
(j)		
MAV	2.16	1.90
RMS	2.33	1.83
SD	9.20	5.06

noted to have a decrease in muscle activity during the 1-week period. The analysis of time domain features indicate the helpful of traction treatment in the reduction of neck pain.

5 Conclusion

This study concluded that the time domain parameters determined. These features can be used efficiently used to determine the MAV, RMS, and SD evaluation of muscle weakness. The traction therapy is valuable to be included in the program of treatment of cervical patients. Following 1 week, traction therapy was more efficient in reducing pain and disability.

Acknowledgements I would like to thank Dr. O. P. Singh, Physiotherapy Department in Sir Sundarlal Hospital Varanasi for their valuable guidance in this study.

References

1. Peterson, D.H., Bergman, T.F.: Chiropractic Technique: Principles and Procedures, United States of America, Mosby (2002)
2. Pan, P.J.; Tsai, P.H., Tsai, et al.: Clinical response and autonomic modulation as seen in heart rate variability in mechanical intermittent cervical traction: a pilot study. J. Rehabil. Med. **44**(3), 229–234 (2012). https://doi.org/10.2340/16501977-0927

3. Harris, P.R.: Cervical traction: review of literature and treatment guideline. Phys. Ther. **57**, 910–914 (1997)
4. Angela Tao, N.G., Arora, R.A.L.: Effectiveness of cervical traction on pain and disability in cervical radiculopathy. Int. J. Recent Sci. Res. **6**(4), 3609–3611 (2015)
5. Hoseinpour, S., Amanollahi, A., Sobhani, V., Mohseni, A., Arazi, E.: Comparison of neck and shoulder strengthening exercises with weights, traction plus physiotherapy, and acupuncture in the treatment of patients with chronic cervical disk herniation. Int. J. Sci. Res. Knowl. **3**(4), 114 (2015). https://doi.org/10.12983/ijsrk2015
6. Dawood, R.S., Kattabei, O.M., Nasef, S.A., et al.: Effectiveness of kinesio taping versus cervical traction on mechanical neck dysfunction. Int. J. Ther. Rehabil. Res. **2**(2) (2013). https://doi.org/10.5455/ijtrr.00000019
7. Sharma, H., Patel, N.: Effectiveness of tens versus intermittent cervical traction in patients with cervical radiculopathy. Int. J. Phys. Res. **2**, 787–792 (2014). https://doi.org/10.16965/ijpr.2014.693
8. Rai, SC., Alith, S,, Bhagvan, KR., Pinto, D.: Cervical traction reduces pain and disability in patients with unilateral cervical radiculopathy. Int. J. Curr. Res. **5**(7) (2013)
9. Akinbo, S.R., Noronha, C.C., Okanlawon, A.O., Danesi, M.A.: Effects of different cervical traction weights on neck pain and mobility. Niger. Postgrad. Med. J. **13**(3), 230–235 (2006)
10. Atteya, A.A.A.: Biofeedback traction versus conventional traction in cervical radiculopathy. Neurosciences **9**(2), 91–93 (2004)
11. Nanno, M.: Effects of intermittent cervical traction on muscle pain. Flowmetric and electromyographic studies of the cervical paraspinal muscles. Nihon Ika Daigaku zasshi **61** (2), 137–147 (1994). https://doi.org/10.1272/jmms1923.61.137
12. Murphy, M.J.: Effects of cervical traction on muscle activity. J. Orthop. Sports Phys. Ther. (May 1991). https://doi.org/10.2519/jospt.1991.13.5.220
13. De Lacerda, F.G.: Effect of angle of traction pull on upper trapezius muscle activity. J. Orthop. Sports Phys. Ther. **1**, 205–209 (1980). https://doi.org/10.2519/jospt.1980.1.4.205
14. Goen, A., Tiwari, D.C.: Review of surface electromyogram signals: its analysis and applications. Int. J. Electr. Comput. Energ. Electr. Commun. Eng. **7**(11) (2013)
15. De Luca, C.J., Adam, A., Wotiz, R., Gilmore, et al.: Decomposition of surface EMG signals. J. Neurophysiol. **96**, 1646–1657 (2006). https://doi.org/10.1152/jn.00009.2006
16. Jia, W., Zhou, R.: The analysis of muscle fatigue based on SEMG. In: Mechatronic Sciences, Electric Engineer and Computer (MEC). IEEE (2013). https://doi.org/10.1109/mec.2013.6885179
17. Phinyomark, A., Phukpattaranont, P., Limsakul, C.: Feature reduction and selection for EMG signal classification. Expert Syst. Appl. **39**(8), 7420–7431 (2012). https://doi.org/10.1016/j.eswa.2012.01.102
18. Yang, J.-D., Tam, K.-W., Huang, T.-W., Huang, S.-W., Liou, T.-H., Chen, H.-C.: Intermittent cervical traction for treating neck pain: a meta-analysis of randomized controlled trials. Spine **42**(13), 959–965 (2017). https://doi.org/10.1097/BRS.0000000000001948
19. Bukhari, S.R.I., Shakil-ur-Rehman, S., Ahmad, S., Naeem, A.: Comparison between effectiveness of mechanical and manual traction combined with mobilization and exercise therapy in patients with cervical radiculopathy. Pak. J. Med. Sci. **32**(1), 31 (2016). https://doi.org/10.12669/pjms.321.8923
20. Elnaggar, I.M., Elhabashy, H.R., Abd El-Menam, E.: Influence of spinal traction in treatment of cervical radiculopathy.Egypt. J. Neurol. Psychiat Neurosurg. **46**, 455–460 (2009)

Part II
Hardware

Simulation Performance of Conventional IDMA System with DPSK Modulation and Modern Fisher–Yates Interleaving Schemes

Vinod Shokeen, Manish Yadav and Pramod Kumar Singhal

Abstract In communication, interleaving is used to perform burst error control. Random interleaving is more advantageous over many other existing interleaving approaches. Recently, the possibility of using modern Fisher–Yates random data shuffle algorithm, which is also named as Durstenfeld's random data shuffle algorithm, in one of such advanced multiple access systems that has been explored. This facilitates controlling of burst errors and various interferences without compromising the system complexity and bit error rate (BER) performance. In this paper, simulation performance of conventional interleave-division multiple access (CIDMA) system in the presence of differential phase-shift keying (DPSK) modulation and modern Fisher–Yates shuffle based random interleaving (MFYI) schemes is inquired. The results show that both MFYI and DPSK are compatible with CIDMA system and have performance equivalence similar to the most of the other popular interleavers.

Keywords Modern Fisher–Yates interleavers (MFYI) · Conventional interleave-division multiple access (CIDMA) · Durstenfeld's random interleaver (DRI)

V. Shokeen
ASE, Amity University, Noida, India
e-mail: vshokeen@amity.edu

M. Yadav (✉)
ASET, Amity University, Noida, India
e-mail: ymaniish@gmail.com

P. K. Singhal
Madhav Institute of Technology and Science (MITS), Gwalior, India
e-mail: pks_65@yahoo.com

A. K. Luhach et al. (eds.), *Smart Computational Strategies: Theoretical and Practical Aspects*, https://doi.org/10.1007/978-981-13-6295-8_14

1 Introduction

Conventional interleave-division multiple access (CIDMA) systems, simply known as IDMA systems, are interleaving-based multiple access systems. These systems employ a modified code division multiple access (CDMA) system approach in which each user receives a unique interleaving pattern instead of unique pseudo noise (PN) sequence [1, 2]. So, their recognition within the network is based on the unique interleaving pattern. CIDMA system mitigates the issues of multiuser interference (MUI) and multiple access interferences (MAI) very efficiently. This facilitates the better overall performance of CIDMA as compared to CDMA. Therefore, CIDMA can easily outperform CDMA, if some system parameters are chosen amicably [3]. These parameters are modulation scheme, forward error correcting code, and type of interleaver used. Proper selection of these parameters results in improved bit error rate (BER) of CIDMA system [4–6]. These parameters also affect the overall system complexity, memory requirements, and speed of response of the system [7, 8].

In this paper, simulation performance of CIDMA system with differential phase-shift keying (DPSK) modulation, simple repetition coding and modern Fisher–Yates shuffle based random interleaver (MFYI) has been analyzed. The remaining part of the paper is divided into the following sections: Sect. 2 discusses about some previous work done in the related area. CIDMA system model is presented in Sect. 3. Section 4 offers a brief introduction to DPSK and MFYI schemes. In Sect. 5, simulation results and discussion on results are covered. Section 6 draws the conclusion.

2 Previous Work

Interleaving was emerged as statistical random shuffling method, which could be applied on a set of data. Initially, its role was limited up to statistical and computer applications only. However, with the time its applications in radio (mobile), data communications, data security, and information sciences also emerged [9–14, 16]. In radio communication, random data shuffle allows the loss of small segments from the multiple frames rather than the complete loss of a single frame. Thus, burst errors distribution among the frames works here. After the code-division multiple access (CDMA) system development, several researchers, in parallel, worked toward developing an equally efficient system, which could distinguish among all the users using a single spreading sequence throughout the system. This led to the development of interleave-division multiple access (IDMA) system which is now also known as conventional IDMA (CIDMA) system. CIDMA system is a modified form of CDMA system with some inherent enhanced features such as burst errors control and interference mitigation put together [19–21].

Though random interleavers are excellent interleavers and are often chosen in CIDMA system; but to enhance the performance of CIDMA systems, several other types of interleavers have also been proposed in the literature [8, 13–16]. For example, user specific chip interleaver, i.e., power interleaver, semi-random, tree, Gaussian, golden dithered, and Rayleigh interleaver. Most of these are derived from random interleavers only and therefore categorized under derived interleaver category [16, 18, 19, 27–29]. Two of such very recently developed interleavers are: Modern Fisher–Yates shuffle based random interleavers (MFYIs) and flip left–right approach based inverse tree interleavers (FLRITIs) [17, 24]. Both interleavers can be categorized under derived interleaver category. CIDMA system can be modified further for performance enhancement. This is achieved by introducing the concept of Integrated IDMA (IIDMA) or grouped IDMA (GIDMA). Thus, IIDMA or GIDMA is the examples of modified CIDMA. However, discussion on IIDMA/GIDMA is beyond the scope of this paper.

Along with different forward error correction (FEC) and interleaving techniques, CIDMA and its variant, i.e., IIDMA has also been explored with different modulation schemes such as binary phase-shift keying (BPSK), quadrature phase-shift keying (QPSK), quadrature amplitude modulation (QAM), etc. In particular, BPSK and QPSK are the modulation techniques extensively employed by the researchers in CIDMA analysis in various contexts. Therefore, to provide the novelty to the work, the idea of using some other modulation technique, e.g., DPSK emerges.

3 Conventional IDMA System Model

CIDMA system model for *k*th user comprises of an information source followed by encoder/spreader and interleaving blocks at the transmitting end. Similarly, at the receiving end, de-interleaving device, a dedicated decoder and/or de-spreader are the main elements. Elementary (received) signal estimation is achieved through a common elementary signal estimator (ESE) device arranged with other blocks in iterative mode. Thus, receiving section of CIDMA system is the replica of an iterative decoding system [1–7]. Figure 1 shows CIDMA block model with minimum essential basic components. Here, multiple access channel (MAC) affected with additive white Gaussian noise (AWGN) is assumed. Total '*K*' users can access the MAC simultaneously.

Assume that the sequence from the *k*th user is $b_k = \{b_1, b_2, \ldots, b(n)\}^{\mathrm{T}}$. After spreading and encoding, it becomes a coded sequence $c_k = \{c_1, c_2, \ldots, c(J)\}^{\mathrm{T}}$; J = frame length. Now, c_k is interleaved with a pattern π_k. This provides a sequence $g_k = \{g_1, g_2, \ldots, g(J)\}^{\mathrm{T}}$ which is modulated and then transmitted over the MAC.

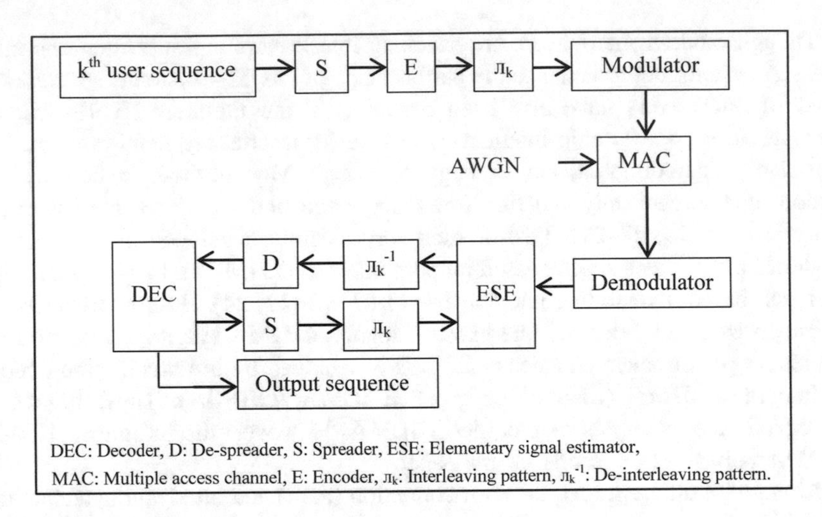

Fig. 1 CIDMA system model

The general output of MAC is given as

$$y(j) = \sum_{k=1}^{K} h_k.g_k(j) + \mathbb{N}(j);\ j = 1, 2, \ldots, J \tag{1}$$

where h_k is channel coefficient for user k and $\mathbb{N}(j)$ are AWGN samples with variance $\sigma^2 = No/2$ respectively. The sequence $y(j)$ is demodulated and then chip-by-chip detection is carried out with one sample of $y(j)$ at a time. Therefore,

$$y(j) = h_k.g_k(j) + \beta_k(j);\ j = 1, 2, \ldots, J \tag{2}$$

where $\beta_k(j)$ is distortion, i.e., noise plus interference component in $y(j)$. Mathematically,

$$\beta_k(j) = y(j) - h_k.g_k(j) = \sum_{k' \neq k} h_{k'}.g_{k'}(j) + \mathbb{N}(j) \tag{3}$$

From central limit theorem, $\beta_k(j)$ can be approximated as a Gaussian random variable and therefore, $y(j)$ can be approximated as a conditional Gaussian probability density function. Estimation of interference mean and variance and log-likelihood ratio (LLR) generation is achieved through chip-by-chip detection algorithm inside elementary signal estimator (ESE). The equation for generated LLR is given as follows:

$$e_{\text{ESE}}\{g_k(j)\} = 2h_k.\frac{y(j) - E(\beta_k(j))}{\text{Var}(\beta_k(j))} \tag{4}$$

Now, iteration process takes place between ESE and kth user decoder (DEC) to LLR exchange. In this process, de-spreading/spreading and de-interleaving/interleaving is performed multiple times to obtain soft-decoded outputs. The final outputs of ESE and DEC are in form of LLRs:

$$e_{\mathrm{ESE}} \text{ or } E_{\mathrm{DEC}}\{g_k(j)\} = \log \frac{p(y(j)|g_k(j) = +1)}{p(y(j)|g_k(j) = -1)} \quad \forall k, j \tag{5}$$

In the final iteration, a hard decision is made considering all the previous soft-decoded outputs.

4 Differential Phase-Shift Keying and Modern Fisher–Yates for Interleavers

DPSK is one of the versions of BPSK modulation scheme. In this, phases of current and previously received bits are compared. There is no additional reference phase. However, the phase of the transmitted sequence bits itself can be treated as a reference phase for this purpose. Unlike BPSK, DPSK modulation supports non-coherent demodulation of the received sequence which is an advantageous feature as CIDMA system uses iterative decoding [21–23, 25–27, 30, 31].

MFYI is random interleaver with computational complexity, time and memory requirement in order of N, i.e.,$\mathcal{O}(N)$; where N is the size of input data set. Hence, it is efficient than ordinary Fisher–Yates random shuffle (interleaving) method and has applications in computer communication. MFYI is beneficial when larger data is to be interleaved. This is most often the case in radio communication. A procedure *SCRAMBLE* for sequence $b_k[i]$; where i = 1, 2, … n is interleaved. The procedure *SCRAMBLE* shuffles the order of $b_k[i]$ randomly by producing a random pattern index for $b_k[i]$. So, an array b_k is declared in proper agreement with 'a'. The algorithm is given as follows:

```
procedure SCRAMBLE (bk, n, rand);
value n;   integer n;
real procedure rand;
integer array bk ;
begin
integer N, m;  real a;
for N : = n step - 1 until 2 do
   begin m : = enter (N X rand + 1);
         a : = bk [N];
      bk[N] : = bk[m];
      bk[m]: = a
   end loop N
end SCRAMBLE
```

Hence, $b_k(i$ = 1, 2 … $n)$ is randomly shuffled through MFYI.

5 Simulation Results and Discussion

Figures 2, 3, and 4 show bit error rate (BER) versus bit error to noise ratio (E_b/N_o) performance of CIDMA system under different cases presented here. All the simulation results shown here are obtained using MATLAB. For maintaining the simplicity and flexibility, repetition codes are employed as spreading codes and no forward error correction performed to focus on modulation and interleaving schemes only. Since flat-fading AWGN MAC is assumed. Therefore, $h_k = 1; \forall k$. Here, in all the simulation results, block size of 5, data length 600, and spread length 12 is used. A total of six iterations have been performed before making a hard decision.

Figure 2 shows the performance of CIDMA system with DPSK modulation and MFYI schemes for 6, 10, and 14 users. BER values are decreasing with increasing BSNR. It is clear from the figure that BER value as low as 10^{-4} can be obtained if bit energy is kept sufficiently high. Even when the No. of users are increased, BER is coming satisfactorily down which is a confirmatory sign that the combination of DPSK and MFYI is compatible with CIDMA system and can be recommended to employ in CIDMA system for different applications.

Figure 3 illustrates the comparison of MFYI and random interleaving schemes for 50 and 80 simultaneous users' cases assuming DPSK modulation in CIDMA system. Since random interleaving is a preferred interleaving scheme, so any interleaving scheme which produces the similar performance can be an acceptable choice. From Fig. 3, it is clear that both random and MFYI results almost overlaps equal No. of users. Therefore, this result also confirms the acceptability of MFYI in CIDMA system. Here, also the performance is significantly improving with increasing E_b/N_o.

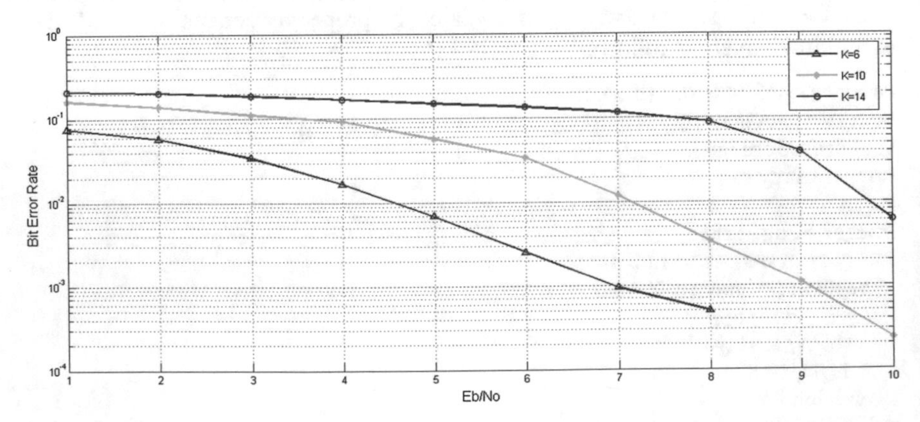

Fig. 2 Performance of CIDMA system with DPSK modulation and MFYI for $K = 6$, 10, and 14 users

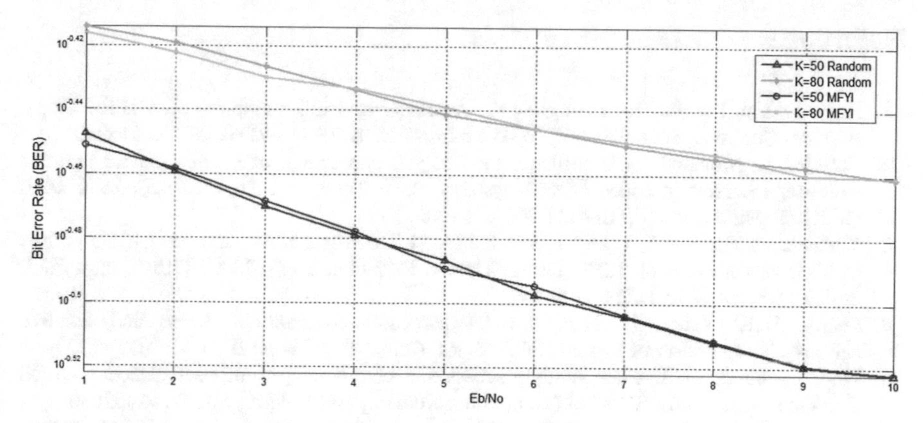

Fig. 3 Comparison of two interleaving techniques in CIDMA system with DPSK modulation

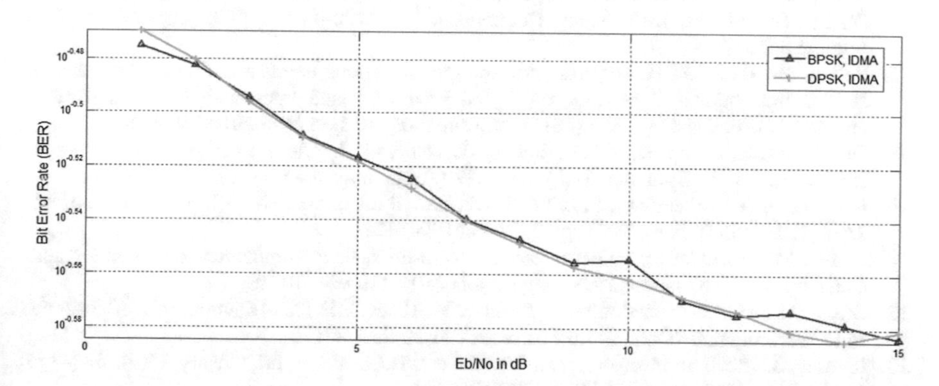

Fig. 4 Comparison of BPSK and DPSK modulation schemes in CIDMA system with MFYI

Figure 4 demonstrates the performance comparison of BPSK and DPSK modulation schemes in CIDMA system considering the presence of MFYI for 40 users. This result also verifies that both BPSK and DPSK are suitable schemes for CIDMA.

6 Conclusion

The performance of CIDMA system is acceptable with both BPSK and DPSK schemes. The two schemes provide almost similar performance in CIDMA system with MFYI in its core functioning. MFYI and general random interleaving approaches are very much comparable and can be a good alternate to each other in CIDMA systems. In future, this research work can be extended considering more advanced modulation schemes and different complex channel scenarios.

References

1. Ping, L., Liu, L., Wu, K., Leung, W.K.: Interleave-division multiple-access. IEEE Trans. Wireless Commun. **5**(4), 938–947 (2005). https://doi.org/10.1109/TWC.2006.1618943
2. Tarable, A., Montorsi, G., Benedetto, S.: Analysis and design of interleavers for iterative multiuser receivers in coded CDMA systems. IEEE Trans. Inf. Theory **51**(5), 1650–1666 (2005). https://doi.org/10.1109/TIT.2005.846848
3. Kusume, K., Bauch, G., Utschick, W.: IDMA vs. CDMA: analysis and comparison of two multiple access schemes. IEEE Trans. Wireless Commun. **11**(1), 78–87 (2012). https://doi.org/10.1109/twc.2011.111211.100954
4. Leung, W.K., Lihai, L., Ping, L.: Interleaving-based multiple access and iterative chip-by-chip multiuser detection. IEICE Trans. Commun. **E86-B**(12), 3634–3637 (2003)
5. Ping, L.: Interleave-division multiple access and chip-by-chip multi-user detection. IEEE Commun. Mag. **43**(6), S19–S23 (2005). https://doi.org/10.1109/MCOM.2005.1452830
6. Wu, H., Ping, L., Perotti, A.: User-specific chip-level interleaver design for IDMA system. IEE Electron. Lett. **42**(4), 233–234 (2006). https://doi.org/10.1049/el:20063770
7. Bruck, S., Sorger, U., Gilgorevic, S., Stolte, N.: Interleaving for outer convolutional codes in DS-CDMA systems. IEEE Trans. Commun. **48**(7), 1100–1107 (2000). https://doi.org/10.1109/26.855517
8. Yadav, M., Banerjee, P.: Bit error rate analysis of various interleavers for IDMA scheme. In: 3rd International Conference on Signal Processing and Integrated Networks (SPIN), pp. 89–94. IEEE, Noida, India (2016). https://doi.org/10.1109/spin.2016.7566668
9. Sadjadpour, H.R., Sloane, N.J.A., Salehi, M., Nebe, G.: Interleaver design for turbo codes. IEEE J. Sel. Areas Commun. **19**(5), 831–837 (2001). https://doi.org/10.1109/49.924867
10. Barbulescu, A.S., Pietrobon, S.S.: Interleaver design for turbo codes. Electron. Lett. **30**(25), 2107–2108 (1994). https://doi.org/10.1049/el:19941434
11. Moher, M.: An iterative multiuser decoder for near-capacity communications. IEEE Trans. Commun. **46**(7), 870–880 (1998). https://doi.org/10.1109/26.701309
12. Black, P.E.: Fisher-Yates shuffle. In: Pieterse, V., Black, P.E. (eds.) Dictionary of Algorithms and Data Structures (Online). Last Accessed on 20 Sept 2017
13. Ramsey, J.: Realization of optimum interleavers. IEEE Trans. Inf. Theory **16**(3), 338–345 (1970). https://doi.org/10.1109/TIT.1970.1054443
14. Durstenfeld, R.: Algorithm 235: random permutation. Commun. ACM **7**(7), 420 (1964). https://doi.org/10.1145/364520.364540
15. Yadav, M., Shokeen, V., Singhal, P.K.: BER versus BSNR analysis of conventional IDMA and OFDM-IDMA systems with tree interleaving. In: 2nd International Conference on Advances in Computing, Communication and Automation (ICACCA) (Fall), pp. 1–6. IEEE, Bareilly, India (2016). https://doi.org/10.1109/icaccaf.2016.7748973
16. Pupeza, I., Kavcic, A., Ping, L.: Efficient generation of interleavers for IDMA. In: International Conference on Communications (ICC), pp. 1508–1513. IEEE, Istanbul (2006). https://doi.org/10.1109/icc.2006.255024
17. Yadav, M., Gautam, P.R., Shokeen, V., Singhal, P.K.: Modern Fisher-Yates shuffling based random interleaver design for SCFDMA-IDMA systems. Wireless Pers. Commun. Springer. **97**, 1–11 (2017). https://doi.org/10.1007/s11277-017-4492-9
18. Nguyen, T.T.T., Lanante, L., Nagao, Y., Kurosaki, M., Yoshizawa, S., Ochi, H.: Low latency interleave division multiple access system. In: International Conference on Information Networking (ICOIN), pp. 7–12. IEEE, Da Nang (2017). https://doi.org/10.1109/icoin.2017.7899444
19. Agiwal, M., Roy, A., Saxena, N.: Next generation 5G wireless networks: a comprehensive survey. IEEE Commun. Surv. Tutor. **18**(3), 1617–1655 (2016). https://doi.org/10.1109/comst.2016.2532458

20. Elnoubi, S.M., Sourour, E., Elshamly, A.: Performance of multicarrier CDMA with DPSK modulation and differential detection in fading multipath channels. IEEE Trans. Veh. Technol. **51**(3), 526–536 (2002). https://doi.org/10.1109/TVT.2002.1002501
21. Wetzker, G., Dukek, M., Ernst, H., Jondral, F.: Multi-carrier modulation schemes for frequency-selective fading channels. In: International Conference on Universal Personal Communications (ICUPC), vol. 2, pp. 939–943. IEEE, Florence (1998). https://doi.org/10.1109/icupc.1998.733647
22. Islam, M., Ahmed, N., Ali, S., Aljunid, S.A., Ahmed, B.B., Sayeed, S.: Design new hybrid system using differential phase shift keying (DPSK) modulation technique for optical access network. In: 3rd International Conference on Electronic Design (ICED), pp. 395–399. Phuket, Thailand (2016). https://doi.org/10.1109/iced.2016.7804676
23. Ninacs, T., Matuz, B., Liva, G., Colavolpe, G.: Non-binary LDPC coded DPSK modulation for phase noise channels. In: International Conference on Communications (ICC), pp. 1–6. IEEE, Paris, (2017). https://doi.org/10.1109/icc.2017.7996748
24. Yadav, M., Shokeen, V., Singhal, P.K.: Flip left-right approach based novel inverse tree interleavers for IDMA scheme. AEU Int. J. Electron. Commun. Elsevier. **81**, 182–191 (2017). https://doi.org/10.1016/j.aeue.2017.07.025
25. Elaiyarani, K., Kannadasan, K., Sivarajan, R.: Study of BER performance of BCJR algorithm for turbo code with BPSK modulation under various fading models. In: International Conference on Communication and Signal Processing (ICCSP), pp. 2323–2326. IEEE, Melmaruvathur, India (2016). https://doi.org/10.1109/iccsp.2016.7754111
26. Madankar, M.M., Ashtankar, P.S.: Performance analysis of BPSK modulation scheme for different channel conditions. In: Students' Conference on Electrical, Electronics and Computer Science (SCEECS), pp. 1–5. IEEE, Bhopal, India (2016). https://doi.org/10.1109/sceecs.2016.7509290
27. Abderrahmane, L.H., Chellali, S.: Performance comparison between Gaussian interleaver, Rayleigh interleaver and dithered golden interleaver. Ann. Telecommun. Springer. **63**, 449–452 (2008). https://doi.org/10.1007/s12243-008-0037-2
28. Ren, D., Ge, J., Li, J.: Modified collision-free interleavers for high speed turbo decoding. Wireless Pers. Commun. Springer. **68**, 939–948 (2013). https://doi.org/10.1007/s11277-011-0491-4
29. Shukla, M., Srivastava, V.K., Tiwari, S.: Analysis and design of optimum interleaver for iterative receivers in IDMA scheme. Wiley J. Wirel. Commun. Mob. Comput. **9**(10), 1312–1317 (2008). https://doi.org/10.1002/wcm.710
30. Chen, Y., Schaich, F., Wild, T.: Multiple access and waveforms for 5G: IDMA and universal filtered multi-carrier (2015). https://doi.org/10.1109/vtcspring.2014.7022995
31. Yadav, M., Shokeen, V., Singhal, P.K.: Testing of Durstenfeld's algorithm based optimal random interleavers in OFDM-IDMA systems. In: 3rd International Conference on Advances in Computing, Communication and Automation. IEEE, Dehradun, India (2017)

Defected Ground Structure Integrated Rectangular Microstrip Patch Antenna on Semi-insulating Substrate for Improved Polarization Purity

Abhijyoti Ghosh and Banani Basu

Abstract A simple rectangular microstrip antenna with defected ground structure on semi-insulating substrate (GaAs) has been studied to improve the polarization purity (co-polarization to cross-polarization isolation) in principle *H*-plane. Wide bandgap with high dielectric constant makes GaAs a very good substrate for Monolithic Microwave Integrated Circuits (MMIC). Unlike the earlier reports, the present one can exhibit much better radiation performance with the simple configuration. In the present paper, initially the effect of dielectric constant of a substrate on radiation performance of a simple conventional patch configuration has been studied and documented. Based on the investigation, defected ground structure has been employed for obtaining much better radiation characteristics. Around 69% improvement in cross-polarization isolation is revealed from the present configuration compared to classical patch antenna. Furthermore, such improvement of polarization purity is maintained over a wide elevation angle. Along with the improvement in polarization purity, the present structure also reveals a broad and stable beam pattern at its design frequency.

Keywords Microstrip antenna · Defected ground structure (DGS) · Gallium arsenide · Polarization purity

1 Introduction

Rectangular microstrip patch antenna (RMA) is the most common and suitable microwave radiator that find potential applications in modern wireless systems due to its several advantages like lightweight, tininess, and easy fabrication process.

A. Ghosh (✉)
Department of ECE, Mizoram University, 796004 Aizawl, Mizoram, India
e-mail: abhijyoti_engineer@yahoo.co.in

B. Basu
Department of ECE, National Institute of Technology-Silchar, 788010
Silchar, Assam, India

A. K. Luhach et al. (eds.), *Smart Computational Strategies: Theoretical and Practical Aspects*, https://doi.org/10.1007/978-981-13-6295-8_15

Apart from all these advantages, conventional RMA suffers low gain, narrow bandwidth, and poor co-polarization to cross-polarization isolation [1, 2]. Conventional RMA radiates some degree of cross-polarization (XP) radiation in its higher order orthogonal mode, while radiating linearly polarized (Co-polarization (CP)) electrical fields along the broadside direction in its fundamental TM_{10} mode. The cross-polarized (XP) radiation from first higher order orthogonal mode (i.e., TM_{02}) is much severe and affects the performance of conventional RMA in some applications where polarization purity is the primary requirement [3]. In probe-feed design of RMA, the XP is more prominent when the thickness increases as well as the dielectric constant of the substrate decreases [4]. Thus, achieving high CP-XP isolation (polarization purity) becomes a challenging topic for antenna research community.

Different techniques like modification of feed structure, aperture-coupled dual polarization have been reported to improve the polarization purity of RMA. Cross-polarization level below −20 dB in both planes has been reported using meander strip line feeding [5] and half wavelength probe-feed suspended patch antenna [6]. Stacked patch structure [7] and stacked patch structure with the "mirrored pair" [8] feeding technique have been employed to obtain CP-XP isolation of 15–20 dB. Nevertheless, the reported feeding structures are very complex. 23 dB of CP-XP isolation is reported using dual-polarized aperture-coupled microstrip patch antenna [9], while the same is −40 dB only in broadside using aperture-coupled microstrip antenna with T-shaped feed [10]. However, all these structures require very complex feeding mechanism. Around 23 dB of CP-XP isolation using simple slot type and dumbbell-shaped defected ground structure is reported in [11, 12]. One very recently reported article [13] has dealt with cross-type defected ground structure in RMA to attain a polarization purity of 24 dB over a wide elevation angle around the broadside direction in *H*-plane. Cross-headed dumbbelled defected patch and circular arc-defected patch surfaces have been used in [14, 15] to achieve polarization purity of 28 and 25 dB, respectively. All the above research work has been done with a single-layer dielectric substrate where dielectric constant is 2.33 only.

In order to alleviate the complexity of the earlier reported structures, in the present investigation, first, a conventional RMA on semi-insulating substrate has been investigated. Gallium arsenide (GaAs) has been used as a substrate for conventional RMA. In fact, some properties of GaAs like larger band gap, higher carrier mobility, etc. compared to silicon are very promising to the electronic device manufacturing industry. The larger bandgap helps manufacturing industry to design devices which can be operated at higher temperatures than Si [16]. Finally, higher carrier mobility of GaAs compared to silicon helps semiconductor manufacturer to design ultrahigh-frequency devices as is clearly indicated in the same report. Moreover, the GaAs substrate is less vulnerable to heating effect as well as noise due to its higher band gap at high frequency. So, this is used to develop microwave

Gunn sources. Therefore, integrating the antenna with other microwave components on same substrate GaAs has been preferred. It is apparent that the use of such substrate improves the RMA radiation characteristics specifically, and the polarization purity of RMA improves dramatically with such substrate.

In the second step of the present investigation, an “I”-shaped DGS has been employed to improve its polarization characteristic to higher degree.

2 Parametric Studies and Proposed Structure

2.1 *Conventional RMA on Different Substrates*

In the first part of the investigation, a conventional RMA with different substrates has been studied. Substrates like RT-DURIOD ($\varepsilon_r = 2.33$), FR4 ($\varepsilon_r = 4.4$), glass ($\varepsilon_r = 5.5$), and GaAs ($\varepsilon_r = 12.9$) have been considered for the investigation. The XP performance of different substrate-integrated RMAs is shown in Fig. 1. It can be seen from the figure that as the value of dielectric constant increases from 2.33 to 12.9, XP performances improve and for GaAs it is the almost around −28 dB which is much better than other materials. Therefore, it may be concluded that the polarization purity improves with high dielectric constant substrate at the cost of lower gain. However, gain may be enhanced with the help of composite substrate [17] or with a substrate having air cavity [18].

2.2 *DGS Integrated RMA on GaAs Substrate*

According to the cavity model analysis of conventional RMA, top (patch) and bottom (ground plane) metals form electric walls, while four open sides are considered to be magnetic walls. Any defect on the ground plane perturbs the

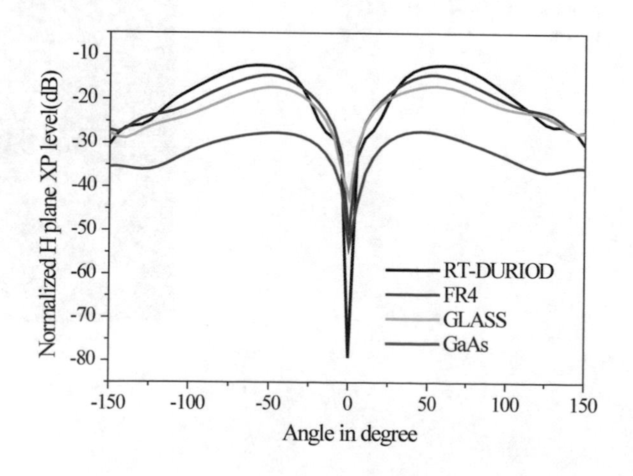

Fig. 1 Simulated normalized *H*-plane XP pattern for different substrate-integrated conventional RMAs

electromagnetic fields below the patch and consequently affects the radiation characteristics. Electric fields correspond to the first higher order orthogonal mode (TM_{02}) which are situated near the non-radiating edges which are the major factor that contributes to XP radiation. Therefore, to reduce the effect of electric fields near the non-radiating edges, an "I"-shaped defect (Fig. 2) has been placed just below the patch such a way that it hampers the radiations from the non-radiating edges. The optimized structure of the defect is achieved by detailed parametric study.

At first, the effect of width of the vertical slot (*sw*) on the XP performance has been studied keeping $a = 1.6$ mm and $l = 4$ mm fixed. Figure 3 shows the normalized *H*-plane XP radiation pattern. It is clear from Fig. 3 that $sw = 1.5$ mm provides uniform XP pattern below −35 dB over whole elevation angle. The minimum CP-XP isolation for the variation of "*sw*" from 1 to 2.5 mm is shown in Fig. 4. Minimum CP-XP isolation of 35 dB is achieved with $sw = 1.5$ mm. So, the optimum value of *sw* is kept at 1.5 mm which provides −35 dB of CP-XP isolation.

Now the effect of "*a*" keeping $sw = 1.5$ mm (optimum) and $l = 4$ mm on XP performance has been investigated and presented in Fig. 5. The figure clearly shows that with the variation of "*a*", XP performance improves. The best XP performance is obtained with $a = 1.6$ mm which is below −36 dB over whole elevation angle between ±150°. Hence, optimum value of "*a*" is fixed at 1.6 mm.

Next, keeping $sw = 1.5$ mm (optimum) and $a = 1.6$ mm (optimum), the normalized *H*-plane XP performance of proposed structure as a function of "*l*" has been presented in Fig. 6. It is clearly evident from the figure that as the value of "*l*" is increased XP performance improves and is best with $l = 6$ mm where CP-XP isolation is approximately 37 dB. Further increment of "*l*" may introduce perturbation to the field at radiating edges which in turn may hamper the dominant mode radiation of the proposed structure. Therefore, we refrain to increase the value of "*l*" above 6 mm (Table 1).

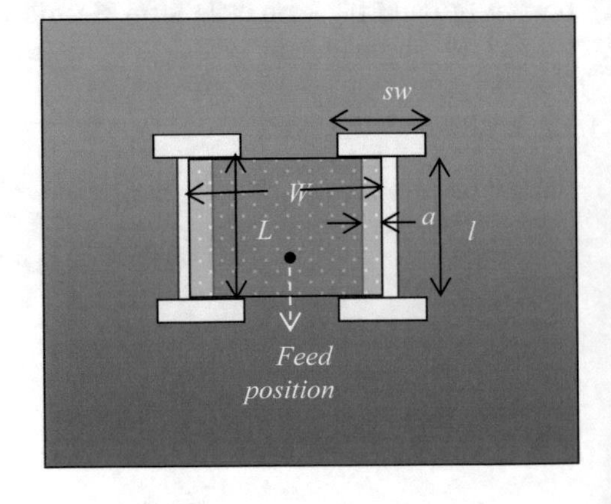

Fig. 2 Schematic representation (top view) of proposed IDGS-integrated RMA

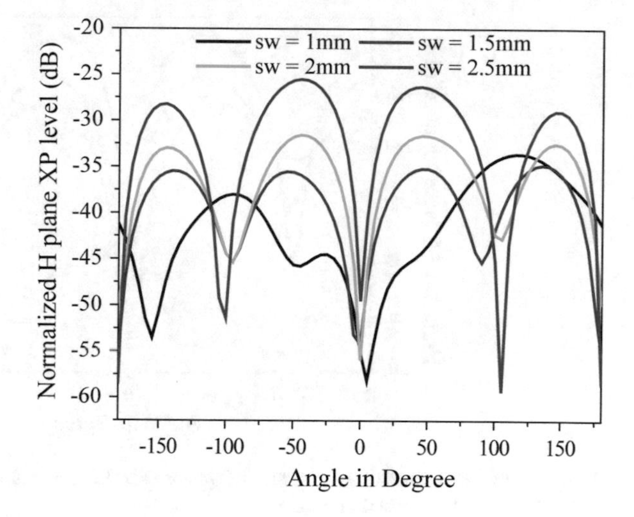

Fig. 3 Simulated normalized *H*-plane XP pattern for different values of "*sw*" with fix "*a*" and "*l*" of proposed structure

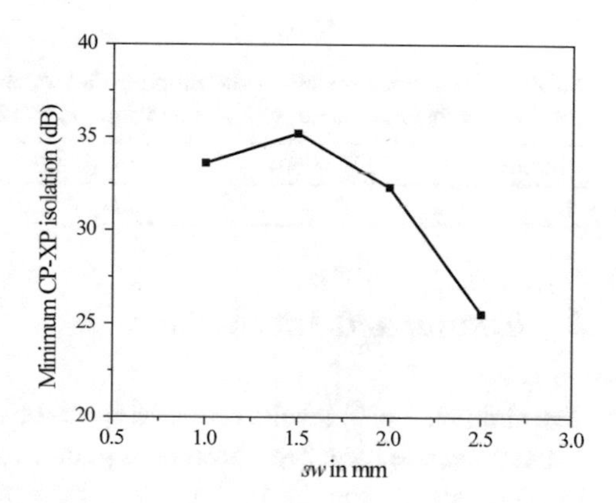

Fig. 4 Variation of CP-XP isolation as a function of "*sw*" keeping $a = 1.6$ mm (fixed) and $l = 4$ mm (fixed)

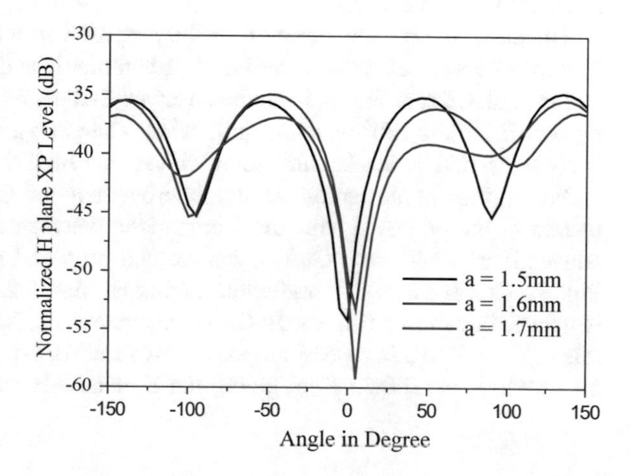

Fig. 5 Simulated normalized *H*-plane XP pattern for different values of "*a*" with $l = 4$ mm (fixed) and $sw = 1.5$ mm (optimum) of proposed structure

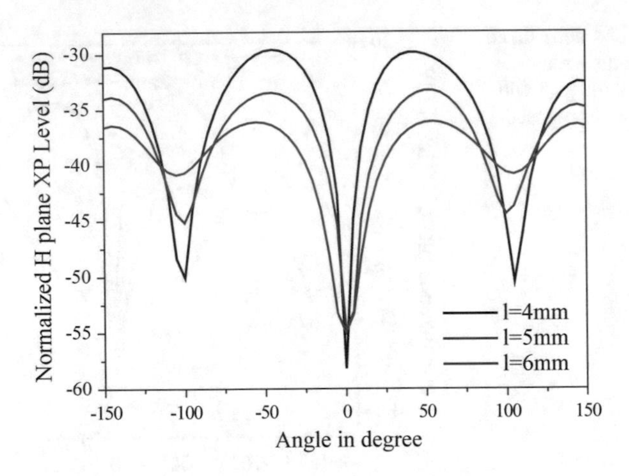

Fig. 6 Simulated normalized *H*-plane XP pattern for different values of "*l*" with optimum value of "*sw*" and "*a*" of proposed structure

Table 1 Detail parameters of the proposed RMA on 70 mm × 70 mm ground plane with GaAs substrate. Substrate thickness h = 1.575 mm, ε_r = 12.9

L (mm)	*W* (mm)	*sw* (mm)	*a* (mm)	*l* (mm)
8	12	1.5	1.6	6

3 Results and Discussions

The simulated [19] results obtained from the conventional and proposed RMA with GaAs substrate have been presented in this section. The reflection coefficient profile is presented in Fig. 7. It is evident from the figure that both the antennas are resonating near 4.5 GHz.

The complete radiation pattern of the proposed structure is presented in Fig. 8. The co-polarized radiation pattern is quite symmetric and stable in both principle *E*- and *H*-planes. From Fig. 8, it can be observed that the 3 dB beamwidth, in principle *E*-plane is almost 130° which is quite wide. The peak gain of the proposed antenna is around 6 dBi which is quite good with such a high dielectric constant substrate.

For further confirmation of the improvement of cross-polarization radiation pattern of the proposed structure, comparison between the XP performance of the conventional RMA with GaAs substrate and proposed structure is documented in Fig. 9. The *H*-plane XP radiation improves significantly in case of proposed structure. For simple RMA with GaAs substrate peak, XP level is at −28 dB, while it is below −37 dB in case of proposed structure. As *E*-plane XP radiation is used to be very low, we refrain from giving the *E*-plane XP pattern for brevity.

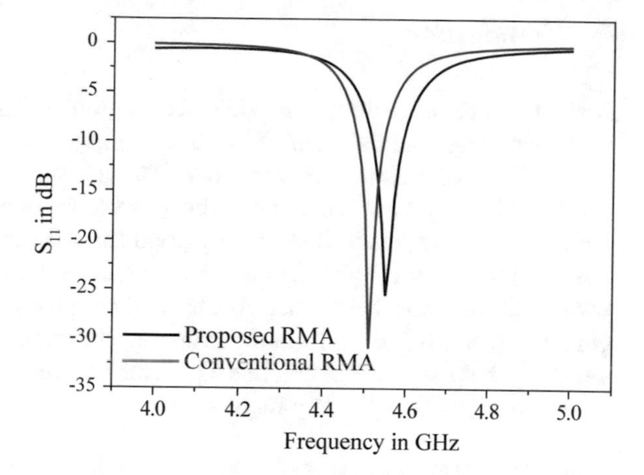

Fig. 7 Reflection coefficient profile of conventional and proposed IDGS-integrated RMA

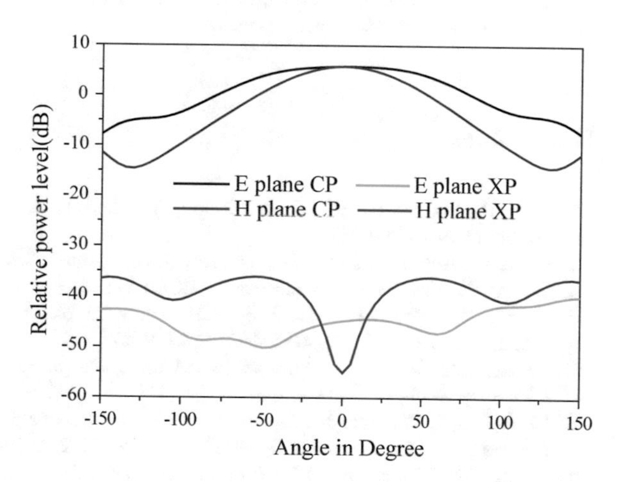

Fig. 8 Radiation pattern both *E*- and *H*-planes of proposed RMA at center frequency

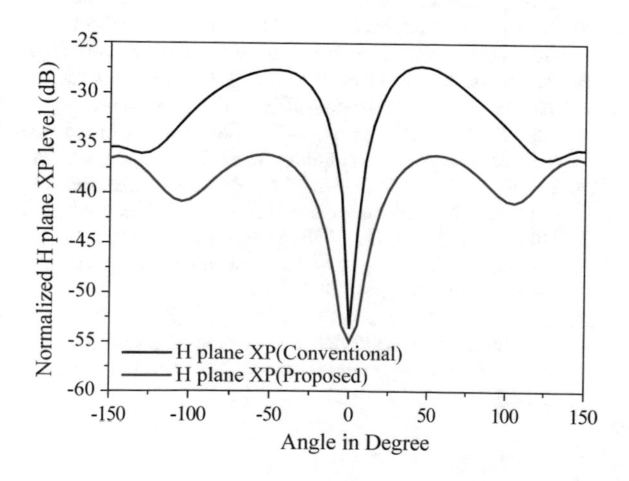

Fig. 9 Radiation pattern (*H*-plane) of conventional and proposed RMA at corresponding center frequency

4 Conclusions

A simple single-element "I"-shaped defected ground structure integrated RMA with semi-insulating material (GaAs) as substrate is proposed for significant improvement of CP-XP isolation performance. The use of semi-insulating material as a substrate is very new technique. The properties, like wide bandgap with high dielectric constant, make GaAs a very good substrate for MMIC. More than 37 dB polarization purity (CP-XP isolation) over a wide elevation angle ($\pm 150^{\circ}$) along with wide and stable beam width is obtained with the proposed structure. Moderate gain of 6 dBi is achieved from the proposed structure. The proposed structure will definitely helpful for the wireless applications, where moderate gain with high polarization purity is the key requirement.

Acknowledgements The authors would like to thank Dr. Sudipta Chattopadhyay, Mizoram University, India, Subhradeep Chakraborty, CSIR-CEERI, Pilani, India for their support by which the whole fact is established in the present form.

References

1. Garg, R., Bhartia, P., Bahl, I., Ittipiboon, A.: Microstrip Antenna Design Handbook. Artech House, Norwood (2001)
2. Guha, D., Antar, Y.M.M. (eds.): Microstrip and Printed Antennas—New Trends, Techniques and Applications. Wiley, Chichester, UK (2011)
3. Huynh, T., Lee, K.F., Lee, R.Q.: Cross-polarization characteristics of rectangular patch antennas. Electron. Lett. **24**(8), 463–464 (1988)
4. Petosa, A.I., Gagnon, N.: Suppression of unwanted probe radiation in wide band probe-fed microstrip patches. Electron. Lett. **35**(5), 355–357 (1999)
5. Li, P., Lai, H.W., Luk, K.M., Lau, K.L.: A wideband patch antenna with cross-polarization suppression. IEEE Antennas Wirel. Propag. Lett. **3**, 211–214 (2004)
6. Chen, Z.N., Chia, M.Y.W.: Broad-band suspended probe-fed plate antenna with low cross-polarization level. IEEE Trans. Antennas Propag. **51**, 345–346 (2003)
7. Loffler, D., Wiesbeck, W.: Low-cost X-polarised broadband PCS antenna with low cross-polarisation. Electron. Lett. **35**(20), 1689–1691 (1999)
8. Granholm, J., Woelders, K.: Dual polarization stacked microstrip patch antenna array with very low cross-polarization. IEEE Trans. Antennas Propag. **49**, 1393–1402 (2001)
9. Gao, S., Li, L.W., Leong, M.S., Yeo, T.S.: A broad-band dual-polarized microstrip patch antenna with aperture coupling. IEEE Trans. Antennas Propag. **51**, 898–900 (2003)
10. Lai, C.H., Han, T.Y., Chen, T.R.: Broadband aperture coupled microstrip antennas with low cross polarization and back radiation. Prog. Electromagn. Res. Lett. **5**, 187–197 (2008)
11. Ghosh, A., Ghosh, D., Chattopadhyay, S., Singh, L.L.K.: Rectangular microstrip antenna on slot type defected ground for reduced cross polarized radiation. IEEE Antennas Wirel. Propag. Lett. **14**, 324–328 (2015)
12. Ghosh, A., Chakraborty, S., Chattopadhyay, S., Basu, B.: Rectangular microstrip antenna with dumbbell shaped defected ground structure for improved cross polarised radiation in wide elevation angle and its theoretical analysis. IET Microw. Antennas Propag. **10**, 68–78 (2016)

13. Ghosh, A., Chattopadhyay, S., Chakraborty, S., Basu, B.: Cross type defected ground structure integrated microstrip antenna: a novel perspective for broad banding and augmenting polarization purity. J. Electromagn. Waves Appl. **31**(5), 461–476 (2017)
14. Ghosh, A., et al.: Rectangular microstrip antenna with cross headed dumbbell defected patch surface for improved polarization purity. In: International Conference on Microwave and Photonics, Dhanbad, India (2015)
15. Ghosh, A., et al.: Rectangular microstrip antenna with defected patch surface for improved polarization purity. In: International Conference on Computer, Communication, Control and Information Technology, Hooghly, India (2015)
16. Kohl, F., Gutberlet, F.G.V.: Gallium-arsenide-the material and its application. Microelectron. J. **12**(3), 5–8 (1981)
17. Chattopadhyay, S., Siddiqui, J.Y., Guha, D.: Rectangular microstrip patch on a composite dielectric substrate for high-gain wide-beam radiation patterns. IEEE Trans. Antennas Propag. **57**, 3324–3327 (2009)
18. Malakar, K., et al.: Rectangular microstrip antenna with air cavity for gain and improved front to back ratio. J. Electromagn. Anal. Appl. **3**, 368–371 (2011)
19. HFSS: High frequency structure simulator, Ver. 14. Ansoft Corp., USA

Design of SIW-Fed Broadband Microstrip Patch Antenna for E-Band Wireless Communication

Maharana Pratap Singh, Rahul Priyadarshi and Prashant Garg

Abstract A microstrip broadband antenna with center frequency 86 GHz is proposed for wireless communication using proximity-coupled feed method which is composed by a thin circular slot etched on broader wall of surface-integrated waveguide (SIW). SIW consists of short-circuited metallic via inserted inside the ring slots which acts as sidewalls of SIW. In the proposed antenna, SIW is designed on 0.5-mm-thick Roger 5880 material, and a patch surface is placed above another substrate of lower refractive index than SIW substrate. The results obtained by simulation show the satisfactory response of designed antenna with return loss of approximately 39 dB, active VSWR < 1 dB at resonant frequency, and 10 dB return loss bandwidth from 82.4 to 100.2 GHz, that is, 20.69% of the fractional bandwidth, which shows that it is broadband antenna. Hence, this antenna is very useful for E-band with very high-speed wireless communication.

Keywords Microstrip patch antenna · Broadband antenna · E-band · SIW · Circular slot · Proximity-coupled feed

1 Introduction

The E-band of frequencies spectrum that covers the range of 71–76, 81–86, and 92–95 GHz offers a huge bandwidth of 13 GHz. Also, atmospheric absorption in free space is very less (<1 dB per kilometer) for this frequency band. These band of frequencies are very suitable for the users as they are used in a wide range of

M. P. Singh · R. Priyadarshi (✉) · P. Garg
Electronics and Communication Engineering Department,
National Institute of Technology, Hamirpur, India
e-mail: rahul.glorious91@gmail.com

M. P. Singh
e-mail: maharanapratap33@gmail.com

P. Garg
e-mail: Prash.garg@gmail.com

A. K. Luhach et al. (eds.), *Smart Computational Strategies: Theoretical and Practical Aspects*, https://doi.org/10.1007/978-981-13-6295-8_16

innovative products and services such as very high-speed point-to-point WLAN, broadband Internet, etc. Potential mobile communications for next generation have also maturated in E-band. To access abovementioned services related to wireless communication system, antenna is one the key elements which establishes a point-to-point, line-of-sight, or high-speed wireless communication link between transceiver units. These band of frequencies which are useful in the technology that support high-speed data transfer across network because of large bandwidth can be an alternative to fiber optics. Nowadays, microstrip patch antenna is precisely fascinated for the scholars because of its attractive features. In simulated microstrip antenna, we have proposed substrate-integrated waveguide (SIW) [1] as feed line, which is very compact and hence can be effortlessly assimilated with planer circuit [2]. In SIW, cylindrical vias are working as vertical walls of the waveguide having well-defined diameter "d" and parting gap "p".

We can calculate active breadth (a_s) of the structure equivalent to rectangular waveguide breadth (a_d) as presented in Fig. 1.

$$a_s = a_d + \frac{d^2}{0.95p} \tag{1}$$

where

$$a_d = \frac{c}{2\sqrt{\varepsilon_r} f_c} \tag{2}$$

and

$$f_c = \frac{c}{2\pi}\sqrt{\left(\frac{m\pi}{a}\right)^2 + \left(\frac{n\pi}{b}\right)^2} \tag{3}$$

This SIW structure is a novel form of transmission line to produce extraordinary results from precise compacted planar circuits. In few years, many scholars have proposed planar circuits for frequencies below 12 GHz only. But with latest VLSI techniques we can integrate circuits in micron range [3]; that is why, it is also

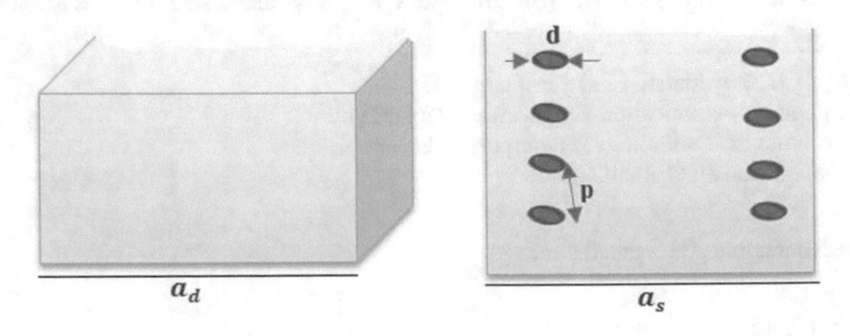

Fig. 1 Rectangular W/G versus SIW

possible to design SIW [4] antenna in E-band frequency range. Here, in case of SIW, diameter of the individual vias and separation between two vias will affect the amount of energy flowing through SIW and bandwidth of the waveguide. The structure as shown in Fig. 2 contains conducting cylinder-shaped vias having diameter "d" and separation "P", which is acting like vertical walls of the waveguide. There are several benefits of using SIW in planar circuits but there are few losses involved. Largely, three types of losses happen in SIW which are conductor loss, dielectric loss, and radiation leakage loss. Conductor and dielectric losses are identical in SIW and in rectangular waveguide; however, the radiation leakage is a feature of SIW only. One main advantage of SIW is that the amount of metal that carries the signal is far greater than that would be in stripline. Therefore, conductor loss is minimum. But dielectric and radiation leakage losses can be substantial. Dielectric loss is proportional to frequency; hence, it increases as we go for higher frequency. Radiation leakage depends on how close the vias are spaced. In order to reduce leakage losses, condition given in Eq. (4) has to be satisfied.

$$d < \frac{\lambda_g}{5} \text{ and } P < 2d \tag{4}$$

where

$$\lambda_g = \frac{2\pi}{\sqrt{\frac{\varepsilon_R (2\pi f)^2}{c^2} - \left(\frac{\pi}{a}\right)^2}} \tag{5}$$

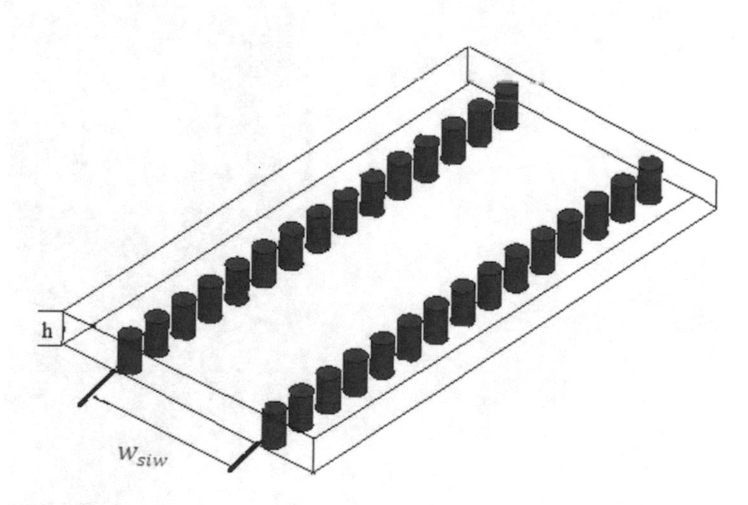

Fig. 2 SIW layout

2 Antenna Configuration

The antenna structure is shown in Fig. 3. It consists of SIW as a feed line [5] which is having dielectric constant $\varepsilon_r = 2.2$ and loss tangent $\tan\delta = 0.0009$ (Rogers RT/duroid 5880 (tm)) and thickness $h_1 = 0.508\,\text{mm}$. Circular slot is engraved on the broad wall of the SIW outer diameter 1 mm, and we have considered two inner diameters of 0.8 and 0.85 mm, respectively, and compared the results in order to find the effect on antenna characteristics for different slot dimensions. From the simulated results as shown in Figs. 4 and 5, we can see that return loss is better for smaller slot width but application bandwidth reduces. Slot is positioned in such a way that it couples maximum energy to the patch surface. Another substrate "Neltec NY9208 (IM) (tm)" having dielectric constant $\varepsilon_r = 2.08$, loss tangent $\tan\,\delta = 0.0006$, and thickness $h_2 = 0.6\,\text{mm}$ is placed above the SIW, which is the second substrate between patch surface and SIW feed line. Patch is placed above the second substrate, and it is centered above the circular slot [6] as shown in Fig. 3. Width of SIW is taken as 2.3 mm; for this antenna simulation, we have considered diameter and separation of metallic vias 0.25 and 0.5 mm, respectively. Amount of coupling from SIW to patch surface [7] depends upon slot width, patch dimensions, and its position with respect to slot. The characteristics impedance of feedline is 50 Ω.

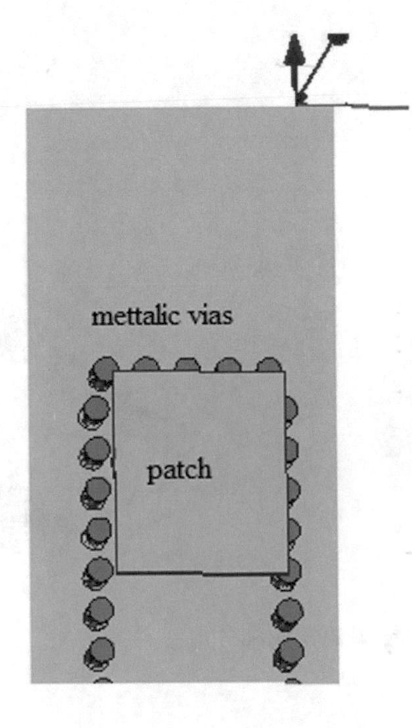

Fig. 3 Configuration of proposed antenna

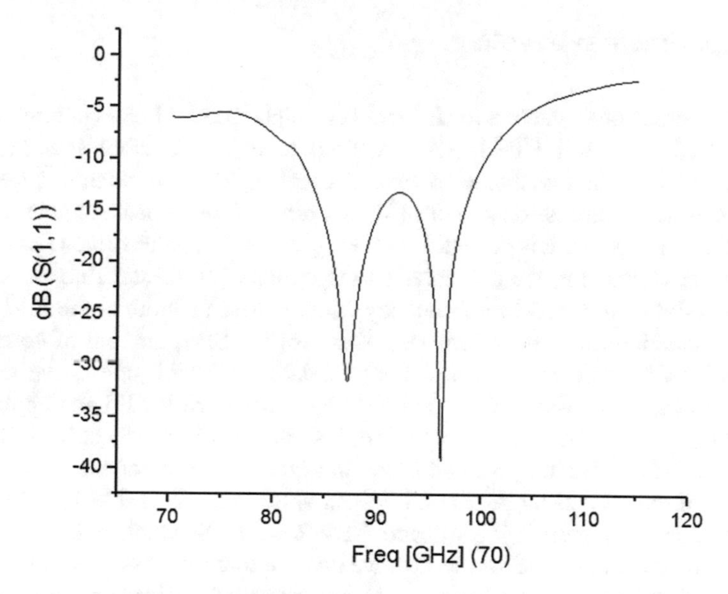

Fig. 4 Simulated S_{11} for slot width = 0.2 mm

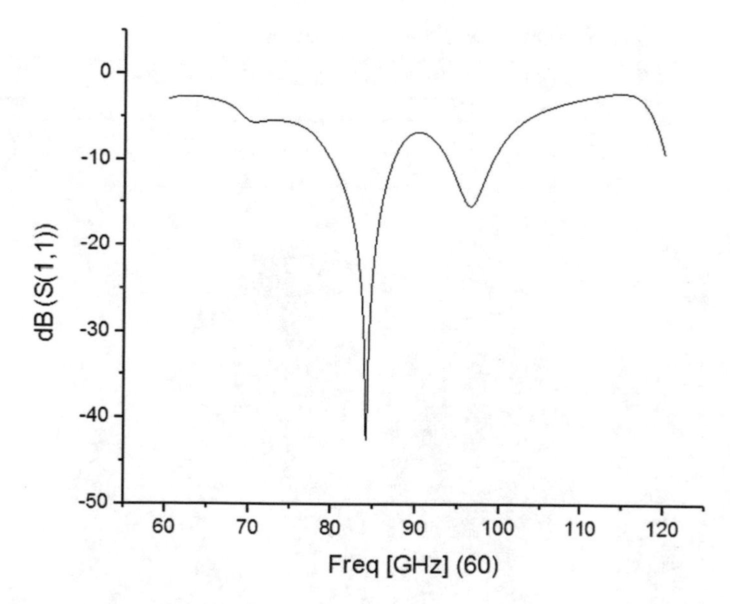

Fig. 5 Simulated S_{11} for slot width = 0.15 mm

3 Experimental Results

Microstrip broadband antenna for E-band very high-speed wireless communication is simulated in Ansoft HFSS (13.0 version) to examine return loss S_{11}, gain, VSWR, and its radiation characteristics. Center frequency of antenna depends on the sizes of its numerous constituents [8], i.e., dimensions of patch structure, width, and slot position where it is placed. By altering slot width, patch dimensions, and its position above the slot, we can enhance the outputs [9]. Center frequency of the antenna is shifted toward lower frequency with increase in width of the SIW. As we keep slot position away from center of Y-axis of the SIW, amount of coupling to patch increases considerably because net cancelation of field is nonzero as we move away from center of Y-axis. Proposed antenna is designed for E-band applications [10]. The experimental results show return loss S_{11}, nearly 33 dB at 87.1 GHz and 39.2 dB at 96.1 GHz for slot thickness 0.2 mm, whereas for slot thickness of 0.15 mm, resonant peak shifted at 84.6 GHz valued 45 dB (Table 1). At resonant frequency, antenna gain is 7.8 dB and VSWR 0.32 dB which is less than 1 dB required for good design. From Fig. 8, we can see that the field is confined within the SIW boundary and also coupling of energy to patch through circular slot is maximum [11, 12] (Figs. 6, 7, 8, and 9).

Table 1 Comparison of results for two different slot widths

Slot width (mm)	Res. freq. (f_0) (GHz)	−10 dB bandwidth (GHz)	Return loss (s_{11}) (dB)	VSWR (dB)
0.2	87.1, 96.1	82.4–100.2	33, 39.2	0.9, 0.78
0.15	84.6	79.6–87.2	45	0.32

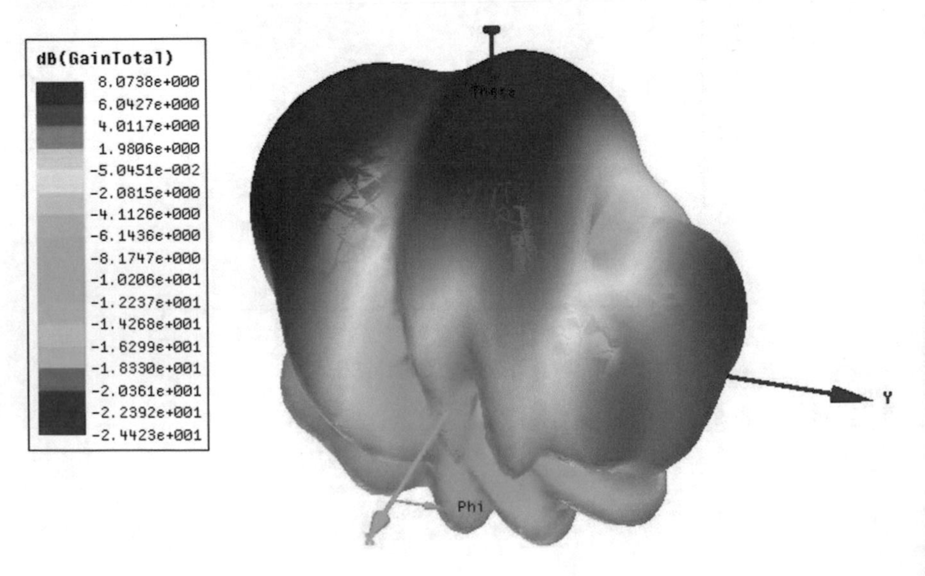

Fig. 6 Simulated gain

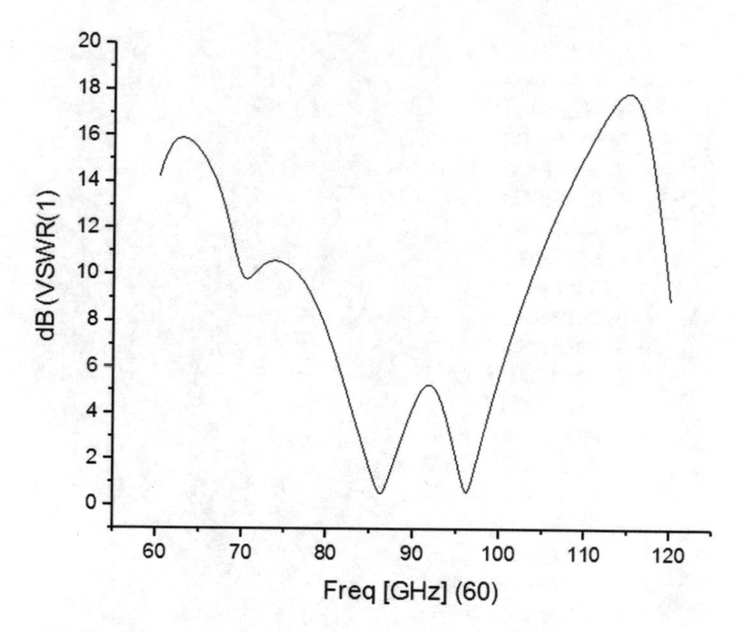

Fig. 7 Simulated VSWR for slot width = 0.2 mm

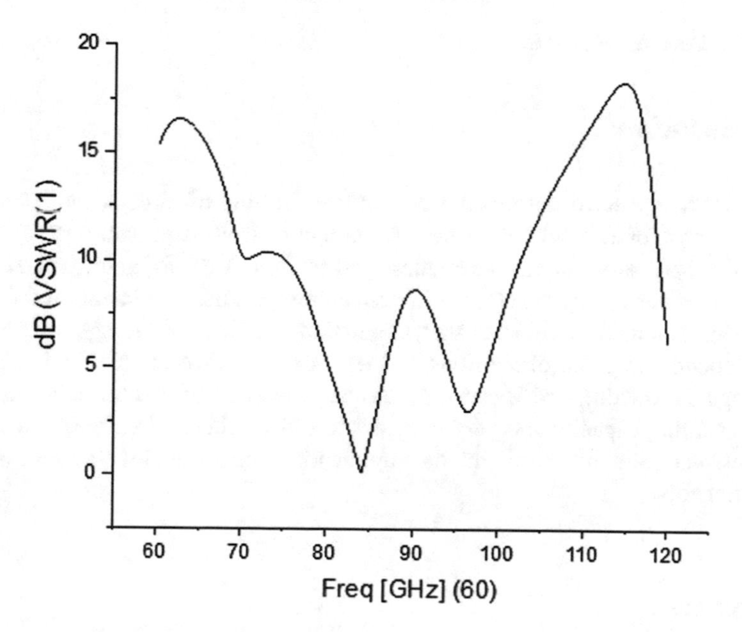

Fig. 8 Simulated VSWR for slot width = 0.15 mm

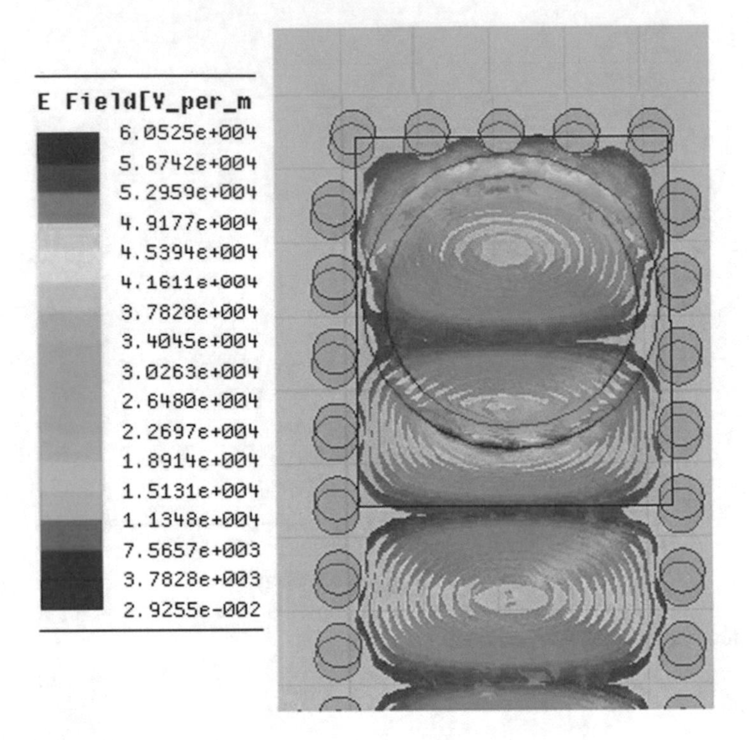

Fig. 9 Field overlays of E-field

4 Conclusion

In this work, we have described the simulated results of circular slot microstrip patch antenna for E-band with a new technique of feed using substrate-integrated waveguide. All antenna characteristics (return loss, VSWR, gain, and radiation pattern) have been analyzed at two different slot widths that can be inferred from the above simulation. It is detected that amount of coupling of energy from SIW to patch depends on patch dimensions and its location above the slot. E-band offers very large bandwidth, and it can generate large amount of power. SIW has larger power handling capacity and very compact to satisfy planar circuit criteria. Hence, this antenna can be used in the application of very high-speed wireless communication.

References

1. Menzel, W., Kassner, J.: Millimeter-wave 3D integration techniques using LTCC and related multilayer circuits. In: 2000 30th European Microwave Conference, EuMC 2000 (2000)

2. Deslandes, D., Wu, K.: Design consideration and performance analysis of substrate integrated waveguide components. In: 32nd European Microwave Conference 2002, pp. 1–4 (2002)
3. Hong, W.H.W.: Development of microwave antennas, components and subsystems based on SIW technology. In: 2005 IEEE International Symposium on Microwave, Antenna, Propagation and EMC Technologies for Wireless Communications, vol. 1, pp. P-14–P-17 (2005)
4. Parment, F., Ghiotto, A., Vuong, T.P., Duchamp, J.M., Wu, K.: Air-filled substrate integrated waveguide for low-loss and high power-handling millimeter-wave substrate integrated circuits. IEEE Trans. Microw. Theory Tech. **63**, 1228–1238 (2015)
5. Belen, M.A., Günes, F., Çaliskan, A., Mahouti, P., Demirel, S., Yildirim, A.: Microstrip SIW patch antenna design for X band application. In: 2016 21st International Conference on Microwave, Radar and Wireless Communications, MIKON 2016 (2016)
6. Awida, M.H., Fathy, A.E.: Design guidelines of substrate-integrated cavity backed patch antennas. IET Microw. Antennas Propag. **6**, 151–157 (2012)
7. Iwasaki, H.: A circularly polarized small-size microstrip antenna with a cross slot. IEEE Trans. Antennas Propag. **44**, 1399–1401 (1996)
8. Priyadarshi, R., Singh, M.P., Tripathi, H., Sharma, P.: Design and performance analysis of vivaldi antenna at very high frequency. In: 2017 Fourth International Conference on Image Information Processing (ICIIP), pp. 1–4, Shimla (2017)
9. Singh, M.P., Priyadarshi, R., Sharma, P., Thakur, A.: Small size rectangular microstrip patch antenna with a cross slot using SIW. In: 2017 Fourth International Conference on Image Information Processing (ICIIP), pp. 1–4, Shimla (2017)
10. Tarbouch, M., El Amri, A., Terchoune, H.: Design, realization and measurements of compact CPW-fed microstrip octagonal patch antenna with H slot for WLAN and WIMAX applications. Int. J. Microw. Opt. Technol. **12**, 389–398 (2017)
11. Yang, F., Zhang, X.-X., Ye, X., Rahmat-Samii, Y.: Wide-band E-shaped patch antennas for wireless communications. IEEE Trans. Antennas Propag. **49**, 1094–1100 (2001)
12. Wong, K.L., Hsu, N.H.: A broad-band rectangular patch antenna with a pair of wide slits (2001)

Part III
Networks

Performance Analysis of Various Cryptographic Techniques in VANET

Sandeep Kumar Arora and Sonika

Abstract Vehicular Ad-hoc network (VANET) is a wireless communication among many vehicles. The main motive of VANET security is to not only to provide safety, secure communication and intelligent transportation service but also another service like entertainment, advertisement and offers based on location wise. As all the services related to communication are more important and vulnerable to attacks, hence, it requires security. Securing communications between vehicles and roadside unit is a great challenge. Security is one of the major concerns in VANETs as nodes in VANETs have high mobility. So, it is a challenging task to design an efficient solution for secure communication in VANETs due to high mobility of nodes. Many authentication protocols have been proposed for secure communication in VANETs using Session Initiation Protocol (SIP). SIP is widely used for signalling and establishing communication between different nodes in VANETs. SIP uses the concept of Voice over Internet Protocol (VoIP) for communication between vehicles. It uses Hypertext Transfer Protocol (HTTP) digest for identity authentication between different vehicles during communication. The proposed SIP using ECC, AES, Dynamic Intrusion Detection Protocol Model (DYDOG) have been implemented in this paper for various traffic scenarios which improve the security and Quality of Services.

Keywords VANET · Security · ECC · AES · Throughput · PDR

S. K. Arora (✉) · Sonika
Electronics and Communication Engineering, Lovely Professional University,
Phagwara 144411, Punjab, India
e-mail: sandeep.16930@lpu.co.in

Sonika
e-mail: sonika01991@gmail.com

A. K. Luhach et al. (eds.), *Smart Computational Strategies: Theoretical and Practical Aspects*, https://doi.org/10.1007/978-981-13-6295-8_17

1 Introduction

1.1 Vehicular Ad-Hoc Network

VANET is a special kind of mobile ad hoc network. In that network, communication takes place between the vehicles called V to V, Vehicle-to-Roadside Units (V to RSU) in a range of 100–300 m [1]. In a VANET, the node can move freely in a dynamic topology and communicate with various vehicles. The major significance of VANET is to give comfort to the users and to save the life of driver and passenger. VANET is part of MANET in which every node travels generously in the community [2]. Each node communicates with other nodes in single-hop or multi-hop. In VANET, On-Board Units (OBUs) regularly broadcast routine traffic-related messages with records approximately role, contemporary time, route, speed, acceleration/deceleration, traffic occasions and so on [3]. Since cell nodes keep running on batteries, the handling vitality to be had for every node is restricted since cell nodes run on batteries, the processing energy to be had for each node is limited. Wi-Fi links have an impressively diminished limit when contrasted with a wired connection. Wireless hyperlinks frequently suffer from the outcomes of multiple get right of entry to, fading, noise, and interference situations that anticipate it from handing over a throughput identical to the most throughput. Each mobile node can also have a specific hardware/software program configuration with extraordinary abilities. Designing protocols and algorithms for such heterogeneous, uneven links becomes a complicated process. The opportunities of eavesdropping, spoofing and denial-of-carrier attacks are greater established compared to fixed line networks [4]. The packages related to MANETs variety from people who involve a small number of nodes to those containing tens of thousands of nodes. Scalability is an essential aspect of a successful deployment of ad hoc networks Compared to small networks, the community management algorithms of massive networks have to cope with altogether specific challenges in areas such as addressing, routing, vicinity control, interoperability, protection, mobility, Wi-Fi technologies, etc. [5].

1.2 Architecture of VANET

This part depicts the framework engineering of vehicular specially appointed systems. VANET engineering can be partitioned into various structures in view of an alternate point of view. VANET is utilized for communication between vehicles for the diverse sort of use, for example street safety, stimulation, movement control and so on. VANET give data auspicious to drivers and obliged specialists to give safety to clients. In VANET, there are two sorts of communication.

One is distributed (P2P) referred to as V2V communication as appeared. V2V communication happens between vehicles. Second is the Vehicle-to-roadside unit

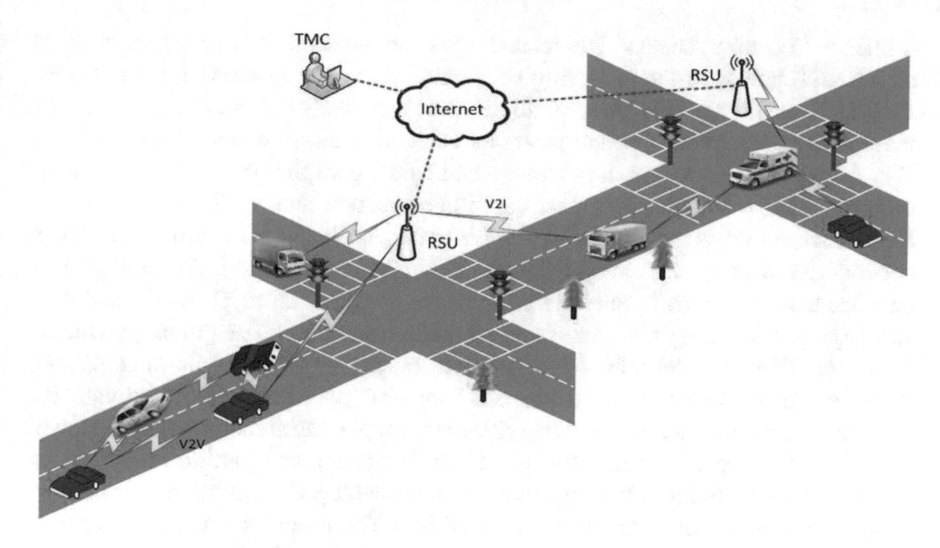

Fig. 1 VANET architecture [2]

(RSU) which is known as V2R, which happens amongst vehicle and RSU. In V2R communication, a vehicle speaks with the closest RSU. The required message is sent by RSU if and just if vehicles are in the scope of those RSU. Something else, RSU makes an impression on the neighbour RSU for communication. Figure 1 shows the architecture of VANET.

In which, communication is finished by taking after three distinctive sorts of process for example,

- Inter-vehicle communication (V2V)
- Vehicle-to-roadside communication (V2I)
- Inter-roadside communication

2 Review of Literature

The accompanying areas will give a short review of some of this exploration, which will help in the examination for this work. It proposed five particular directions of attacks and every radiance is expected to offer higher viewpoints for the VANET assurance. The authors endeavoured to embrace a characterization and a recognizable proof of various attacks in VANET. In top-notch Organize Attacks; assailants can immediately influence diverse vehicles and foundation. These attacks are at the abnormal state of peril on the grounds that those affect the entire system. The assailant is particularly intrigued by changing over substance used in applications and mishandling it for his or her own special endowments. At last, attacks wherein following and observing exercises are refined are laying inside the brilliance Checking Attacks. The related works above caution a disturbing situation of VANET safety. In the following segments, we expect to stress security necessities in VANET, and then present, all the more succinctly the suitable attacks, their comparing countermeasures and promoter each other classification of these attacks [1]. After this, the author has shown validation conspire for session initiation protocol (SIP) is unreliable against user impersonation attacks. This means that a legal user can impersonate other legal users to perpetrate toll fraud and use the VoIP network for making free long-distance calls. Keeping in mind, the end goal to enhance the security and productivity, we have proposed an ECC-based validation and key understanding plan for session initiation protocol. The proposed scheme frustrates different attacks, as well as demonstrates a superior execution contrasted with the related plans. Henceforth, our proposed plan is more appropriate for IP-based communication frameworks [2]. In this paper, the author has demonstrated that Yoon's scheme is vulnerable to offline dictionary and partition attacks. An attacker can recover the correct password in reasonable a time [3–9]. In this review, another plan is proposed for the security of SIP. The proposed technique depends on ECC. Favoured properties of ECC are utilized for verification of session initiation protocol. Along these lines, the article covers the essential components of SIP and security consideration relying upon the test/reaction systems [4–7]. This arrangement fits conveniently in the SIP protocols as depicted in RFC3261. The new protocol has security properties required by SIP standard and requires just negligible changes to the standard. The plan is intended to give data confidentiality, data integrity, authentication, access control and perfect forward secrecy, and it is secure against known-key attacks. Also, NAKE depends on ECC, so it is more proficient and ideal in the applications which require low memory and fast exchanges [5]. The authors recognized distinctive security attacks. Though, an untouchable, who is not confirmed to specifically speak with different individuals from the system, have a restricted ability to play out an attack. They proposed a new group-based protocol for VANET called VANET QOS-OLSR. To diminish the soundness of group, they include the parameters including speed and separation that speak to the versatility measurements to the QOS work. In ideal method, select the MPR nodes with the goal that they communicate the three message to at most extreme 2-jump away nodes [6]. Dynamic and flexible intrusion detection model to detect the intrusion node and prevents the denial-of-service attack. The authors have discussed how the same node will work as detector as well as forward the information to next node and also discussed the symmetric key sharing for this mechanism [7]. In this paper, the author proposed a Vehicular Crash Warning Communication (VCWC) protocol to enhance street safety. In particular, it portrays blockage control routes of action for emergency alerted messages so that a low emergency advised message transport deferral can be proficient and a broad number of harmonizing abnormal vehicles can be supported. This familiarizes a procedure with discard dull emergency advised messages, abusing the normal chain effect of emergency events [8]. We in this way proposed an enhanced protocol that dispenses with the shortcoming. As per our investigation and exchange, the proposed protocol accomplishes shared verification, withstands different attacks, and is more productive [9]. Vulnerabilities have been shown by the authors for Tsai's nonce-based authentication scheme for SIP to various attacks but it did not provide perfect forward secrecy. To solve that problem, they presented the ECC scheme for SIP. As a result, it provides more efficiency and faster speed as well as security to the data [10]. In this paper, the author is discussed that his proposed technique is able to withstand user anonymity and stolen verifier attack as well as other attacks also. In this paper, the author is discussing about elliptic curve cryptography to prevent the attack.

3 Security in VANET

3.1 Dynamic Intrusion Detection Protocol Model (DYDOG)

Wireless Sensor Networks (WSNs) are simply the accumulation—sorting out sensor nodes sent in different physical situations statically or progressively relies on the application. In Wi-fi condition, these sensor nodes are protection less or defenceless against attacks. To take care of this issue, the Intrusion Detection System (IDS) has been utilized and for remote systems, Dynamic Intrusion Detection System (DIDS) has been utilized. Be that as it may, this is not adequate to accomplish the greatest versatility against attacks. Considering the issues, here, a DYDOG technique is implemented to protect the node from various attacks [6].

3.1.1 DYDOG Technique

It is a technique to detect the several attacks like Blackhole attack, Wormhole attack, Sybil attack and selective forwarding attack, by creating the dynamic intruder detection node called Compromised Node (CN) in sensor node. Here, node will be monitored by more than one. In this technique, we are detecting each and every node so that malicious node can be detected. The task of node is not only detecting or monitoring of the node but also responsible for forwarding the node dynamically. The node itself takes decision/action against any kind of attack and sends the update to their neighbour nodes without any cluster head [7]. The behaviour of node as forwarding node in the forwarding list is for a moment, and then becomes idle dynamically, until this node is one hop neighbour. The nodes will become only idle when they are one hop distance, and they are not in the path of forwarding node. The result of this scheme is shown in Fig. 2.

The nodes which are one hop distant from forwarding node will act as Intrusion detection when data transmission taking place and other nodes functionality are stable.

At a single time, one node is monitored by more than one DIDN node. If any malicious activity or attack is traced by any adversary node and other node has also

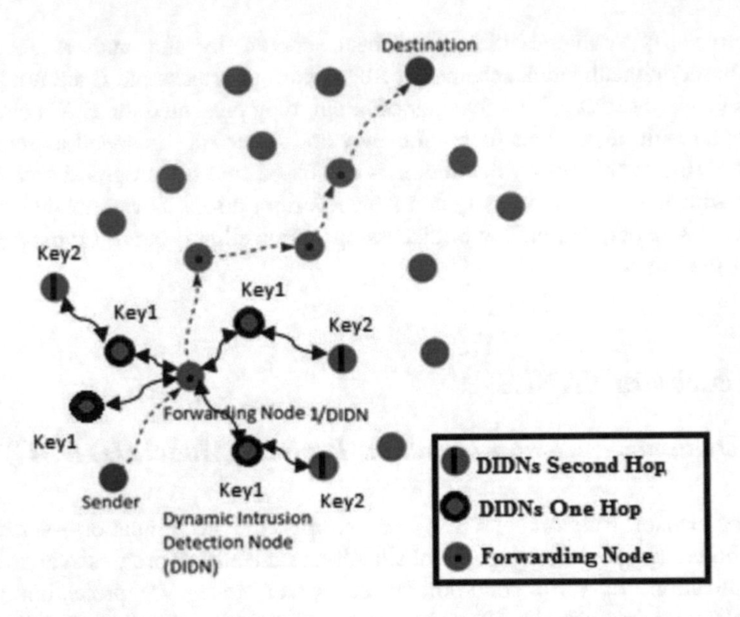

Fig. 2 DIDN and forwarding nodes [7]

sensed this activity, then action will be taken against the malicious node. This is a method in which it is very difficult for intruders to identify and attack on intrusion detection node. We are considering the idle node which is one hop from forwarding node as Dynamic Intrusion Detection Node when these nodes are not coming in their forwarding path. Since monitoring node is two-hop distant from forwarding node, it happens only when higher data is transmitted but there is no need of maintaining two-hop information all time to the forwarding node. During critical situation or higher data rate situation, one-hop distance monitoring nodes will share its own information with forwarding node as its second hop monitoring node dynamically with predefined node shared session key. In this way, we can increase the availability of Intrusion detection node in the worst scenario. We select this Intrusion Detection Node using secure key management and malicious node are also avoided to act as monitoring node. Figure 3 represents the result of secure DIDN selection (Hop-2) with shared secret session key (key 2).

3.1.2 Secret Key Management

For selection of DIDN, we have to use secure way which identify malicious node from DIDN. To achieve this Intrusion detection node should maintain two secretly shared session key. These two IDs are concatenated in forwarding node and EX-ORed in intrusion detection node, and send that key to forwarding node.

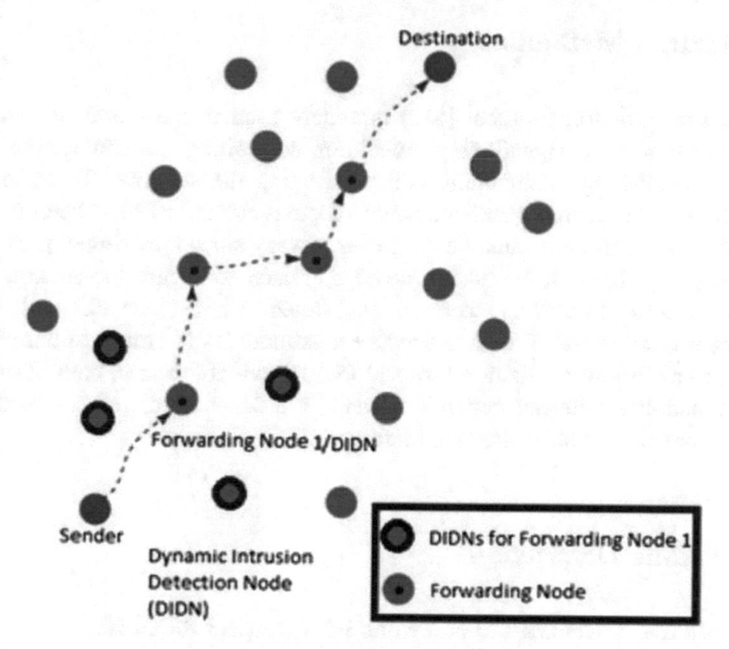

Fig. 3 Secure DIDN selection (Hop-2) with shared secret session key (Key 2)

By using reverse EX-OR operation, monitoring node's ID is checked. In this manner, the security is increased against intruder node. It is very difficult to find the session key within a particular session and using this authentication mechanism, we can avoid malicious node [7].

Decision Key for Decision-Making Dynamic Intrusion Detection (DMDIDN)
In this, we are talking about decision after the attack is identified. When intruder attack is identified, more than one intrusion detection node monitoring forwarding node having one-hop distance from that monitoring node. Each monitoring node should find the attack as much as possible, but when any action is taken against these attacks, Intrusion detection node will select an alternative path and reroute the data via alternatively selected forwarding path after healing the attacked node or infected packet. Even though there is various intrusion detection nodes but only one will take a decision of route change within the intrusion time. Here, Forwarding node will send the decision key to their monitoring nodes and wait for reply from those nodes. The nodes reply with decision-making key which is having TTL field to the forwarding node. The node which has lowest TTL value will be selected as decision-making intrusion detection node. In next step for ensuring authentication, the forwarding node will only send initial portion of data to the selected intrusion detection node. After this, the rest of the data will be sent to authentic node to make sure route selection. Depending on node mobility, decision-making node selection changes [8].

4 Research Methodology

The Session Initiation Protocol (SIP) is widely used in the world of multimedia communication as a signalling protocol for controlling communication on the internet, establishing, maintaining and terminating the sessions. To ensure communication security, many authentication schemes for the SIP has been proposed. However, those schemes cannot ensure user privacy since they cannot provide user anonymity. We have proposed the novel approach to secure the session against various attacks like user anonymity, stolen verifier attack and various denial-of-service attack. We have used various modules to complete our objective as they are as follows: Initiation Protocol (SIP) is widely used in both of the wired network and the wireless network decade. We have used various modules to complete our objective as they are as follows:

4.1 Module Description

4.1.1 Module 1 Generation of Traffic with Respect to Number of Nodes

Nodes are taken randomly and they forward the data to each other. We do not know how to transmit the packets and basic needs of network topology. It can be wired or wireless or ad-hoc networks. Building a simple topology with minimum number of nodes, for example seven nodes will be deployed in the network. The flowchart of research methodology is shown in Fig. 4.

4.1.2 Module 2 Implementation of AODV

In this module, we are going to implement the routing protocol to the node. We have taken Ad-hoc On-Demand Distance Vector protocol (AODV). We are going to build a complex topology with more number of nodes and same configurations will be set as above. In this, we will be giving the mobility to all nodes with different timings. We will take the values of QOS parameters like average delay, packet delivery ratio, throughput at the corresponding time like 2, 4, 6, 8, 10. The values are inserted in a system generated trace file and graph will be schemed for each parameter.

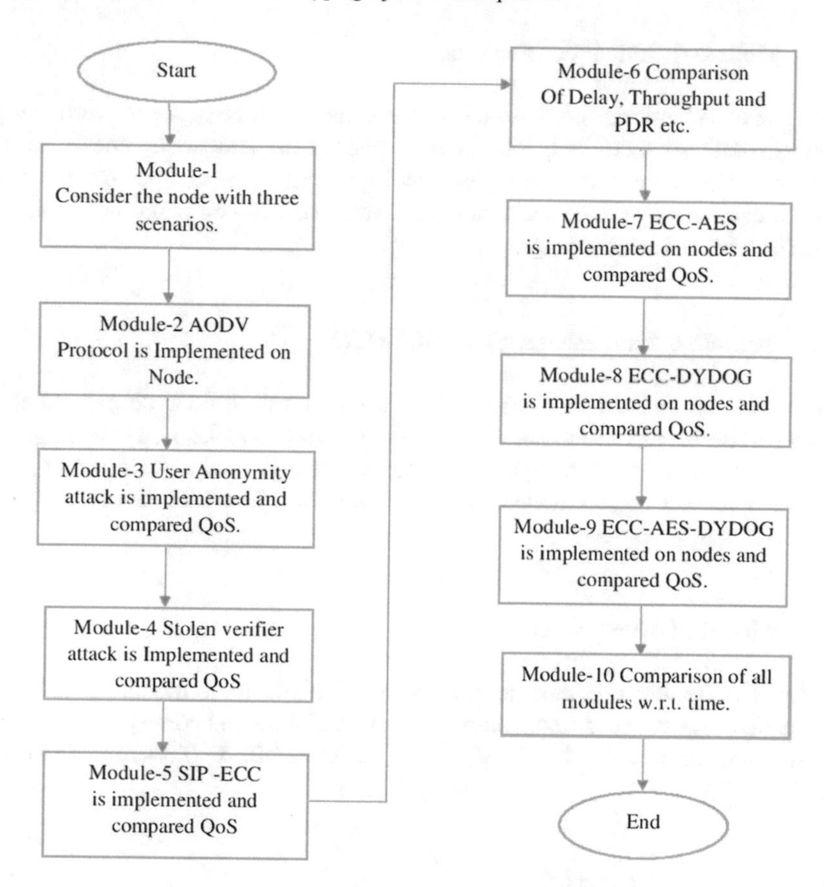

Fig. 4 Flowchart of research methodology

4.1.3 Module 3 User Anonymity Attack

User anonymity means that the adversary cannot get the user's identity from the message transmitted in the login and authentication phase. In this, we will be giving the mobility to all nodes with different timings. We will take the values of QOS parameters like end-to-end delay, packet delivery ratio, throughput at the corresponding time like 2, 4, 6, 8, 10. The values are inserted in a system generated trace file and graph will be schematic for each parameter. The attacker is trying to access the user identity and he/she is able to get the users identity by hook or crook. Now, there will be some data loss when the communication takes place among nodes. Here, the network performance will be degraded.

4.1.4 Module 4 Stolen Verifier Attack

When the nodes are transmitting their packets, this intruder steal password and get access to data so there is a loss in data transmitted and nodes energy is also increasing. We have built a complex topology with more number of nodes and same configurations will be set as above. In this, we will be giving the mobility to all nodes with different timings.

4.1.5 Module 5 Implementation of SIP-ECC

In this section, we have implemented the communicating node scenario. Let the intruder node be mixed among communicating node and he is trying to get the user's identity from the message transmitted in the login and authentication phase. The user sends the protected identity [9]. Then, he will face with the computational Diffie–Hellman problem.

4.1.6 Module 6 Comparison

The QOS values taken in module 2, module 3, module 4 and module 5 will added in a single user generated trace file and graph will be plotted correspondingly. This is just Compare module of AODV, USER_ANONYMITY, STOLEN VERIFIER and SIP ECC.

4.1.7 Module 7 ECC-AES

In this module, we have applied the novel technique (Advance Encryption Standard and Elliptic Curve Cryptography) on the different traffic scenario like 10, 30 and 50 nodes to protect the communication session from attack. AES is more secure than ECC as it uses 10, 12 and 14 rounds of processing for 128,192 and 256-bit key length respectively. Intruder node will try to get the packet but he/she is unable to get it since AES is used for authentication and node will ask for key shared between nodes.

4.1.8 Module 8 ECC-DYDOG

In this module, we have applied the ECC-Dydog encryption mechanism for securing the communication from malicious node so that we can protect our node from attacks. ECC-Dydog is implemented on different traffic scenarios like low, medium and high. We have calculated the QoS parameter, and after that, we have compared these parameters with ECC implemented QoS parameter and found that ECC-DYDOG gives better results. Rounds of processing for 128, 192 and 256-bit key length, respectively.

4.1.9 Module 9 ECC-AES-DYDOG

This is a module in which we have applied ECC-AES-DYDOG for encrypting the session among vehicles. This technique is implemented in low, medium and high traffic scenarios. Attacker node will drop the packet and this behaviour is detected by watchdog node. We are using AES for authentication of node. Hence, we are reducing the traffic and prevented attack in the network.

Proposed Technique

Step-1: We have taken three scenarios like 10, 30 and 50 nodes and considered 50 node scenarios for analysing proposed technique. Moreover, four more nodes are considered, i.e. node 51, 52, 53 and 5 which behaves as attacker node.
Step-2: To find the packet drop we have used Dydog technique.
Step-2.1: In Dynamic Intrusion Detection Protocol Model (DYDOG) technique, all the normal node will behave like watchdog. We will observe the behaviour of every node.
Step-2.2: When source node will try to send packet to destination, attacker node will come forward in between and will receive the packet and drop it.
Step-2.3: The normal node behaving as watchdog will observe pattern of traffic flow from source to destination. When malicious node will drop packet, the watchdog node will detect this malicious node.
Step-3: Packet drop in the network may not be only due to attacker, it can also be possible that there is bandwidth blockage or link congestion. We will detect through AES to get to know whether these nodes are authentic node or not.
Step-4: Node, which has observed the node 51, 52, 53 and 54 is dropping packet in the network, will ask their key. Node will reply with their key "XYZ" but the original key used in the network is "ABC".
Step-5: Since there is a mismatch in the key so observer node will consider that these nodes are not authentic and will confirm as malicious node.
Step-6: Now observer node will broadcast the malicious node ID in the network so that no other node will communicate with these malicious nodes.

4.1.10 Module 10 Final Comparison

In this module, we have done final comparison of all modules in a single graph window and find out the average delay, packet delivery ratio and average throughput with respect to time as well as number of nodes.

5 Simulation and Results

5.1 Implementation

In this section, we are going to implement the base paper model assumption and try to get the results closest to base paper result. In base paper, the authors have taken the scenario of user anonymity and many more attacks. But we have implemented two scenarios only, first is user anonymity and second stolen verifier attack on network simulator 2. We will use the proposed scheme (Module 9) to overcome attacks.

5.2 Simulator

According to Shannon, simulation is "the process of designing a model of a real system and conducting experiments with this model for the purpose of understanding the behaviour of the system and/or evaluating various strategies for the operation of the system" (Table 1).

5.3 QOS Parameter

We have used the following parameters for our study.

5.3.1 Delay

The delay of a network specifies how long it takes for a bit of data to travel across the network from one node or endpoint to another. It is typically measured in multiples or fractions of seconds.

Table 1 Simulation parameter

Simulation parameter	Value
No. of nodes	10, 30, 50
Propagation model	Two ray ground
Antenna type	Omnidirectional
Routing protocol	AODV
MAC	802.11
Packet size	200
Simulation area	500 * 500

5.3.2 Packet Delivery Ratio (PDR)

The calculation of Packet Delivery Ratio (PDR) is based on the received and generated packets as recorded in the trace file. In general, PDR is defined as the ratio between the received packets by the destination and the generated packets by the source [10].

$$\text{Packet delivery ratio} = (\text{Received packets/generated packets}) * 100$$

5.3.3 Throughput

In data transmission, network throughput is the amount of data moved successfully from one place to another in a given time period, and typically measured in bits per second (bps), as in megabits per second (Mbps) or gigabits per second (Gbps).

5.4 Low Traffic Scenario (10 Nodes)

We are considering the low traffic in which only 10 nodes are communicating among them. Nodes are transmitting their data to the interested user using Ad-hoc on-demand distance vector protocol. Here, we are considering only user anonymity attack and stolen verifier attack [10].

5.4.1 Average Delay

In simple AODV the delay found to be less but for user anonymity the delay increased. Increment in the delay during attack is attributed to the fact that the attackers do not forward the all the other modules tested. This result shows in Fig. 5.

5.4.2 Average Throughput

In case of low traffic, the average throughput of the various nodes is shown in Fig. 6. When nodes are in communication using AODV protocol, the throughput is initially increasing and after that, it slightly starts decreasing as time increases. It is because of congestion over the path increase as time passes. But in case of attacks, throughput is less in comparison to the previous case with the fact that very less packets reach to the destination. To improve the throughput, we have encrypted using AES when nodes wants to get packet, the observer node will detect the

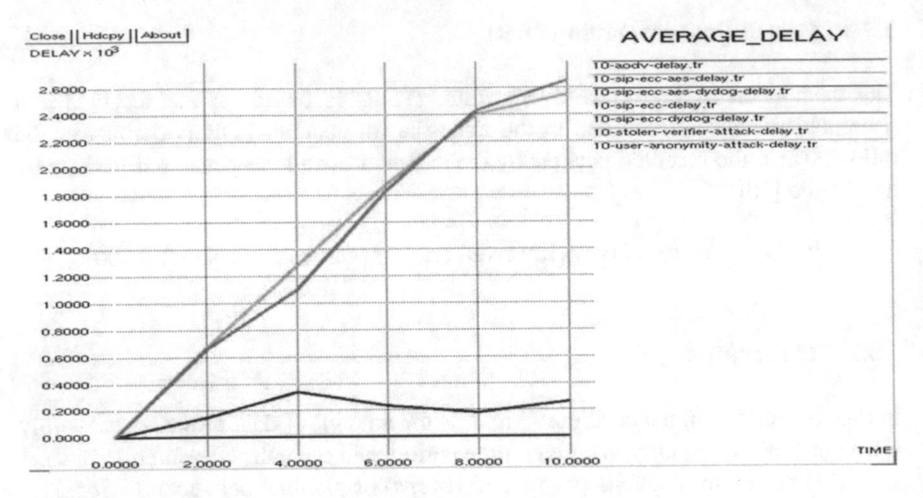

Fig. 5 Comparison of average delay (10 Nodes)

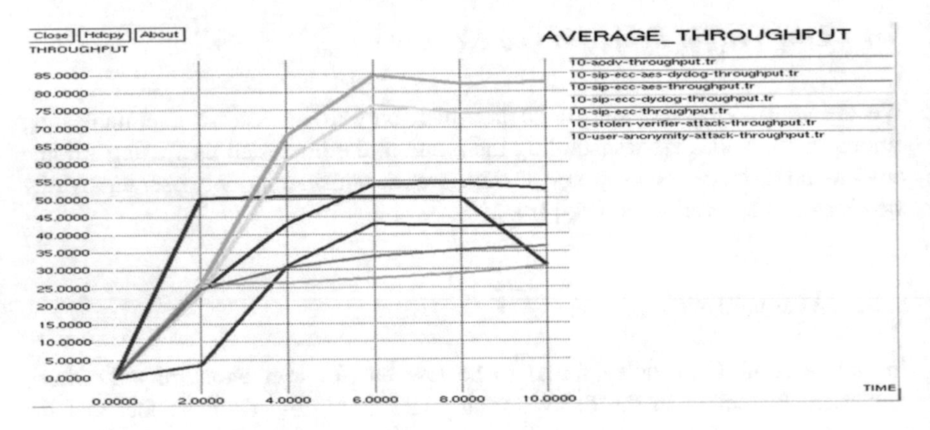

Fig. 6 Comparison of average throughput (10 Nodes)

attacker node and broadcast their ID in network traffic. Rest of the node in the network will not communicate with these nodes.

When nodes are in communication using AODV protocol, the throughput is initially increasing, and after that, it slightly starts decreasing as time increases. It is because of congestion previous case with the fact that very less packets reach to the destination. To improve the throughput, we have encrypted using AES.

5.4.3 Packet Delivery Ratio (PDR)

In case of low traffic, the packet delivery ratio is shown in Fig. 7. By observing the graph, we can say that as time passes, the PDR is normal in case of AODV but in

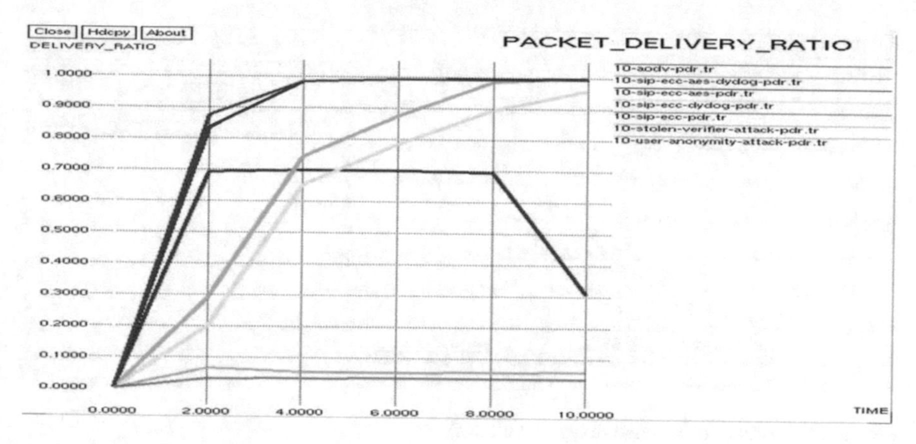

Fig. 7 Comparison of PDR (10 Nodes)

case of attacks, the PDR is very less since most of data traffic is created by attacker so there is a drop of packets but when our proposed technique is implemented in the network, only authorized person is allowed to send so there will be no loss of data leading to increase in the PDR.

5.5 *High Traffic Scenario (50 Nodes)*

We are considering the high traffic in which 30 and 50 nodes are communicating among themselves. We have analysed the impact of the number of nodes over the proposed scheme. It has been found that as the number of nodes increases in the network, the value for the packet delivery ratio goes down a bit.

5.5.1 Average Delay

Delay of nodes when simple AODV protocol is implemented, it is less but when attackers are activated in the communication the delay is increasing in case of both user anonymity as well as stolen verifier attack the increment in the delay during attack is attributed to the fact that the attackers do not forward the packet to the destination as a normal node. The result is shown in Fig. 8.

5.5.2 Average Throughput

In case of high traffic, the average throughput of the various nodes is shown in Fig. 9. When nodes are in communication using AODV protocol, the throughput is initially increasing, and after that, it slightly starts decreasing as time increases.

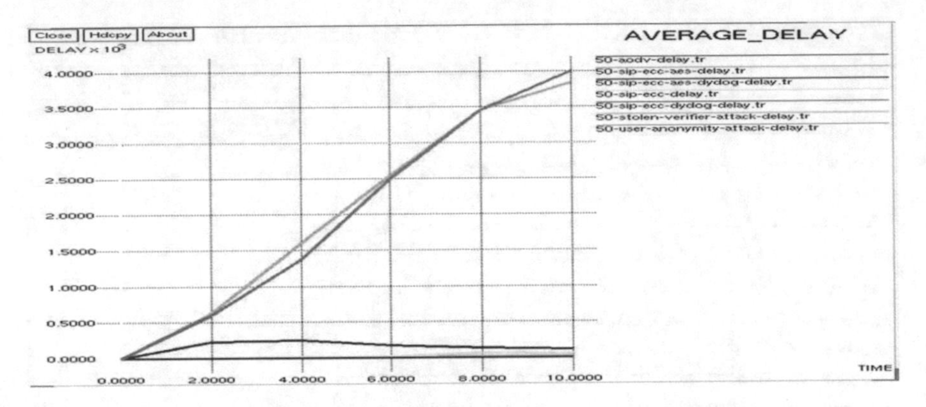

Fig. 8 Comparison of average delay (50 Nodes)

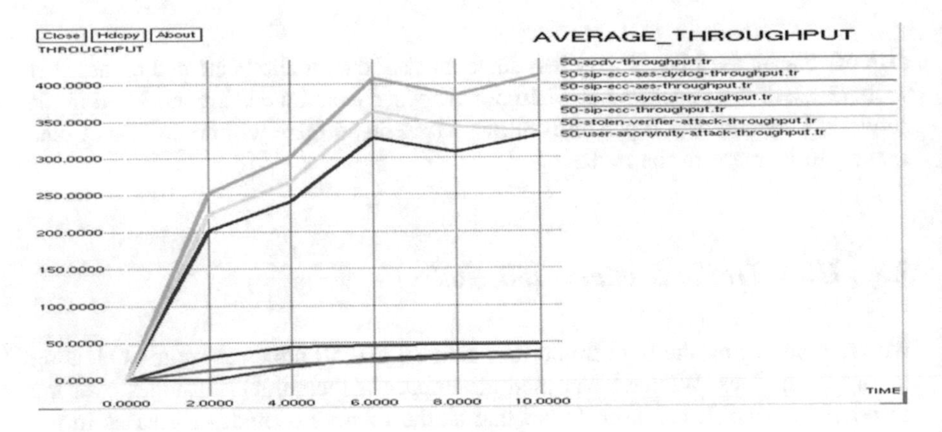

Fig. 9 Comparison of average throughput (50 Nodes)

In Fig. 9 throughput is improved but still lesser than low traffic scenario. As the time elapsed the congestion over path increased. Due to increase in node, it leads to high traffic in the network. To improve the throughput, we have encrypted using AES when nodes wants to get packet, the observer node will detect the attacker node and broadcast their ID in network traffic. The rest of the node in the network will not communicate with these nodes.

5.5.3 Packet Delivery Ratio (PDR)

In case of high traffic, the packet delivery ratio is shown in Fig. 10. By observing the graph, we can say that as time passes the PDR is normal in case of AODV but in case of attacks, the PDR is very less since most of data traffic is created by attacker so there is a drop of packets but when our proposed technique is implemented in the

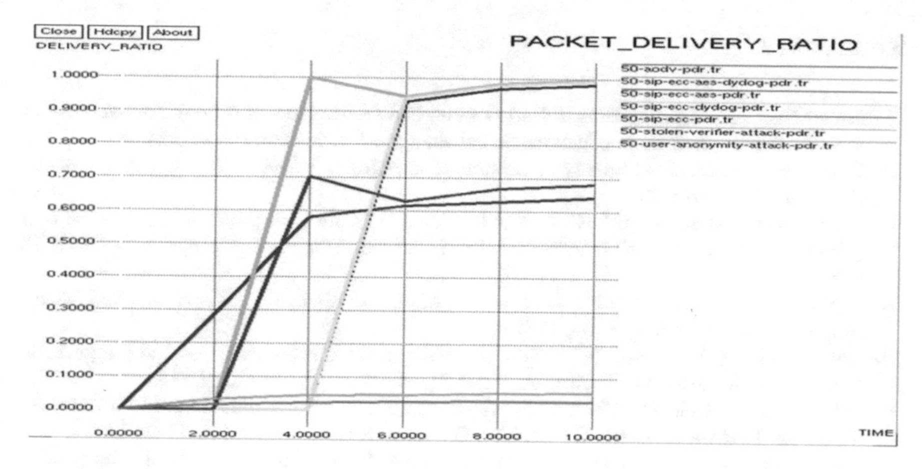

Fig. 10 Comparison of PDR (50 NODES)

network, only authorised person is allowed to send the information. The node has to pass the authenticity of 256 bit AES verification. So, there will be no loss of data but due to congestion PDR is little bit low compared to low traffic case.

6 Conclusion and Future Scope

We have proposed a novel technique ECC-AES-DYDOG to overcome the user anonymity and stolen verifier attack, and compared the QOS parameters in different traffic scenarios like Low (10 Nodes), Medium (30 Nodes) and High (50 Nodes). It can be also used for prevention of various Denial-of-Service attack (DOS) such as selective forwarding attacks, Blackhole attack, Wormhole attack, Sybil attack, jamming attacks. The proposed scheme has shown better values because in DYDOG mechanism, all the nodes (except the attackers) perform the function of observing their neighbourhoods. This leads to better and quick detection of the malicious nodes in the network thus leading to lower delays and high packet delivery ratio. These DYDOG nodes will identify the packet dropping nodes and perform their authentication as quickly as the other schemes. In future, we can enhance the parameters of this novel technique by reducing the payload of the packet. Moreover, we can use other crypto techniques to overcome the attacks and compare it with ECC.

References

1. Jeroen Hoebeke, Ingrid Moerman, Bart Dhoedt, and Piet Demeester (2004). An overview of mobile ad hoc networks: applications and challenges. J. Commun. Netw. **3**(3)
2. Dhamgaye, Chavhan, N.: Survey on security challenges in VANET. Comput. Sci. **2**, 88–96 (2013). ISSN 2277-5420
3. He, D., Khan, M.K., Kumar, N.: A new handover authentication protocol based on bilinear pairing functions for wireless networks. Int. J. Ad Hoc Ubiquitous Comput. (IJAHUC) **18** (2015)
4. Durlanik, Sogukpinar, I.: SIP authentication scheme using ECDH. World Inf. Soc. Trans. Eng. Comput. Technol. **8**, 350–353 (2005)
5. Wu, L., Zhang, Y., Wang, F.: A new provably secure authentication and key agreement protocol for SIP using ECC. Comput. Stand Interfaces **31**(2), 286–291 (2009)
6. Wahab, O.A., Otrok, H., Mourad, A.: A cooperative watchdog model based on dempster shafer for detecting misbehaving vehicles. Comput. Commun. **41**(2014), 43–54 (2014)
7. Janakiraman, S., Rajasoundaran, S., Narayanasamy, P.: The Model—Dynamic and flexible intrusion detection protocol for high error rate wireless sensor networks based on data flow. In: International Conference on Computing, Communication and Applications, pp. 1–6 (2012)
8. Gollan, L., Gollan, I.L., Meinel, C.: Digital signatures for automobiles. In: Systemic, Cybernetics and Informatics, SCI, Citeseer (2002)
9. Schneier, B.: Applied Cryptography. Protocols, Algorithms, and Source Code in C. Wiley, New York (1996)
10. Mejri, M.N., Ben-Othman, J., Hamdi, M.: Survey on VANET security challenges and possible cryptographic solutions

Network Selection Techniques Using Multiple-Criteria Decision-Making for Heterogeneous Cognitive Radio Networks with User Preferences

Krishan Kumar and Mani Shekhar Gupta

Abstract It is a vision that in next-generation cognitive radio networks deployment, there will be different heterogeneous cognitive radio networks with no overlapping spectrum pool owned by different network operators. These heterogeneous cognitive radio networks may coexist together in the same coverage area. In such environment, cognitive radio may have multiple options to select different networks which may face different traffic conditioned generated by the primary user. Hence, the mobile subscriber with various interfaces will be in position to select the networks of their choice which are best fulfill for his needs. This paper shows the analysis of different network selection techniques in heterogeneous cognitive radio networks as multiple-criteria decision-making. Three main multiple-criteria decision algorithms compared with inputs are taken from the user as per requirement of the user. The main parameter for comparison is the user satisfaction achieved by that algorithm. Every algorithm is simulated for the given scenario and their performance is compared on the basis of the above-specified parameter.

Keywords Network selection · Cognitive radio network · User satisfaction index

1 Introduction

The current trend in wireless communication networks has hoisted the requirement of more frequency bands. Presently, the wireless networks are limited by static spectrum allocation scheme, which is being controlled by the authorized government bodies. The current static spectrum allocation scheme used by the authorized

K. Kumar (✉) · M. S. Gupta
Electronics and Communication Engineering Department,
National Institute of Technology, Hamirpur, India
e-mail: krishan_rathod@nith.ac.in

M. S. Gupta
e-mail: mi_sr87@yahoo.com

A. K. Luhach et al. (eds.), *Smart Computational Strategies: Theoretical and Practical Aspects*, https://doi.org/10.1007/978-981-13-6295-8_18

government bodies is insufficient to accommodate the increasing requirement of the wireless networks. In fact, few assigned bands are less occupied like some part of TV UHF and VHF band also known as TV white space [1]. According to Federal Communication Commission (FCC), generally, assigned spectrum is partially utilized due to assignment of static frequency bands. This creates the large space between recent increased demand of more frequency bands and static spectrum allocation scheme. To combat this space, FCC has advised to practice the partially utilized spectrums and spectrum parts that are not being utilized are called white spaces and this is called as Dynamic Spectrum Access (DSA) [2, 3]. This creates the new direction for research in Cognitive Radio (CR) networks that is an emerging technology to realize DSA [4]. The active research is going on the selection of network in wireless heterogeneous networks. The wireless heterogeneous networks equipped with CRNs are referred to as Heterogeneous Cognitive Radio Networks (Het CRNs). Hence, Het CRNs faces multiple challenges such as the time and location varying spectrum availability, heterogeneous types of node, and best network selection out of available networks. Therefore, spectrum selection in heterogeneous cognitive radio networks is a natural way to meet out the above. This type of complex challenges is rarely explored in the research community [5]. Figure 1 shows spectrum hole concept which can be identified with the help of cognitive capability by which cognitive radio can configure itself. It is a vision that in next-generation cognitive radio networks implementation, there will be different heterogeneous cognitive radio networks with no overlapping spectrum pool owned by several network providers as shown above. These heterogeneous cognitive radio networks may coexist together in the same coverage area [6]. Hence, the mobile subscriber with different interfaces will have the ability to adopt one of these available networks options that are best suitable for his needs.

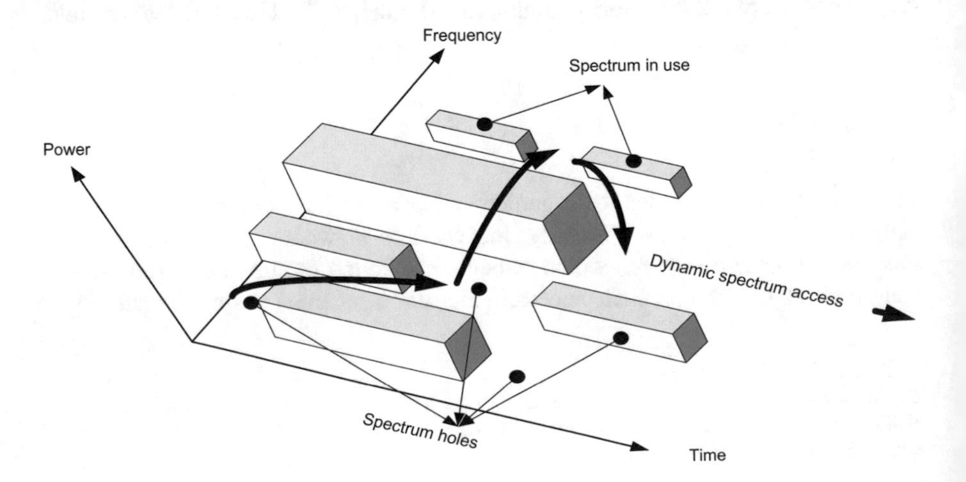

Fig. 1 Spectrum hole at various bands

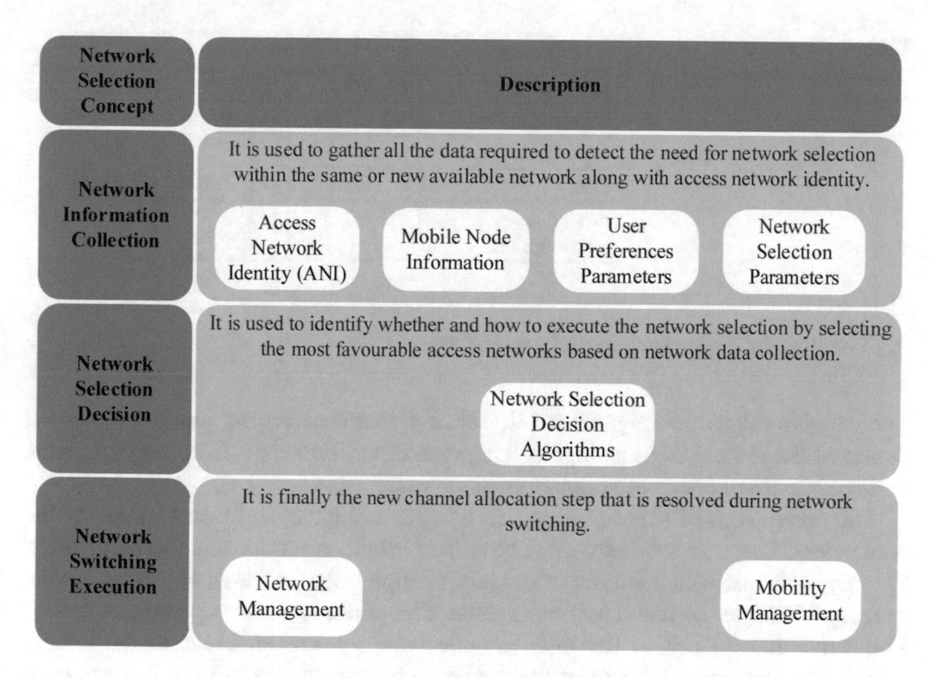

Fig. 2 Network selection steps

This paper shows the analysis of different network selection techniques in heterogeneous cognitive radio networks as multiple-criteria decision-making [7]. Figure 2 depicts the self-explanatory the network selection steps [8] for heterogeneous cognitive radio networks with user preferences.

2 Different Network Selection Algorithms

There are many network selection algorithms in the literature. Selection of the network deals with selecting an optimum candidate network. This selection of network is based on the several criteria so it is a referred as Multi-Criteria Decision-Making (MCDM) problem [9]. Here, we are taking three different network selection schemes. Simple Additive weighing (SAW), Technique for Order Preference by Similarity to Ideal Solution (TOPSIS) [10, 11], and Cost function-based algorithm [12]. For instance, suppose there are three candidate networks available with the given specification in Table 1.

For comparing performance of these algorithms, we need to generate weights of the handover decision parameters by a weight calculation algorithm.

Table 1 Parameter specification of candidate networks

Network code	1	2	3
Network type	WLAN	WLAN	UMTS
Bandwidth	~2 Mbps	~1 Mbps	~384 kbps
Security level	~1	~2	~7
Cost	~3 units	~2 units	~5 units
Power consumption	~3	~2	~1

2.1 Weight Distribution Algorithm

The weights calculation algorithm [10] takes preferences of user and the measured power of the mobile node as inputs. This creates utility/weight factors for different vertical handover decision parameters as outputs.

The preferences of user are specified by user and have significant levels for the parameters related to the network. These user-specified parameters are bandwidth, security, and cost which is used in monetary. High, medium, low, and none are the four significant levels that a user can define. The preset level for a parameter is low. The remaining battery of the mobile node is also taken an input to the weight calculation algorithm. Weight of the available bandwidth, security, monetary cost, and power consumption is calculated by the described method as:

The assumptions made for this method are the following:

- The remaining battery power of the mobile node is s_u, where $0 < s_u \leq 1$, $s_u = 0$ represents the empty state of battery and $s_u = 1$ represents full state of battery.
- The utility/weights factor for parameters bandwidth, cost, security, and power level are u_{B}, u_{C}, u_{S}, and u_{P}, respectively.

where $u_{\text{P}} = 1 - s_w$ and $u_{\text{B}} + u_{\text{C}} + u_{\text{S}} + u_{\text{P}} = 1$.

- The factors with respect to importance level of high, medium, low, and none are k_{H}, k_{M}, k_{L} and 0, respectively. The mobile system designer described these values.
- The user specified different importance levels m_{H}, m_{M}, m_{L}, and m_{N}, respectively, where $m_{\text{H}} + m_{\text{M}} + m_{\text{L}} + m_{\text{N}} = 3$ (since the user specified network parameters are three).
- The utility/weight factors of the four significant levels, after considered to subscriber preferences and power level of battery are $u_{k_{\text{H}}}$,$u_{k_{\text{M}}}$, $u_{k_{\text{L}}}$, and $u_{k_{\text{N}}}$, respectively.

To obtain the values of u_{B}, u_{C}, u_{S} and u_{P}, based on preference of user and mobile node battery power level, the following equations are considered:

$$m_{\text{H}} \times u_{k_{\text{H}}} + m_{\text{M}} \times u_{k_{\text{M}}} + m_{\text{L}} \times u_{k_{\text{L}}} + m_{\text{N}} \times u_{k_{\text{N}}} = s_u \tag{2.1}$$

$$u_{k_{\mathrm{M}}} = u_{k_{\mathrm{H}}} \times \frac{k_{\mathrm{M}}}{k_{\mathrm{H}}} \tag{2.2}$$

$$u_{k_L} = u_{k_H} \times \frac{k_L}{k_H} \tag{2.3}$$

$$u_{k_{\mathrm{N}}} = 0 \tag{2.4}$$

Putting the values of $u_{k_{\mathrm{M}}}$, $u_{k_{\mathrm{L}}}$ and $u_{k_{\mathrm{N}}}$ in Eq. (2.1), then

$$m_{\mathrm{H}} \times u_{k_{\mathrm{H}}} + m_{\mathrm{M}} \times u_{k_{\mathrm{H}}} \times \frac{k_{\mathrm{M}}}{k_{\mathrm{H}}} + m_{\mathrm{L}} \times u_{k_{\mathrm{H}}} \times \frac{k_{\mathrm{L}}}{k_{\mathrm{H}}} = s_u \tag{2.5}$$

So, the utility/weights of four different significant levels are determined by the given equations as

$$u_{k_{\mathrm{H}}} = \frac{k_{\mathrm{H}} s_u}{m_{\mathrm{H}} k_{\mathrm{H}} + m_{\mathrm{M}} k_{\mathrm{M}} + m_{\mathrm{L}} k_{\mathrm{L}}} \tag{2.6a}$$

$$u_{k_{\mathrm{M}}} = \frac{k_{\mathrm{M}} s_u}{m_{\mathrm{H}} k_{\mathrm{H}} + m_{\mathrm{M}} k_{\mathrm{M}} + m_{\mathrm{L}} k_{\mathrm{L}}} \tag{2.6b}$$

$$u_{k_{\mathrm{L}}} = \frac{k_{\mathrm{L}} s_u}{m_{\mathrm{H}} k_{\mathrm{H}} + m_{\mathrm{M}} k_{\mathrm{M}} + m_{\mathrm{L}} k_{\mathrm{L}}} \tag{2.6c}$$

$$u_{k_{\mathrm{N}}} = 0 \tag{2.6d}$$

With the help of above equations, mobile node can allocate utility/weights to four different parameters according to the preferences of user and power level of battery.

After assigning the weights to the related parameters, the second task chooses the best available network. For this purpose, we will compare four different algorithms, SAW, TOPSIS, and cost function-based method.

Before applying these methods, we can form decision matrix D from Table 1 and utility/weight matrix U.

$$D = \begin{bmatrix} 2000 & 3 & 1 & 3 \\ 1000 & 2 & 2 & 2 \\ 384 & 5 & 7 & 1 \end{bmatrix} \tag{2.7}$$

$$U = \begin{bmatrix} u_{\mathrm{B}} & u_{\mathrm{S}} & u_{\mathrm{C}} & u_{\mathrm{P}} \end{bmatrix} \tag{2.8}$$

2.2 Simple Additive Weighing

SAW [10, 11] is a renounced and generally used technique. In this, the total value of an options/alternative is calculated as the sum of all weight for all possible attribute values. Simplicity and ease of understanding included in this technique.

It needs a comparable measurement scale for all terms in the matrix used for decision. If an assumed criterion is advantageous, i.e., the larger, comparable scale and better is determined by using Eqs. (2.9) and (2.10) is applied for criterion smaller the better.

$$r_{ij} = x_{ij}/x_j^{\max} \quad i = 1,\ldots,3 \; j = 1,\ldots,4 \tag{2.9}$$

$$r_{ij} = x_j^{\min}/x_{ij} \quad i = 1,\ldots,3 \quad j = 1,\ldots,4 \tag{2.10}$$

After scaling, we will get normalized decision matrix then after applying weight factors, we will get ranking matrix. The network with highest ranking score will be selected as a target network.

2.3 TOPSIS (Technique for Order Preference by Similarity to Ideal Solution)

In TOPSIS [10, 11], the options/alternative is selected on the basis of distance. It should have the least separation from the ideal solution and the maximum separation from the practical solution. This scheme is also very simple and understandable.

The decision matrix can be normalized in TOPSIS by using the equation in the first step as

$$p_{ij} = q_{ij}/\sqrt{\sum_{i=1}^{3} q_{ij}} \quad i = 1,\ldots,3, \; j = 1,\ldots,4 \tag{2.11}$$

The decision matrix is weighted with the help of utility/weighting factors from (2.8), and gives normalized weighted matrix is M in second step.

The ideal solutions C^* is determined in third step and the practical solutions C^-. They are represented in (2.12) and (2.13) where Q is related with the advantageous criteria and Q' is related with the cost and power consumption criteria.

$$C^* = [v_1^* \quad v_2^* \quad v_3^* \quad v_4^*] = \left\{ \left(\max_i v_{ij} | q \in Q \right), \left(\min_i v_{ij} | q \in Q' \right) \middle| \begin{matrix} i = 1,\ldots,3 \\ j = 1,\ldots,4 \end{matrix} \right\} \tag{2.12}$$

$$C^- = [v_1^- \quad v_2^- \quad v_3^- \quad v_4^-] = \left\{ \left(\min_i v_{ij} | q \in Q \right), \left(\max_i v_{ij} | q \in Q' \right) \middle| \begin{matrix} i = 1, \ldots, 3 \\ j = 1, \ldots, 4 \end{matrix} \right\} \tag{2.13}$$

The separation of each alternative from the ideal solution and practical solution is calculated in fourth step by using the formula as in (2.14) and (2.15):

$$F_{i^*} = \sqrt{\sum_{j=1}^{4} \left(v_{ij} - v_j^* \right)^2} \quad i = 1, \ldots, 3 \text{ and } j = 1, \ldots, 4 \tag{2.14}$$

$$F_{i^-} = \sqrt{\sum_{j=1}^{4} \left(v_{ij} - v_j^- \right)^2} \quad i = 1, \ldots, 3 \text{ and } j = 1, \ldots, 4 \tag{2.15}$$

In fifth step, the closeness which is relative to the ideal solution is determined by using the formula:

$$R_{i^*} = F_{i^-} / (F_{i^-} + F_{i^*}) \quad i = 1, \ldots, 3 \tag{2.16}$$

The network with the highest ranking score will be selected as target network.

2.4 Cost Function-Based Algorithm

The algorithm used for cost factor calculation (CFC) [10] determines involved cost for performing a handover to one of the available candidate networks with the help of cost function. The handover target network is selected on the basis of lowest cost factor. The cost factor G_i provides calculated value of cost for ith network, which is determined as the following function:

$$G_i = G(u_B B_i, u_S S_i, u_C C_i, u_P P_i) \tag{2.17}$$

where $G_i(.)$ is the cost function, B_i, S_i, C_i ,and P_i represents for available bandwidth (in Mbps), security level (ranging from 1 to 10 scale, from least to maximum), cost used in monetary per minute (in cents) and power consumption level (ranging from 1 to 10 scale, from least to maximum), and u_B, u_S, u_C, and u_P are their utility/weights calculated from the WD algorithm.

The cost factor can be normalized for the n candidate networks re given as

$$G_i = \frac{u_B(1/B_i)}{\max((1/B_1), \ldots, (1/B_i))} + \frac{u_C C_1}{\max(C_1, \ldots, C_n)} + \frac{u_S(1/S_i)}{\max((1/S_i), \ldots, (1/S_1))} + \frac{u_P p_i}{\max(p_1, \ldots, p_n)} \tag{2.18}$$

Finally, the candidate network that has a cost factor of $\min(G_1, \ldots.., G_i)$ is identified as target network.

3 Evaluation of Performance for Different Network Selection Algorithms

MATLAB is used for the evaluation of performance of these network selection decision algorithms. The different set of user preferences, importance level for each parameter (BW, security, cost, and power consumption) is generated.

A performance evaluation parameter "User's satisfaction index," [12] is introduced for the purpose of comparison and is given by

$$\text{User's Satisfaction index} = \frac{preferred\_bw - actual\_bw}{actual\_bw} + \frac{preferred\_cst - actual\_cst}{actual\_cst} + \frac{preferred\_SEC - actual\_SEC}{actual\_SEC} \tag{3.1}$$

where *preferred_bw*, *preferred_cst*, and *preferred_SEC* are the measurement of the network parameters defined by the mobile subscriber, *actual_bw, actual_cst*, and *actual_SEC* are the measurement of the networks parameters of the target network selected by the above algorithms. Figures 3, 4, and 5 show the user satisfaction generated by applying cst function-based method, TOPSIS method, and SAW method.

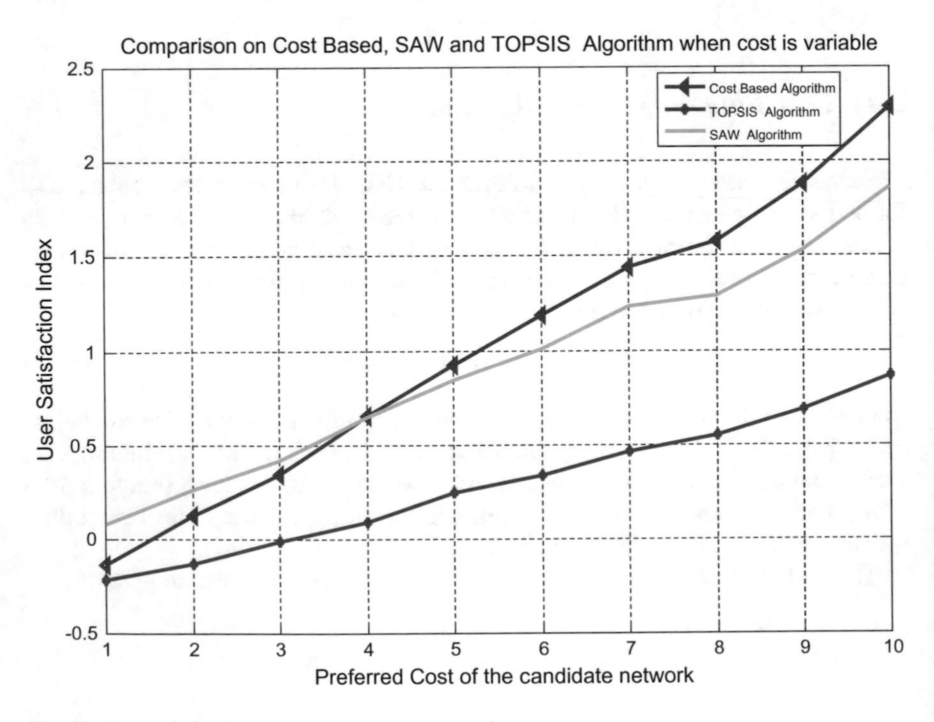

Fig. 3 User satisfaction index when cost is varied

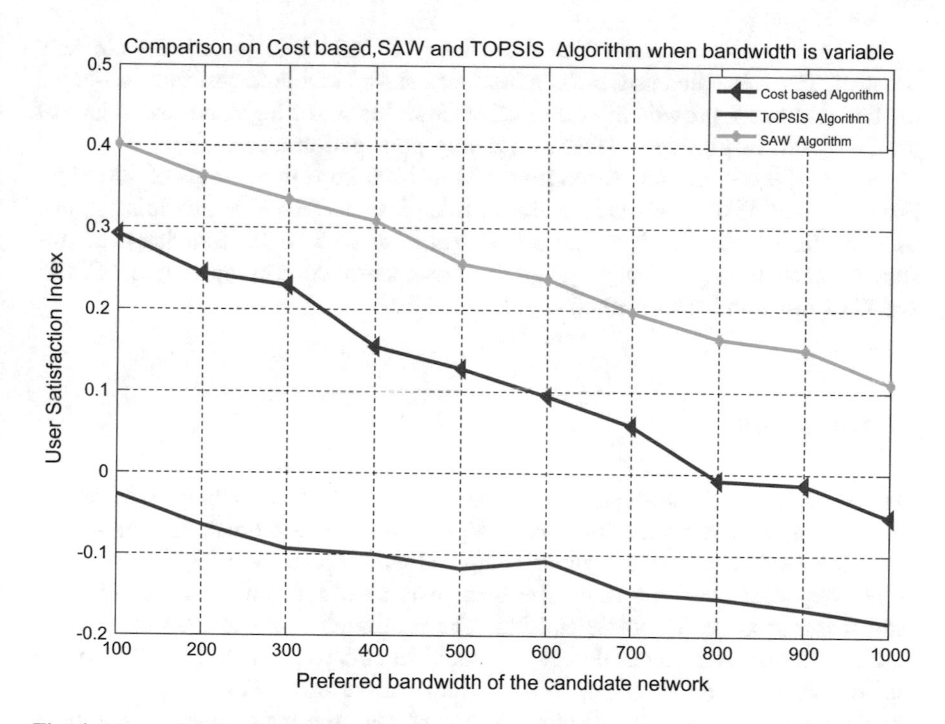

Fig. 4 User satisfaction index when bandwidth is varied

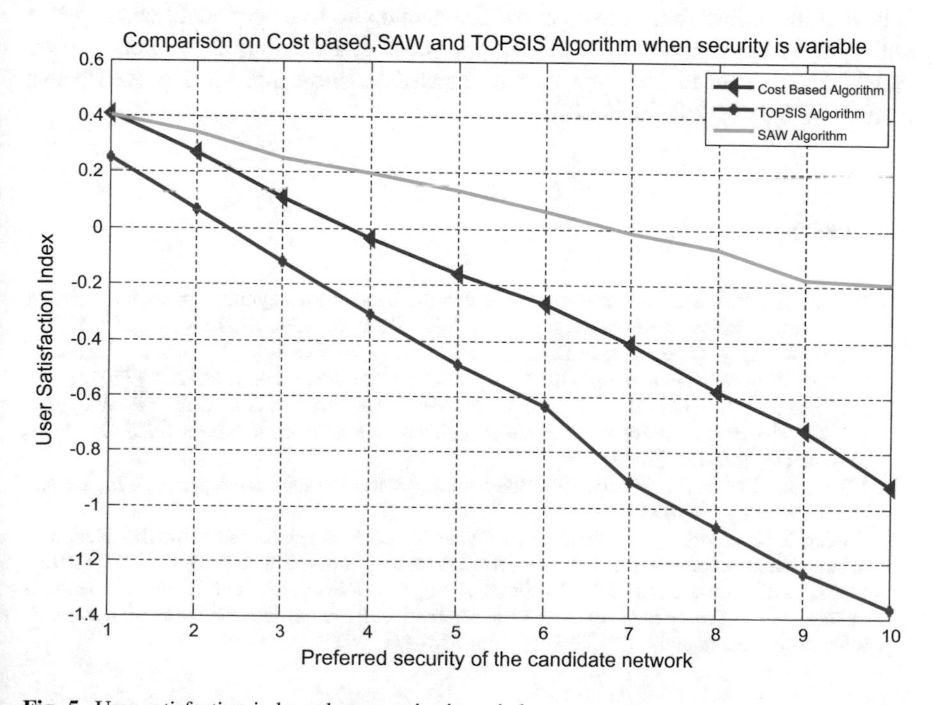

Fig. 5 User satisfaction index when security is varied

In Fig. 3, *preferred_cst* varied from 1 to 10. All other parameters are kept constant. The calculated user satisfaction index shows that both cost function-based method and SAW provide high user satisfaction index for high and low values of *preferred_cst*, respectively. TOPSIS has very poor performance.

In Fig. 4, *preferred_bw* varied from 100 to 1000 kbps in the steps of 100 kbps. From the plot of user satisfaction index, it is clear that SAW is providing higher user satisfaction. In Fig. 5, *preferred_sec* varied from 1 to 10. Simulation results shows that SAW is providing high grade of user satisfaction comparing to TOPSIS and Cost function based methods.

4 Conclusion

This paper shows the analysis of different network selection techniques in heterogeneous cognitive radio networks as multiple-criteria decision-making. Three main multiple-criteria decision algorithms compared with inputs are taken from the user as per requirement of the user. The main parameter for comparison is the user satisfaction achieved by that algorithm. The results represent that SAW provides high amount of user satisfaction as compared to cost function and TOPSIS-based method when priority is given to bandwidth and security. From cognitive radio networks point of view, bandwidth and security are the main constraints. Hence, SAW algorithm is preferred over the TOPSIS and cost-based function. This method is having the ability to maximize users' satisfaction up to nearly 50%, as compared with the existing methods TOPSIS and cost function-based method. When weightage is given to cost, and then cost-based function method is proved better than SAW and TOPSIS methods.

References

1. Kumar, K., Prakash, A., Tripathi, R.: Spectrum handoff in cognitive radio networks: a classification and comprehensive survey. J. Netw. Comput. Appl. **61**, 161–181 (2016)
2. Jouini, W., Moy, C., Palicot, J.: Decision making for cognitive radio equipment: analysis of the first 10 years of exploration. EURASIP J. Wirel. Commun. Netw. **2012**(26) (2012)
3. Stevenson, C.R., Chouinard, G., Lei, Z., Hu, W., Shellhammer, S.J., Caldwell, W.: IEEE 802.22: The first cognitive radio wireless regional area network standard. IEEE Commun. Mag. **47**(1), 130–138 (2009)
4. Mitola, J., Maguire, G.Q.: Cognitive radio: making software radios more personal. IEEE Pers. Commun. **6**(4), 13–18 (1999)
5. Haldar, K.L., Ghosh, C., Agrawal, D.P.: Dynamic spectrum access and network selection in heterogeneous cognitive wireless networks. Pervasive Mob. Comput. **9**(4), 484–497 (2013)
6. Lai, J., Dutkiewicz, E., Liu, R.P., Vesilo, R., Fang, G.: Network selection in cooperative radio networks. In: Proceedings of the 11th international symposium on communication & information technologies (ISCIT 2011), pp. 378–381 (2011)

7. Ozturk, Z.: A review of multi criteria decision making with dependency between criteria. In: Proceedings of MCDM 2006, Chania, Greece (2006)
8. Akyildiz, I.F., Wang, W.: A dynamic location management scheme for Next-Generation multitier PCS systems. IEEE Trans. Wirel. Commun. **1**(1), 178–189 (2002)
9. Hwang, C.-L., Yoon, K.: Multiple Attribute Decision Making. Springer, Berlin (1981)
10. Zhang, W.: Handover decision using fuzzy MADM in heterogeneous networks: In: Proceedings of IEEE Wireless Communications and Networking Conference (WCNC 2004)., vol. 2, pp. 653–658 (2004)
11. Kumar, K., Prakash, A., Tripathi, R.: A spectrum handoff scheme for optimal network selection in cognitive radio vehicular networks: a game theoretic auction theory approach. Phys. Commun. **24**, 19–33 (2017)
12. Hasswa, A., Nasser, N., Hassanein, H.: A context aware cross-layer architecture for next generation heterogeneous wireless networks. In: Proceedings of the 2006 IEEE International Conference on Communications (ICC'06), pp. 240–245, Istanbul, Turkey (2006)

Energy Detection-Based Spectrum Sensing Scheme Using Compel Cooperation in Cognitive Radio Networks

Krishan Kumar, Hitesh Tripathi and Mani Shekhar Gupta

Abstract Cognitive radio network is a next-generation technology to address the spectrum scarcity issue. One important operation of cognitive radio is sensing of the available spectrum. To utilize an unused licensed network's spectrum, cognitive radio needs to sense and identify the frequency bands and detection of the primary users. Thus, an efficient spectrum sensing is a primary need of cognitive radio for detection of free channels and licensed users. Several schemes have been introduced for spectrum sensing. In this paper, benefits of compel cooperation are utilized for efficient spectrum sensing in cognitive radio. Hence, this paper proposed an efficient spectrum sensing scheme, which reduces the detection time and improved activity gain by the permission of the cognitive radio in the same band to compel cooperate with each other. Numerical results reveal that the introduced scheme is well suited for spectrum sensing in a scenario of two users/multiusers with reduced delay and enhanced activity gain.

Keywords Cognitive radio network · Secondary user · Compel cooperation · Probability of false alarm · Activity gain

1 Introduction

To fulfill the demand of ever increasing wireless services, the Cognitive Radio (CR) network has been introduced. The basic concept behind CR networks is that the CR can access the unused spectrum of licensed networks by detecting spectrum

K. Kumar (✉) · H. Tripathi · M. S. Gupta
Electronics and Communication Engineering Department,
National Institute of Technology, Hamirpur, India
e-mail: krishan_rathod@nith.ac.in

H. Tripathi
e-mail: tripathihitesh80@gmail.com

M. S. Gupta
e-mail: mi_sr87@yahoo.com

A. K. Luhach et al. (eds.), *Smart Computational Strategies: Theoretical and Practical Aspects*, https://doi.org/10.1007/978-981-13-6295-8_19

holes [1–3]. CR nodes do not cause any interference to Primary Users (PUs), which are also referred to as the licensed users. Spectrum sensing is one of the most important tasks to be implemented in CR networks. Various popular spectrum sensing methods like energy detection [4], matched filter detection [5], and cyclostationary feature detection [6] have been already briefly surveyed by researchers. However, the problems like noise levels, interference and fading make spectrum sensing inaccurate. Thus, spectrum sensing performance is undesirable, which leads to low detection probability and high false alarm probability. Hence, to overcome these problems, cooperative spectrum sensing [7, 8] is presented. In cooperative spectrum sensing, different CR nodes communicate with each other to minimize receiver uncertainty and fading [9]. The cooperative sensing can be categorized into two approaches named as centralized and distributed [10, 11]. In centralized approach, a CR base station collects all the sensing data from different CR nodes and accordingly makes the decision for spectrum sensing. In distributed approach, each CR node provides coordination to other CR nodes in the network to provide high networks throughput. The cooperation among users to improve diversity is discussed in [12, 13]. In [13], a cooperative protocol for Code-Division Multiple Access (CDMA) network is analyzed. In [14], the Amplify-and-Forward (AF) protocol has been considered for a two-user network scenario, in which a signal is sent by the transmitter and retransmitted by the relay.

In this paper, a spectrum sensing scheme using compel cooperation is introduced for CR networks. The impact of AF on the spectrum sensing performance is also considered. In Sect. 2, a PU detection hypothesis is formulated for a simple two CR nodes and energy detector is used to identify the absence or presence of the PU signal using compel cooperation approach. In Sect. 3, activity gain for two CR nodes is analyzed. In Sect. 4, the results show the reduction in sensing time and improved activity gain. Finally, the conclusion is drawn in Sect. 5.

2 Formulation of the Detection Problem

In this part, a channel model [15] is portrayed and used throughout the paper. This defines the PU detection and implement a two CR nodes compel cooperation scheme in CR networks and thus improving activity gain of the CR networks.

A. Channel Model

Rayleigh fading has been considered for all the channels and channels are assumed to be independent for different CR nodes. If the input signal sent is s, and r is the received signal, then the channel model is calculated as

$$r = as + u$$

where u is the Additive White Gaussian Noise (AWGN) with zero mean and unit variance and a is the fading coefficient. It is also assumed that CR nodes communicate with the centralized CR base station throughout the paper.

B. Compel Cooperation Scheme

CR nodes should always monitor the spectrum to detect the presence of a PU. Consider a network with two CR nodes (CR1 and CR2), sending the signal to a shared receiver, as shown in Fig. 1. As soon as the PU starts using the spectrum, the two CR nodes should vacate it. Now, CR1 takes a long time to sense the presence of PU because it is some distant from it and signal which is received by CR1, is weak to identify the presence of PU. So, in this case, compel cooperation can be used between the two CR nodes to reduce the time of detection of 'far away' node and thereby increasing the activity gain of the CR networks. Both the CR node, CR1 and CR2 cooperate with each other and CR2 act as a relay for CR1. In Fig. 2, the data is being transmitted to a shared receiver by CR1 and CR2 in a frequency band in successive time intervals.

In T_1 time interval, node CR1 transmits the signal, CR2 listens and in T_2 time interval, the signal is relayed by node CR2. The PU has the highest priority for occupying the band. Therefore, the presence of PU should be detected quickly. In T_1 time interval, the received signal by CR2 from CR1 is calculated as

$$r_2 = \eta g_{p2} + b g_{12} + u_2 \tag{1}$$

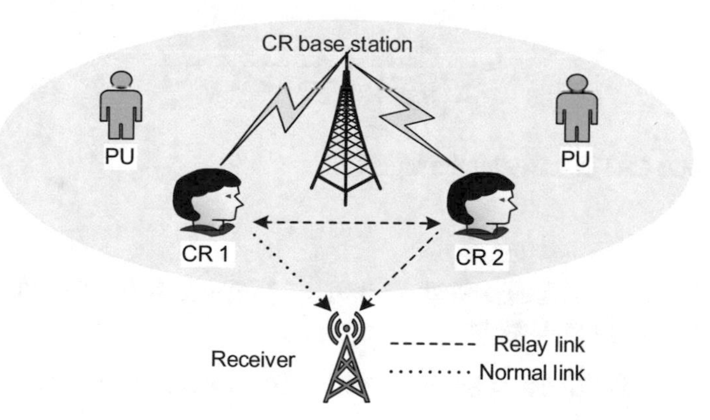

Fig. 1 Compel cooperation in CR networks

Fig. 2 Relay protocol

CR1 T_x	CR2 Relay	CR2 T_x	CR1 Relay	…………….

time

where g_{pi} is the instantaneous channel gain between CR nodes and PU, g_{12} is the instantaneous channel gain between CR1 and CR2, u_2 is the AWGN, b is the signal sent from CR1, and η indicates the PU, i.e., if $\eta = 1$, it means that PU is present and if $\eta = 0$, it means that PU is absent. Let transmit power of CR1 be S then,

$$E\left[|bg_{12}|^2\right] = SC_{12}$$

where $C_{12} = E\left\{|g_{12}|^2\right\}$, is the channel gain between nodes CR1 and CR2. All the parameters are considered independent, then it becomes:

$$E\left[|r_2|^2\right] = \eta^2 S_2 + SC_{12} + 1$$

where $S_i = E\left\{\left|g_{pi}\right|^2\right\}$, is the power signal received at particular CR node from PU. In T_2 time interval, CR2 (act as a relay node), the data is relayed from CR1 to a shared receiver. The maximum power constraint of relay node is $\acute{S}$. In T_2 time interval, when CR2 relays the data of CR1 to the receiver, node CR1 hears its own message. The received signal by CR1 from CR2 is calculated as

$$\begin{aligned} r_1 &= \sqrt{\gamma_1} r_2 g_{12} + \eta g_{p1} + u_1 \\ &= \sqrt{\gamma_1} h_{12}\left(\eta g_{p2} + b g_{12} + u_2\right) + \eta g_{p1} + u_1 \end{aligned} \tag{2}$$

where γ_1 is the scaling factor for relaying the information to the shared receiver used by CR2. The value of γ_1 is calculated as

$$\gamma_1 = \frac{\acute{S}}{E\left\{|r_2|^2\right\}} = \frac{\acute{S}}{\eta^2 S_2 + SC_{12} + 1}$$

The node CR1 is left with the signal as

$$R = \eta G + U \tag{3}$$

where $G = g_{p1} + \sqrt{\gamma_1} g_{12} g_{p2}$ and $U = u_1 + \sqrt{\gamma_1} g_{12} u_2$. Now, the PU detection problem can be formulated as

$$\begin{aligned} &H_1 : R = \eta G + U \quad \text{when PU signal is present} \\ &\text{or} \\ &H_0 : R = U \qquad\quad\;\; \text{when PU signal is absent} \end{aligned}$$

This is a common detection problem. Various detectors can be utilized to solve this detection problem. In this paper, Energy Detector (ED) is considered for testing compel cooperation approach between two CR nodes.

C. **Energy Detector**

The basic concept behind Energy Detector (ED) [16] scheme is to calculate the received energy of the signal and compare it with a predefined threshold. If the received signal is greater than the predefined threshold, then it indicates the presence of PU and the spectrum cannot be used by CR nodes, else, PU is absent and spectrum is available to be used by CR nodes. It is considered that the random variable H and U in Eq. (3) are complex normal distributed with zero mean and unit variances,

$$\sigma_g^2 = S_1 + \gamma S_2 g \tag{4}$$

and

$$\sigma_U^2 = 1 + \gamma g \tag{5}$$

where

$$g = \frac{g_{12}}{E\left\{|g_{12}|^2\right\}} = \frac{|g_{12}|^2}{C_{12}}$$

and

$$\gamma = \frac{\acute{S}C_{12}}{\eta^2 S_2 + SC_{12} + 1} \tag{6}$$

where g_{12} is complex Gaussian, the g has probability density function given as

$$f(g) = \begin{cases} e^{-g} & g > 0 \\ 0 & g \leq 0 \end{cases}$$

The ED forms the statistics as

$$D(R) = |R|^2$$

This is compared with a threshold θ, and it can be determined by false alarm probability α,

$$\text{define } \Phi(t; c, d) = \int_0^\infty \mathrm{e}^{-g - \frac{t}{c + dg}} \mathrm{d}g \quad \text{for positive } t,\ c \text{ and } d \tag{7}$$

Let $E_i(t)$ represent the cumulative distributive function of the random variable $D(R)$ where i = 0,1 for H_i hypothesis. From (5), we get

$$E[D(R)|H_0, g] = E[|W|^2 g, \eta = 0] = 1 + \frac{SC_{12}}{\acute{S}C_{12}+1} g$$

For $H_0(\eta = 0)$,

$$E_0(t) = P\big(D(R) > t|H_0\big) = \int_0^\infty P\big(D(R) > t|H_0, g\big) f(g) \mathrm{d}g = \Phi(t; 1, \frac{SC_{12}}{\acute{S}C_{12}+1})$$

Similarly, it is shown that

$$E_1(t) = \Phi(t; S_1 + 1, \gamma(S_2 + 1))$$

where γ is given by Eq. (6). The threshold θ for the given false alarm probability α is calculated as

$$\Phi\left(\theta; 1, \frac{SC_{12}}{\acute{S}C_{12}+1}\right) = \alpha \tag{8}$$

τ can be calculated uniquely as Φ in (7) is decreasing in t. Therefore, the detection probability by CR1, with compel cooperation with CR2, is given by

$$p_{\mathrm{dc}} = \Phi(\theta; S_1 + 1, \gamma(S_2 + 1)) \tag{9}$$

Now, when two CR nodes, follows compel cooperation scheme as described, the overall probability of detection is determined as

$$P_{\mathrm{d}} = p_{\mathrm{dc}} + p_{\mathrm{dn}}^2 - p_{\mathrm{dc}} p_{\mathrm{dn}}^2$$

For the case of noncooperation between the two nodes, γ_1 is zero. The terms p_{dn}^1 and p_{dn}^2, represents the probability of detection for CR1 and CR2, respectively. Similarly, it is calculated as

$$p_{\mathrm{dn}}^1 = \alpha^{\frac{1}{S_1+1}} \tag{10}$$

and

$$p_{\mathrm{dn}}^2 = \alpha^{\frac{1}{S_2+1}} \tag{11}$$

Now, the probability of detection of overall CR networks, i.e., the PU is detected by any of the nodes (either by CR1 or by CR2). For noncooperation, i.e., each node detect PU independently, then overall probability of detection is determined as

$$\mathrm{P_d} = p_{\mathrm{dn}}^1 + p_{\mathrm{dn}}^2 - p_{\mathrm{dn}}^1 p_{\mathrm{dn}}^2$$

3 Activity Gain

In the previous sections, improved detection probability is obtained for the compel cooperation scheme. In this part, the focus shifts to reduce the entire detection time for compel cooperation scheme. The detection time is affected by two protocols, namely Non Cooperative (NC) Protocol and Totally Cooperative (TC) Protocol. In first, all CR nodes detect PU, independently and the first CR node to identify the presence of PU informs rest of the CR nodes through the base station controller. In second, two or more CR nodes cooperate to identify the presence of PU. Activity gain is calculated as the ratio of time taken in TC protocol and time taken in NC protocol for two CR nodes. Mathematically

$$\beta_{n/c} = \frac{T_n}{Tc}$$

where

$$T_n = \frac{2 - \frac{p_{\rm dn}^1 + p_{\rm dn}^2}{2}}{p_{\rm dn}^1 + p_{\rm dn}^2 - p_{\rm dn}^1 p_{\rm dn}^2}$$

and

$$Tc = \frac{2 - \frac{p_{\rm dc} + p_{\rm dn}^2}{2}}{p_{\rm dc} + p_{\rm dn}^2 - p_{\rm dc} p_{\rm dn}^2}$$

4 Results and Discussion

Figure 3 represents the variation in detection probability with respect to received signal power from PU. Set the value of $\alpha = 0.1$ (since it is false alarm probability), detection probability for cooperation and noncooperation (i.e., p_{dc} and $p_{\rm dn}^1$), as a function of S_2 for $S = \acute{S} = 0$ dB and for $S_1 = 0$ dB, 4.5 dB and 8 dB is analyzed. It can be observed that for different values of S_1, the detection probability for compel cooperation scheme is better than noncooperation scheme for definite range of S_2. In Fig. 4, the overall detection probability in CR networks, for compel cooperation and noncooperation plotted for $\alpha = 0.1$ as a function of S_2 with $S_1 = 0$ dB. It can be observed from the graph that the cooperation between the users enhances the overall detection probability.

Figure 5 shows the graph for $\beta_{n/c}$, i.e., activity gain as a function of S_2 for $\alpha = 0.1$ and $S_1 = 0$ dB. From the graph, it is inferred that activity is high. The activity should be high because CR users monitor the spectrum continuously in order to identify the presence of PU.

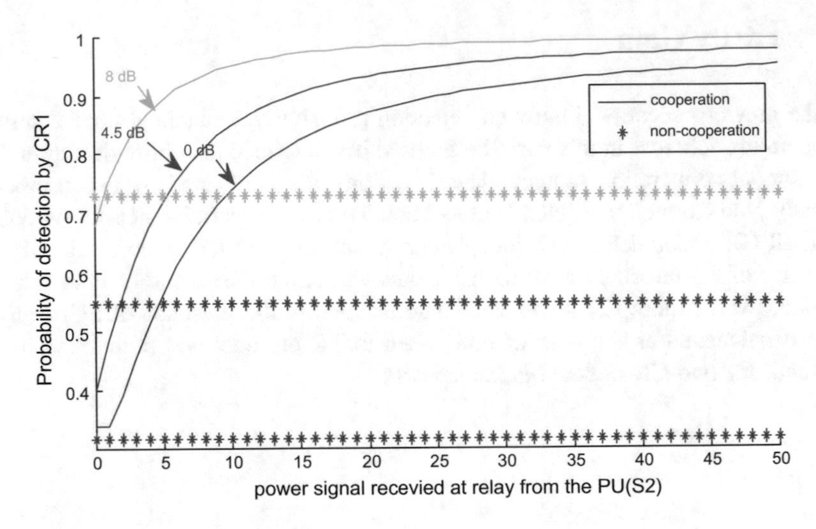

Fig. 3 Cooperation improves detection probability

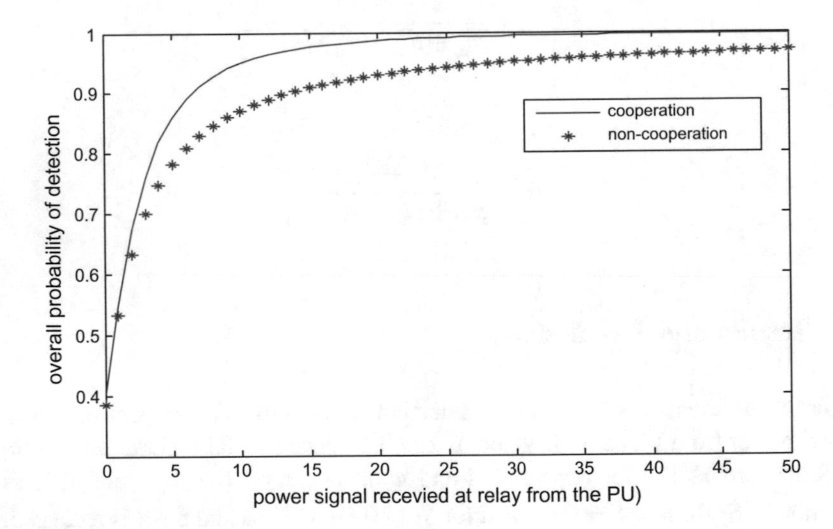

Fig. 4 Overall probability of detection ($S_1 = 0\,\text{dB}$)

Further, in Fig. 6, the graph between the time taken (detection time) to identify the presence of PU as a function of S_2 has been plotted. This gives us the idea that with the help of cooperation, the detection time (i.e., time to detect the presence of PU) is significantly reduced when compared to noncooperation scheme.

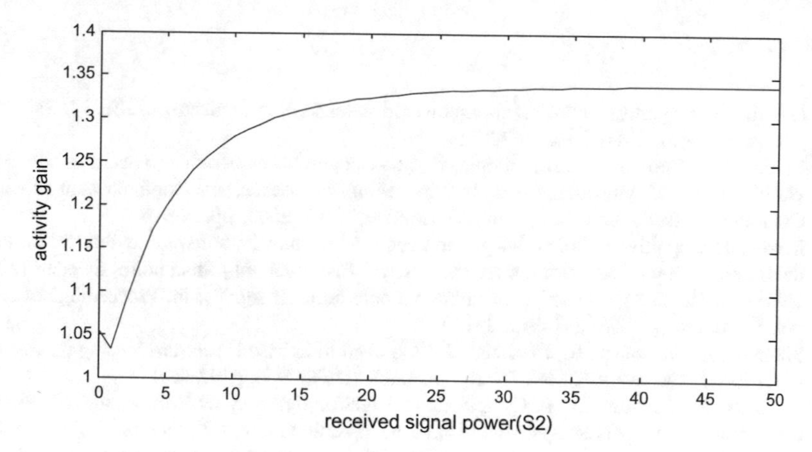

Fig. 5 Activity gain

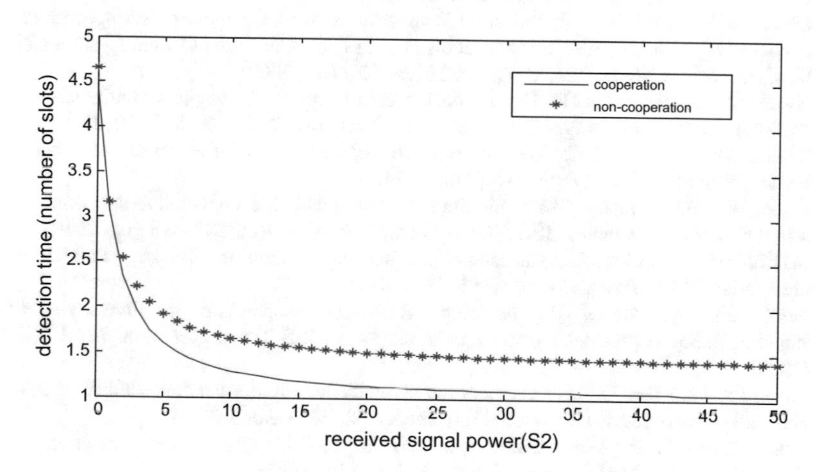

Fig. 6 Time for detection($(S_1 = 0\,\mathrm{dB}, \alpha = 0.1)$

5 Conclusion

In this paper, the two-user compel cooperation scheme has been analyzed, which not only increases the probability of detection of the CR network but also increases activity gain (i.e., reduction in detection time) of the CR network. First, compel cooperation scheme has been shown for which cooperation between the users increases the probability of detection for the CR network when compared to adjacent schemes. Then, the activity gain for two-user CR network has been analyzed and it is shown that the activity of the network increases when we use cooperation scheme. Further, the detection time is also reduced, i.e., CR nodes sense the PU faster when compared to adjacent schemes.

References

1. Haykin, S.: Cognitive radio: brain-empowered wireless communications. IEEE J. Sel. Areas Commun. **23**(2), 201–220 (Feb 2005)
2. Barve, S.S., Deosarkar, S.B., Bhople, S.A.: A cognitive approach to spectrum sensing in virtual unlicensed wireless network. In: Proceedings of International Conference on Advanced Computation Communication, Control, Mumbai, India, 2009, pp. 668–673
3. Mitola, J.: Cognitive radio: an integrated agent architecture for software defined radio. Ph.D. dissertation, Dept. Tele informatics, KTH Royal Inst. Technol., Stockholm, Sweden (2000)
4. Urkowitz, H.: Energy detection of unknown deterministic signals. In: Proceedings of IEEE, vol. 55, no. 4, pp. 523–531 (Apr 1967)
5. Shobana, S., Saravanan, R., Muthaiah, R.: Matched filter based spectrum sensing on cognitive radio for OFDM WLANs. Int. J. Eng. Technol. (IJET) **5**(1), p142 (2013)
6. Enserink, S., Cochran, D.: A Cyclostationary feature detector. In: Proceeding 28th Asilmar Conference on Signals & Systems, Computers, Pacific Grove, CA, Oct 1994, pp. 806–810
7. Cabric, D., Mishra, S.M., Brodersen, R.W.: Implementation issues in spectrum sensing for cognitive radios. In: Proceedings of 38th Asilomar Conference Signals, Systems, Computation, vol. 1, pp. 772–776 (Nov 2004)
8. Ghasemi, A., Sousa, E.S.: Optimization of spectrum sensing for opportunistic spectrum access in cognitive radio networks. In: Proceedings of 4th IEEE Consumer Communication Network Conference, Las Vegas, NV, USA, pp. 1022–1026 (Jan 2007)
9. Huang, X.-L., Wang, G., Hu, F.: Multitask spectrum sensing in cognitive radio networks via spatiotemporal data mining. IEEE Trans. Veh. Technol. **62**(2), 809–823 (2013)
10. Yildiz, M.E., Aysal, T.C., Barner, K.E.: In-network cooperative spectrum sensing. In: Proceedings of EUSIPCO, pp. 1–5 (Aug 2009)
11. Li, Z., Yu, F.R., Huang, M.: A distributed consensus-based cooperative spectrum sensing scheme in cognitive radios. IEEE Trans. Veh. Technol., **59**(1), 383–393 (Jan 2010)
12. Sendonaris, A., Erkip, E., Aazhang, B.: User cooperation in diversity. Part I: System description. IEEE Trans. Commun. **51**, 1927–1938 (2003)
13. Sendonaris, A., Erkip, E., Aazhang, B.: User cooperation in diversity—Part II: Implementation aspects and performance analysis. IEEE Trans. Commun. **51**, 1939–1948 (2003)
14. Laneman, J.N., Tse, D.N.C.: Cooperative diversity in wireless networks: efficient protocols and outage behaviour. IEEE Trans. Inf. Theory **50**, 3062–3080 (2004)
15. Sklar, Bernard: Rayleigh fading channels in mobile digital communication system Characterization. IEEE Commun. Mag. **35**, 136–146 (1997)
16. Oh, D.C., Lee, Y.H.: Energy detection based spectrum sensing for sensing error minimization in cognitive radio networks. Int. J. Commun. Netw. Inf. Secur. **1** (Apr 2009)

Performance and Comparison Analysis of MIEEP Routing Protocol Over Adapted LEACH Protocol

Rahul Priyadarshi, Soni Yadav and Deepika Bilyan

Abstract Technical innovations in wireless sensor provided a way to the expansion of low-power and low-cost tiny sensor nodes with multiple functions. Routing problems in sensor networks done by network layer. As radio reception and transmission ingests huge quantity of energy, so investigation on power is the main factor. Energy management is consequently a crucial issue in wireless networks. The present study includes scheming routing protocols which need smaller amount of energy throughout the communication as a result prolonging the network lifetime. In various applications, replacement of node or charging the node battery is very expensive work. This paper delivers comparison analysis of different variant of PEGASIS protocol and also different variants of LEACH protocol. Finally, performance analysis of MIEEP routing protocol with adapted LEACH protocol using MATLAB tool has been analyzed which further reflects that MIEEP routing protocol shows better result compared to adapted LEACH protocol.

Keywords Wireless Sensor Network · Nodes · Base station · Routing · Energy

1 Introduction

A wireless sensor network collects environment information by utilizing sensor modules and analyzes collection information with various applications. However, the performance of sensor nodes in sensor networks is limited in terms of energy, computational capability [1–5], communication radius, and storage memory.

R. Priyadarshi (✉) · S. Yadav · D. Bilyan
Electronics and Communication Engineering Department,
National Institute of Technology, Hamirpur, India
e-mail: rahul.glorious91@gmail.com

S. Yadav
e-mail: soni151291@gmail.com

D. Bilyan
e-mail: deepikabilyan@gmail.com

A. K. Luhach et al. (eds.), *Smart Computational Strategies: Theoretical and Practical Aspects*, https://doi.org/10.1007/978-981-13-6295-8_20

Therefore, the algorithms of all wireless sensor network schemes should be studied to minimize energy consumption by utilizing energy efficiently and considering the limited performance of sensor nodes. WSNs are vulnerable to external attacks, as they are constructed mostly in unmanned environments to monitor environmental information or military zones, and sensor nodes use wireless communication to transfer collected data to base stations [6–9]. Due to such characteristics, data can be easily exposed during data transmission. This can be a serious problem in applications that deal with military information or individual privacy. To solve these problems, studies on security schemes have been performed. To configure a safe sensor network, it is possible to combine a routing protocol [10], which is a basic element for information transmission, with an encryption protocol, such as key exchange and authentication. Recently, the various data transmission schemes [11–13] were proposed such as a transmission scheme that periodically collects real and virtual data at the same time based on encryption algorithms and a data transmission scheme that uses reliability levels. However, such schemes are not suitable for wireless sensor nodes with performance limitations, such as energy consumption due to additional data transmission or increased communication between nodes for increasing security. Therefore, it is necessary to study energy-efficient security schemes [14–18] while considering the characteristics of wireless sensor networks. Therefore, this paper aims to provide an algorithm to restrict data analysis of transferred data over a wireless sensor network despite transferred data exposure as well as minimize energy consumption for communication and improve security and energy efficiency [19]. Communication structural design of WSN which is shown in Fig. 1 and hardware structural design of WSN is shown in Fig. 2.

Figure 3 shows the clustering in WSN.

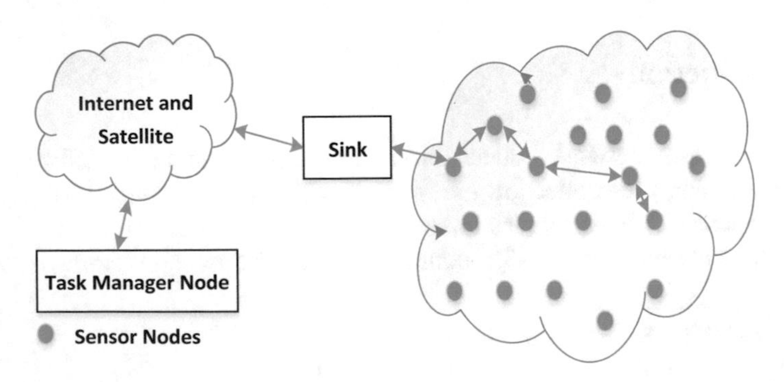

Fig. 1 Communication structural design in WSN

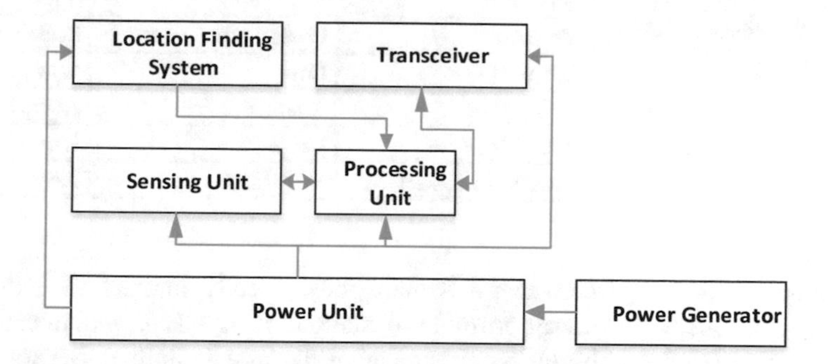

Fig. 2 Hardware structural design in WSN

Fig. 3 Clustering in WSN

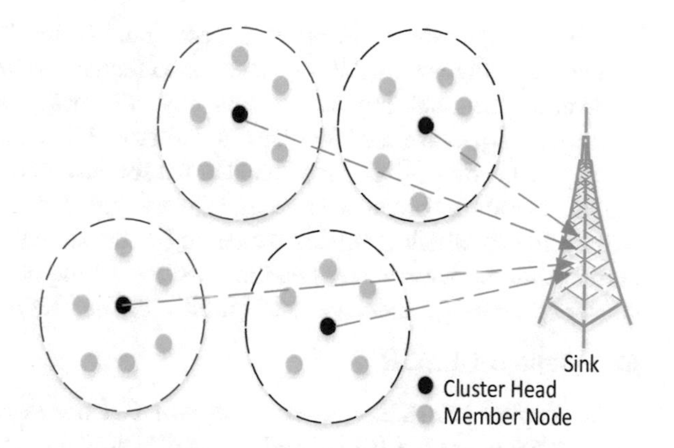

2 Comparison of LEACH Variants

Leach performs the re-clustering function and self-organization after each round. Sensor nodes are combined to form cluster in LEACH routing protocol. In LEACH protocol, all nodes will transmit their information to the CH node, here CH node will do the data aggregation and compression [20–24] and after this, it will transmit this aggregated information to the sink. Because the cluster head will perform the additional task of data aggregation, it will dissipate more energy and if it will remain the cluster head permanently, then it will die quickly. Thus, the ultimate goal of LEACH protocol is to increase the lifetime of whole system by selecting cluster head among more powerful sensor network. The process of implementation of LEACH includes many rounds. There are various variants of LEACH protocol that is listed below:

A. **LEACH-L**

LEACH-L is an advanced energy balanced multi-hop cluster-based routing protocol. LEACH-L relies on the distance only and optimal hop counts are assumed

Table 1 Variants of LEACH protocol

Protocol	Energy efficiency	Scalability
LEACH	Low	Low
LEACH-B	Average	Average
LEACH-L	Large	Large
Adapted LEACH	Very large	Large

in it. When CHs are placed near BS then, they directly interact with the BS. When CH is positioned away from base station, then CH communicates via multi-hop mode and the shortest communication distance is limited. Sensor nodes communicate with BS at different frequencies in LEACH-L [25].

B. **LEACH-B**

It is an improvement in original leach. For cluster formation, LEACH-B uses a decentralized algorithm. All sensor nodes contain information about their location and receiver location but they do not have any information about other nodes. Multiple access techniques are used for data transmission. Each node decides based on formula whether to become (CH) cluster head or not for current round. After this, each elected CH node sends the broadcast message advertising all nodes. After this, member nodes will decide to which it relates depending upon strength of signal and distance for current round. Once cluster head is selected, all the nodes join their corresponding cluster by sending a message to CH informing that it is a member of that cluster [26].

C. **Adapted LEACH**

In adapted LEACH, cluster nodes and CH nodes can move during setup and steady state phase but BS is fixed. Primarily, all the sensor nodes are supposed to be homogeneous and information about their position gets through GPS. The distributed setup phase of original leach has been modified to select the suitable cluster head in adapted LEACH. Selection of CH [27] is done by least mobility and lowermost attenuation energy. If CH moves from its cluster nodes or cluster nodes move away from its CH, another CH is elected in the steady-state phase which results in effective cluster formation. A comparison between LEACH variants has been listed in Table 1.

3 Comparison of PEGASIS Variants

PEGASIS is a sequence-based algorithm where every single node connects with its immediate sensor node neighbors. Creation of sequence basically starts with furthermost sensor node from the BS. Average energy consumed by every single node for each round is diminished and advances lifetime of system which is up to just about 280% as related to LEACH Protocol. There are various variants of PEGASIS protocol that is listed below:

A. **Energy Reduction PEGASIS**

In this algorithm, nodes are organized in a manner where data packets come to the target point over the direct route which further drops overall energy consumption of the network. The sequence structure is altered such that distances among nodes will continually be diminished. Moreover, the data accumulation too diminishes overall network energy consumption [28].

B. **Energy Efficient PEGASIS**

EEPB is a sequence-based protocol that assumes distance threshold when creating sequence, lessening the creation of lengthy links. Subsequently, energy debauchery of nodes is relational to communication distance, and the lead is carefully chosen by bearing in mind the cooperation of remaining energy of sensor nodes and distance among BS and node. Difficulty and ambiguity in threshold are weaknesses of this protocol.

C. **Improved Energy-Efficient PEGASIS**

MIEEP is a multi-sequence prototypical integrating sink movement by this means accomplishing less significant sequences and decreased loads on lead nodes. A mobile sink diminishes energy usage of nodes and also supports in dropping data supply delay for all the sensor nodes. Multi-sequence concept diminishes distance among associated nodes. It drops system overhead subsequently. There is only a smaller number of nodes in the sequences [29–31]. A comparison between PEGASIS variants has been listed in Table 2.

4 Simulation and Results

The performance analysis of adapted LEACH and MIEEP is simulated by MATLAB tool taking the following simulation parameters into consideration with its value which are listed in Table 3.

Figures 4 and 5 show that the number of alive nodes per round in MIEEP protocol is more related to the adapted LEACH protocol which additionally reflects superior network lifetime of MIEEP protocol. The obtained result shows that 200th

Table 2 Variants of PEGASIS protocol

Protocol	Energy efficiency	Scalability
PEGASIS	Good	Good
Energy reduction PEGASIS	Very large	Large
EEPB	Good	Large
MIEEP	Large	Very large

Table 3 Simulation parameter

Simulation parameters	Value
Energy dissipation	50 nj/bit
Area	100 m * 100 m
Initial energy of nodes	0.4 J
Sensor nodes	200
Number of round	3000
Packet length	3000 bits

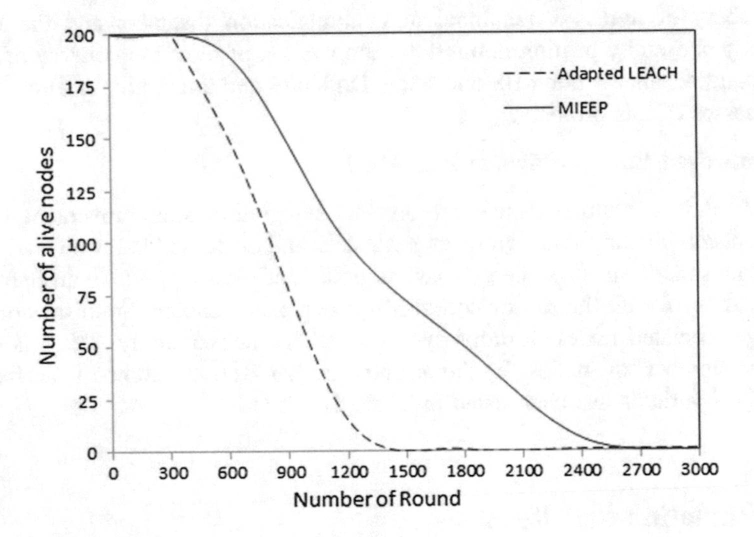

Fig. 4 Number of alive nodes

node of adapted LEACH protocol dies out at 1432 round where 200th node of MIEEP protocol dies out at 2634th round.

Figures 6 and 7 display the plot of remaining energy of nodes. At the starting point, every single node has 0.4 J energy and also overall number of node is considered as 200. As a result, network overall energy is 80 J. MIEEP protocol favors improved results in comparison to the adapted LEACH protocol in reference to energy consumption consideration.

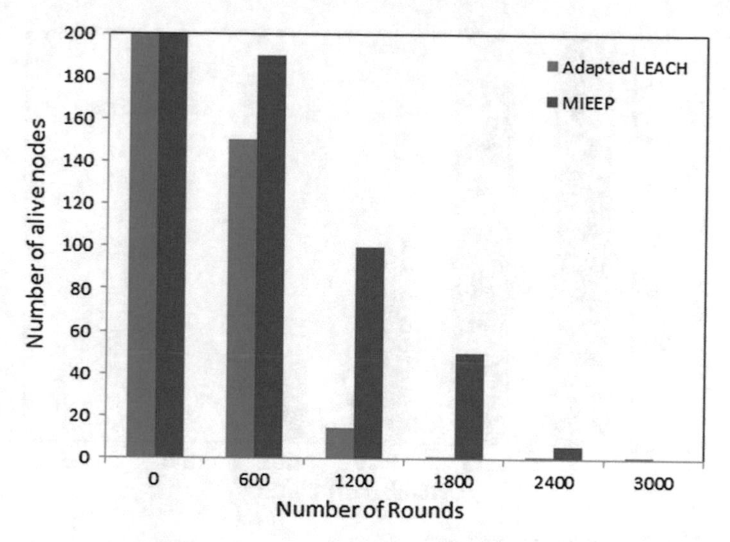

Fig. 5 Number of alive nodes (histogram view)

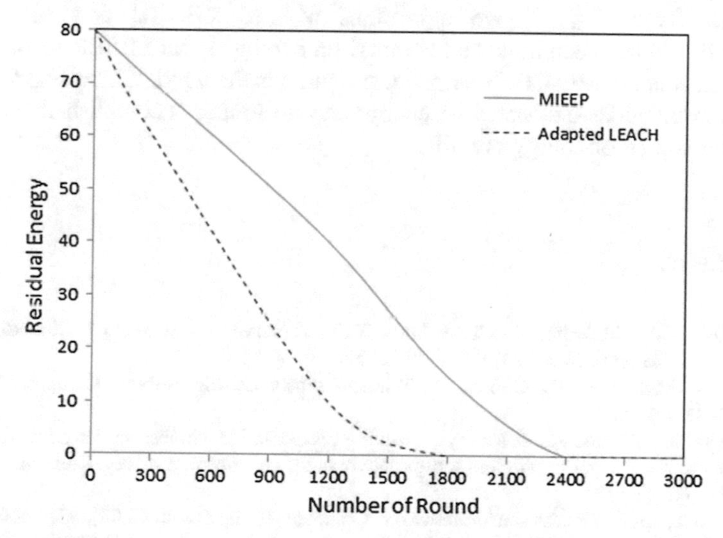

Fig. 6 Residual energy

5 Conclusion

This paper analyzed MIEEP protocol with adapted LEACH protocol and also surveyed about different variants of LEACH and PEGASIS protocols. Protocols planned had better objective in keeping sensor nodes active for a large period of time to realize the application necessities and should come across the scalability

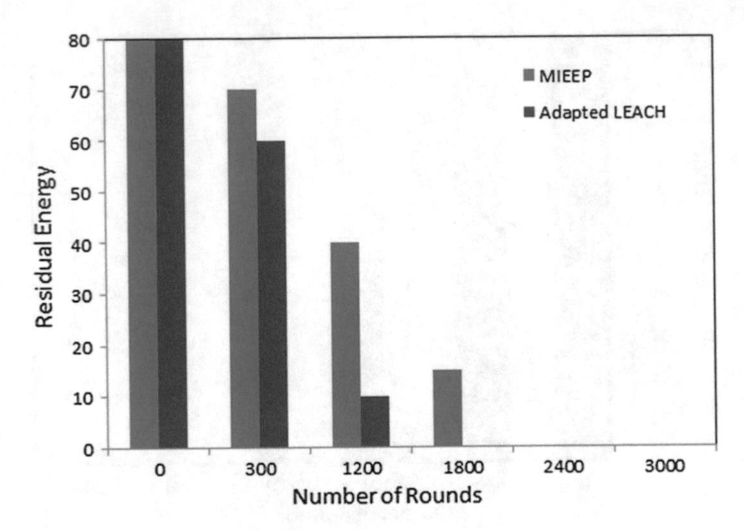

Fig. 7 Residual energy (histogram view)

problems. Simulation of two algorithms adapted LEACH and MIEEP via MATLAB tool has been done and comparison exhibited that MIEEP accomplishes better than adapted LEACH. Energy consumption is the major difficulty handled by WSN. Sensor nodes exhausted of energy can no longer accomplish its character unless the source of energy is refilled.

References

1. Akyildiz, I.F., Melodia, T., Chowdhury, K.R.: A survey on wireless multimedia sensor networks. Comput. Netw. **51**, 921–960 (2007)
2. Yick, J., Mukherjee, B., Ghosal, D.: Wireless sensor network survey. Comput. Netw. **52**, 2292–2330 (2008)
3. Akkaya, K., Younis, M.: A survey on routing protocols for wireless sensor networks (2005)
4. Alemdar, H., Ersoy, C.: Wireless sensor networks for healthcare: a survey. Comput. Netw. **54**, 2688–2710 (2010)
5. Stankovic, J.A.: Wireless sensor networks. Computer (Long Beach Calif.), **41**, 92–95 (2008)
6. Karl, H., Willig, A.: Protocols and Architectures for Wireless Sensor Networks (2006)
7. Akyildiz, I.F., Su, W., Sankarasubramaniam, Y., Cayirci, E.: A survey on sensor networks. IEEE Commun. Mag. **40**, 102–105 (2002)
8. Zhang, C., Zhang, M., Su, Y., Wang, W.: Smart home design based on ZigBee wireless sensor network. In: 7th International Conference on Communications and Networking in China, pp. 463–466 (2012)
9. Yiming, Z., Xianglong, Y., Xishan, G., Mingang, Z., Liren, W.: A design of greenhouse monitoring control system based on ZigBee wireless sensor network. In: International Conference on Wireless Communications, Networking and Mobile Computing, WiCom 2007, pp. 2563–2567 (2007)
10. Stavrou, E., Pitsillides, A.: A survey on secure multipath routing protocols in WSNs. Comput. Netw. **54**, 2215–2238 (2010)

11. Deniz, F., Bagci, H., Korpeoglu, I., Yazici, A.: An adaptive, energy-aware and distributed fault-tolerant topology-control algorithm for heterogeneous wireless sensor networks. Ad Hoc Netw. **44**, 104–117 (2016)
12. Chen, I.R., Speer, A.P., Eltoweissy, M.: Adaptive fault-tolerant QoS control algorithms for maximizing system lifetime of query-based wireless sensor networks. IEEE Trans. Dependable Secure Comput. **8**, 161–176 (2011)
13. Chen, Q., Lam, K.Y., Fan, P.: Comments on "Distributed Bayesian algorithms for fault-tolerant event region detection in wireless sensor networks" (2005)
14. Priyadarshi, R., Soni, S.K., Sharma, P.: An Enhanced GEAR Protocol for Wireless Sensor Networks, pp. 289–297. Springer, Singapore (2019)
15. Priyadarshi, R., Singh, L., Kumar, S., Sharma, I.: A hexagonal network division approach for reducing energy hole issue in WSN. Eur. J. Pure Appl. Math. **118** (2018)
16. Priyadarshi,R., Rawat, P., Nath, V.: Energy dependent cluster formation in heterogeneous wireless sensor network. Microsyst. Technol. 1–9 (2018 September)
17. Priyadarshi, R., Bhardwaj, A.: Node non-uniformity for energy effectual coordination in WSN. Int. J. Inf. Technol. Secur. **9**(4), 3–12 (2017)
18. Priyadarshi, R., Soni, S.K., Nath, V.: Energy efficient cluster head formation in wireless sensor network. Microsyst. Technol. (2018 April)
19. Priyadarshi, R., Singh, L., Sharma, I., Kumar, S.: Energy efficient leach routing in wireless sensor network. Int. J. Pure. Appl. Math. **118**(20), 135–142 (2018)
20. Priyadarshi, R., Tripathi, H., Bhardwaj, A., Thakur, A., Thakur, A.: Performance metric analysis of modified LEACH routing protocol in wireless sensor network. Int. J. Eng. Technol. **7**(1–5), 196 (2017)
21. Wang, N., Zhu, H.: An energy efficient algorithm based on LEACH protocol. In: Proceedings on the 2012 International Conference on Computer Science and Electronics Engineering, ICCSEE 2012, pp. 339–342 (2012)
22. Liu, X.: A survey on clustering routing protocols in wireless sensor networks (2012)
23. Peng, Z., Li, X.: The improvement and simulation of LEACH protocol for WSNs. In: Proceedings 2010 IEEE International Conference on Software Engineering and Service Sciences, ICSESS 2010, pp. 500–503 (2010)
24. Wu, X., Wang, S.: Performance comparison of LEACH and LEACH-C protocols by NS2. In: Proceedings on the 9th International Symposium on Distributed Computing and Applications to Business, Engineering and Science, DCABES 2010, pp. 254–258 (2010)
25. Qing, L., Zhu, Q., Wang, M.: Design of a distributed energy-efficient clustering algorithm for heterogeneous wireless sensor networks. Comput. Commun. **29**, 2230–2237 (2006)
26. Feng, S., Qi, B., Tang, L.: An improved energy-efficient PEGASIS-based protocol in wireless sensor networks. In: Proceedings on the 2011 8th International Conference on Fuzzy Systems and Knowledge Discovery, FSKD 2011, pp. 2230–2233 (2011)
27. Lindsey, S., Raghavendra, C., Sivalingam, K.M.: Data gathering algorithms in sensor networks using energy metrics [PEGASIS]. IEEE Trans. Parallel Distrib. Syst. **13**, 924–935 (2002)
28. Guo, W., Zhang, W., Lu, G.: PEGASIS protocol in wireless sensor network based on an improved ant colony algorithm. In: 2010 Second International Workshop on Education Technology and Computer Science, vol. 3, pp. 64–67 (2010)
29. Wang, Z.M., Basagni, S., Melachrinoudis, E., Petrioli, C.: Exploiting sink mobility for maximizing sensor networks lifetime. In: Proceedings of the 38th Annual Hawaii International Conference System Science, pp. 1–9 (2005)
30. Priyadarshi, R., Singh, L., Singh, A., Thakur, A.: SEEN: stable energy efficient network for wireless sensor network. In: 5th International Conference on Signal Processing and Integrated Networks (SPIN 2018), Noida, Delhi-NCR, India, pp. 338–342 (2018)
31. Priyadarshi, R., Singh, L., Randheer., Singh, A.: A novel HEED protocol for wireless sensor networks. In: 5th International Conference on Signal Processing and Integrated Networks (SPIN 2018), Noida, Delhi-NCR, India, pp. 296–300 (2018)

Performance Analysis of Adapted Selection Based Protocol Over LEACH Protocol

Rahul Priyadarshi, Soni Yadav and Deepika Bilyan

Abstract In wireless sensor network, sensor nodes nearer to sink will drain their energy rapidly in comparison to the sensor nodes that are far-flung from the sink. This is because they will have to forward enormous amount of data in the course of multi-hop communication which further reduces the lifetime of system. One of the main research issues is how to enhance network lifetime. In this paper, an adapted selection established protocol is suggested, judgment of picking cluster heads through sink is totally created on basis of connected extra energy, location of node, and remaining energy of node. Congested link concept has been used to find a shorter path while transmission of data from cluster head to sink. Simulation outcomes display that the proposed routing protocol is superior to present LEACH protocol.

Keywords Wireless sensor network · Nodes · Base station · Routing · Multipath routing · Energy

1 Introduction

A wireless sensor network is one of the core next-generation application fields. It has been applied in a variety of fields, such as ecosystem monitoring, military zone surveillance, and U-City applications [1]. A wireless sensor network consists of multiple sensor nodes that can perform environment information collection, computation, and wireless communication [2]. A wireless sensor network collects

R. Priyadarshi (✉) · S. Yadav · D. Bilyan
Electronics and Communication Engineering Department,
National Institute of Technology, Hamirpur, India
e-mail: rahul.glorious91@gmail.com

S. Yadav
e-mail: soni151291@gmail.com

D. Bilyan
e-mail: deepikabilyan@gmail.com

A. K. Luhach et al. (eds.), *Smart Computational Strategies: Theoretical and Practical Aspects*, https://doi.org/10.1007/978-981-13-6295-8_21

environment information by utilizing sensor modules and analyzes collection information with various applications. However, the performance of sensor nodes in sensor networks is limited in terms of energy, computational capability, communication radius, and storage memory [3–7]. Therefore, the algorithms of all wireless sensor network schemes should be studied to minimize energy consumption by utilizing energy efficiently and considering the limited performance of sensor nodes. Wireless sensor networks are vulnerable to external attacks, as they are constructed mostly in unmanned environments to monitor environmental information or military zones, and sensor nodes use wireless communication to transfer collected data to base stations [8–10]. Due to such characteristics, data can be easily exposed during data transmission. This can be a serious problem in applications that deal with military information or individual privacy. To solve these problems, studies on security schemes have been performed. To configure a safe sensor network, it is possible to combine a routing protocol, which is a basic element for information transmission, with an encryption protocol, such as key exchange and authentication. Recently, the various data transmission schemes [11] were proposed such as a transmission scheme that periodically collects real and virtual data at the same time based on encryption algorithms and a data transmission scheme that uses reliability levels. However, such schemes are not suitable for wireless sensor nodes with performance limitations [12], such as energy consumption due to additional data transmission or increased communication between nodes for increasing security. Therefore, it is necessary to study energy-efficient security schemes while considering the characteristics of wireless sensor networks [13–17]. Therefore, this paper aims to provide an algorithm to restrict data analysis of transferred data over a wireless sensor network despite transferred data exposure as well as minimize energy consumption for communication and improve security and energy efficiency.

A characteristically clustered sensor system is shown in Fig. 1. In this paper, we put forward an Adapted selection based protocol (ASP), which employs the conclusion of picking Cluster Head (CH) by sink, is totally on the basis of linked extra

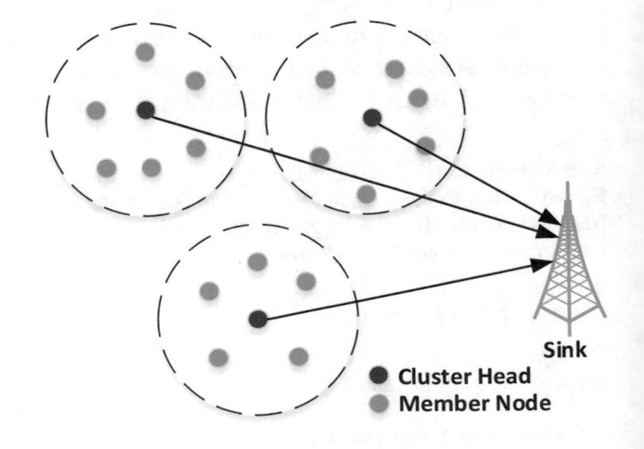

Fig. 1 Cluster sensor network

energy and remaining energy at every sensor node [18]. In this adapted protocol, the CH picks out the straight path to reach the Base Station (BS) using concept of congested link. The main issue that we are talking in this paper is operative energy administration of a gathered routing to get the most out of network lifetime process in the occurrence of undependable nodes that are accountable for packet damage [19–21]. More precisely, we are investigating the optimum quantity of idleness in WSN over and done with which data are directed to a far-flung BS in occurrence of untrustworthy sensor nodes, consequently likelihood to reply operator's inquiry must make the most of while exploiting the network lifetime over multipath routing, there are dual difficulties to resolve [22].

First number of pathways to use and second are what pathways to use. We are focusing on to statement that how many pathways to use to reach BS. We are retaining a protocol for jammed node, as a result, there should be smaller amount of energy preservation by sensor nodes in system [23–26]. Jammed sensor nodes are sensed and pathway over that sensor node is overlooked from dissimilar network. In this paper, we adopt how many pathways to use to bear remaining cooperated nodes, thus enhance network lifetime.

2 Related Work

Low-Energy Adaptive Clustering Hierarchy (LEACH) is a standard protocol for WSN. It uses randomized rotation of CH to uniformly allocate energy in midst of the sensor in the network. Authors appealed, LEACH diminishes communication energy almost eight times related to shortest communication and smallest communication energy routing. Its compensations can be abridged as tracks: First, it can lengthen network lifetime related to new level routing protocol and stationary clustering algorithms. Another, CH fuses data composed from conforming zones, and handovers it to sink node that could effectually advance energy use relation [27].

As a final point, LEACH deals out the duty in the midst of all sensor nodes, dropping the burden of specific nodes. LEACH-C protocol is recommended to deal with shortcomings of LEACH Protocol. It practices a dominant control procedure to form clusters, which allocates CH uniformly all over the network. To make assured energy burden is uniformly circulated between all nodes, BS must computes average amount of sensor node energy. Sensor nodes having energy lower than the average may not be used as CH for the existing round [28]. On the other hand, LEACH protocol has specific shortcomings like

- Protocol agreements no assurance about the location of number of CH nodes.
- If the CH expires in any round, then the entire cluster is incapable to hand over its data to BS till subsequent round. This irregular catastrophe of cluster could be adversity as soon as observing a zone in real-time.

- Specific nodes handover their data to CH over particular hops that is not appropriate for the systems which are too large.

For that reason, advance exploration of this field has been started into some of these concerns. The foremost indication is to make the energy load circulation further uniform between sensors for WSN. Every single node will obtain the data and communicate to adjacent neighbors and yield opportunities being trailblazer for communications to the BS [29].

In [30–32], authors construct a sequence to make sure that all sensor nodes have adjacent neighbors. As soon as a node deceases, the sequence is reassembled to circumvent the deceased node. Authors select CH conferring to a mixture of node remaining energy and a subordinate constraint like node closeness or node degree. There is cooperation among prolonging the time till leading node decease (FND) and the time till the last node deceases (LND).

3 Proposed Model

We are considering the well-known routing protocol in WSN, i.e., geographic routing which is used to forward data from CH to BS lengthwise with multipath routing. As a consequence, in that situation, there is not at all necessity to sustain path info of the network. We should also have info about position of the target node so that we can appropriately direct the packet in the route of it. CH is correspondingly conscious with the position of neighbor CHs lengthways with route in the way of BS. Sensor nodes are nominated as CH by threshold value. Sensor node will communicate their detected data to the BS over CH through single of multi-hop communication. The design of network can be defined as a connectivity graph. We take radio energy model in the expression to communicate one bit span message with a distance d, energy ingesting by radio is specified by Eq. (1) which is valid for multi-hop communication.

$$E(n) = h(n) \times e(n) \times c(n) \tag{1}$$

where

$e(n)$ is consumption of energy while transmitting one bit.
$h(n)$ is average number of hops for communication.
$c(n)$ is number of transmitted bit.

Node generally sends data to the CH in case of value of $h(n)$ should be 1 thus energy ingesting by each node n_i in time interval ‘t’ is specified by

$$E(n_j) = e(n) \times 1 \times c(n_j) \times t \times \mathrm{sr}_j \tag{2}$$

where sr_j stands for sampling rate of sensor node n_j.

CH is accountable for advancing data in cluster. Therefore, energy consumption of CH is specified by

$$E(\mathrm{ch}) = h(\mathrm{ch}) \times \sum_{j=0}^{n} E(n_j) \quad (3)$$

where $h(\mathrm{ch})$ is defined as total number of hops while data transmission from CH to BS and n is number of nodes. In present method, node is nominated as CH based on value of hop count, the similar node will be nominated as CH all the way through network lifetime. Energy consumed by cluster node in time t is given by

$$E_t(\mathrm{ch}) = t \times h(\mathrm{ch}) \times \sum_{j=0}^{n} E(n_j) \quad (4)$$

In this planned method, CH is nominated from time to time based on residual energy of sensor node. Every single node is getting chance to be CH for certain period of time. In t amount of period, every single node is CH for time slot $\left(t - \frac{t}{n}\right)$ or $\frac{t}{n}$ as normal sensor node. Now, consumption of energy by every single node in time slot 't' is given as

$$E(n_j) = f_1 + f_2 \quad (5)$$

where

$$f_1 = e(n_j) \times 1 \times c(n_j) \times \left(t - \frac{t}{n}\right) \times \mathrm{sr}_j$$

And

$$f_2 = h(\mathrm{ch}) \times \frac{t}{n} \times \sum_{j=0}^{n} E(n_j)$$

In our suggested routing protocol, assuming network to be dissimilar or heterogeneous in nature, where total $m\%$ of advanced node have extra energy factor β in this one while comparing with normal nodes. Assuming each node has initial energy E_0. Energy of advanced sensor node in suggested routing protocol is $E_0(1+\beta)$. Now formulation of total energy of the proposed network is computed as Eq. (6).

$$NE_0(1 + m\beta) = nmE_0(1 + \beta) + N(1 - m)E_0 \quad (6)$$

As a result, overall energy is increased by factor $(1 + \beta m)$ times. According to the probability of normal node and advanced node, we enhanced selection technique with the remaining energy of some sensor nodes. Normal node probability equation is given by Eq. (7)

$$P_{\text{normal}} = \frac{P_{\text{opt}}}{1+\beta m} \times \frac{E_{\text{residual}}}{E_0} \tag{7}$$

where P_{opt} optimal percentage of CH and E_0 is preliminary energy of nodes. In the same way, advanced node probability equation is given by Eq. (8)

$$P_{\text{advance}} = \frac{P_{\text{opt}}}{1+\beta m} \times \frac{E_{\text{residual}}}{E_0} \times (1+\beta) \tag{8}$$

Selection of CH is given by Eq. (9)

$$T(n) = \begin{cases} \frac{P}{1-P(r \bmod 1/P)} & \text{if } n \in G \\ 0 & \text{otherwise} \end{cases} \tag{9}$$

where

- P is fraction of choosing CHs.
- r is present round.
- G is group of all the sensor nodes which are not in CHs in $1/P$ rounds.

In every round, node produces random number which belongs to 0 and 1. If generated random number is smaller than $T(n)$, it will be selected as a CH otherwise not. Once the selection of CH is done, then the function $T(n)$ is set to be 0. The defined threshold for advanced and normal node is specified as

$$T(P_{\text{normal}}) = \begin{cases} \frac{P_{\text{normal}}}{1-P_{\text{normal}}(r \bmod 1/P_{\text{normal}})} & n \in G \\ 0 & \text{otherwise} \end{cases} \tag{10}$$

$$T(P_{\text{advanced}}) = \begin{cases} \frac{P_{\text{advanced}}}{1-P_{\text{advanced}}(r \bmod 1/P_{\text{advanced}})} & n \in G \\ 0 & \text{otherwise} \end{cases} \tag{11}$$

4 Simulation and Results

Simulation has been done using MATLAB tool. Simulation parameter and its value are listed in Table 1. Figures 2 and 3 show the number of alive nodes in proposed ASP protocol is more compared to the existing LEACH protocol which further reflects the better lifetime of the proposed ASP protocol.

Figures 4 and 5 reflect plot of residual energy. At the start, each node has 0.4 J energy and overall number of node is considered as 500. Therefore, network energy is 200 J. Proposed ASP protocol approves greater outcomes in contrast of existing LEACH protocol in situation of energy consumption.

Table 1 Simulation parameter

Parameters	Description	Value
E_{elec}	Energy dissipation	50 nJ/bit
E_0	Initial energy of nodes	0.4 J
N	Sensor nodes	500
E_{DA}	Data aggregation	5 nJ/bit
K	Packet Length	3000 bits

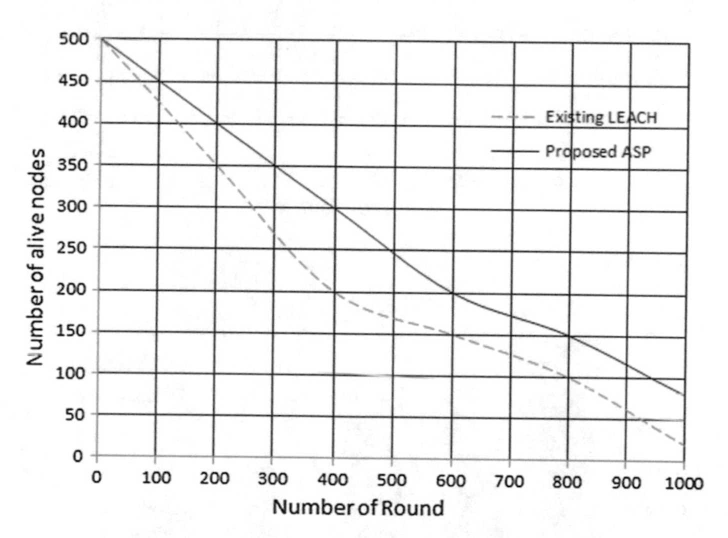

Fig. 2 Number of alive nodes

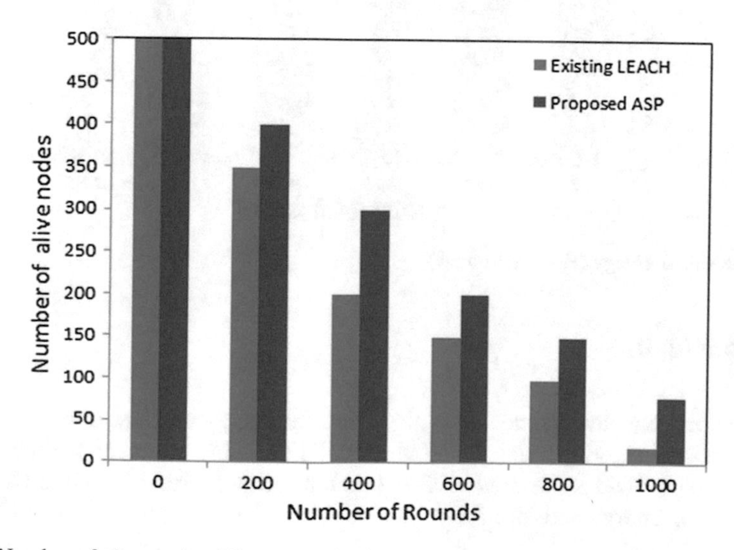

Fig. 3 Number of alive nodes (Histogram view)

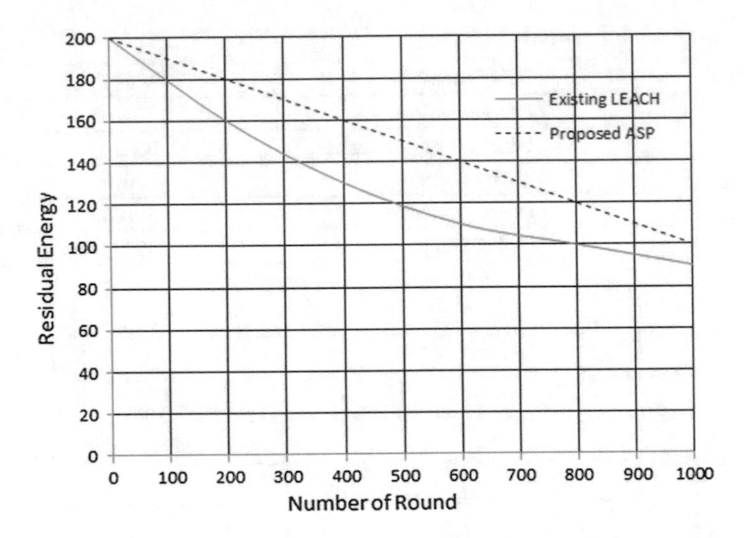

Fig. 4 Residual energy

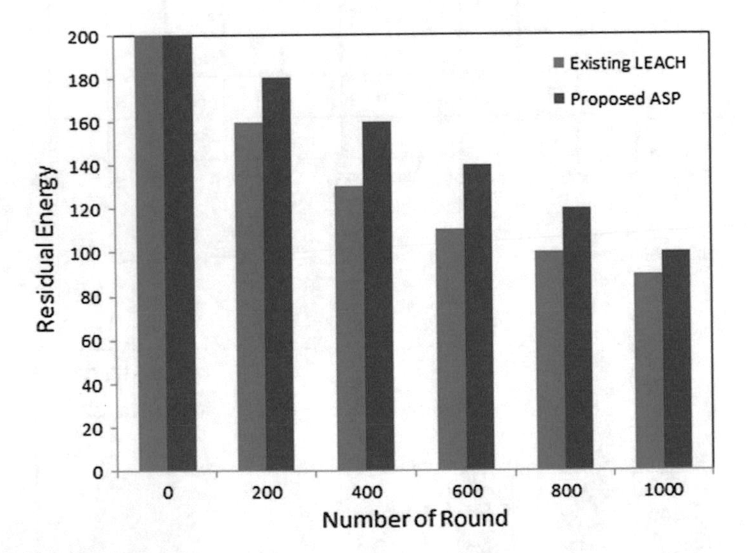

Fig. 5 Residual energy (Histogram view)

5 Conclusion

In this paper, we have defined ASP algorithm for energy efficient routing in WSN. Our proposed algorithm developed a classified routing protocol in hierarchical manner which divides the network first into cluster and further selecting CH based on remaining energy and location of nodes. Also, Congested link concept has been used to find shorter path while transmission of data from CH to the BS. The proposed ASP algorithm displays good performance in reference to energy consumption and extending network lifetime.

References

1. Heinzelman, W.R., Chandrakasan, A., Balakrishnan, H.: Energy-efficient communication protocol for wireless microsensor networks. Proc. Hawaii Int. Conf. Syst. Sci. 1–10 (2000)
2. Heinzelman, W.B., Chandrakasan, A.P., Balakrishnan, H.: An application-specific protocol architecture for wireless microsensor networks. IEEE Trans. Wirel. Commun. **1**, 660–670 (2002)
3. Wang, Y., Wu, H., Nelavelli, R., Tzeng;, N.-F.: Balance-based energy-efficient communication protocols for Wireless sensor networks. In: Distributed Computing Systems Workshops 2006. ICDCS Workshops 2006. 26th IEEE International Conference, p. 85 (2006)
4. Raghunathan, V., Schurgers, C., Park, S., Srivastava, M.B.: Energy-aware wireless microsensor networks. IEEE Signal Process. Mag. **19**, 40–50 (2002)
5. Younis, O., Fahmy, S.: HEED: a hybrid, energy-efficient, distributed clustering approach for ad-hoc sensor networks. Mob. Comput. IEEE Trans. **3**, 366–379 (2004)
6. Qing, L., Zhu, Q., Wang, M.: Design of a distributed energy-efficient clustering algorithm for heterogeneous wireless sensor networks. Comput. Commun. **29**, 2230–2237 (2006)
7. Akyildiz, I.F., Su, W., Sankarasubramaniam, Y., Cayirci, E.: A survey on sensor networks. IEEE Commun. Mag. **40**, 102–105 (2002)
8. Younis O., F.S.: A hybrid, energy-efficient, distributed clustering approach for ad hoc sensor networks. IEEE Trans. Mob. Comput. **3**, 366–379 (2004)
9. Guo, D., Xu, L.: LEACH clustering routing protocol for WSN. In: Lecture Notes in Electrical Engineering, pp. 153–160 (2013)
10. Kumar, D., Aseri, T.C., Patel, R.B.: EEHC: energy efficient heterogeneous clustered scheme for wireless sensor networks. Comput. Commun. **32**, 662–667 (2009)
11. Al-Karaki, J.N., Kamal, A.E.: Routing techniques in wireless sensor networks: a survey. IEEE Wirel. Commun. **11**, 6–28 (2004)
12. Priyadarshi, R., Soni, S.K., Sharma, P.: An Enhanced GEAR Protocol for Wireless Sensor Networks, pp. 289–297. Springer, Singapore (2019)
13. Priyadarshi, R., Singh, L., Kumar, S., Sharma, I.: A hexagonal network division approach for reducing energy hole issue in WSN. Eur. J. Pure Appl. Math. vol. 118 (2018)
14. Priyadarshi, R., Rawat, P., Nath, V.: Energy dependent cluster formation in heterogeneous wireless sensor network. Microsyst. Technol. 1–9 (2018)
15. Priyadarshi, R., Bhardwaj, A.: Node non-uniformity for energy effectual coordination in WSN. Int. J. Inf. Technol. Secur. **9**(4), 3–12 (2017)
16. Priyadarshi, R., Soni, S.K., Nath, V.: Energy efficient cluster head formation in wireless sensor network. Microsyst. Technol. (2018)
17. Priyadarshi, R., Singh, L., Sharma, I., Kumar, S.: Energy efficient leach routing in wireless sensor network. Int. J. Pure. Appl. Math. **118**(20), 135–142 (2018)
18. Priyadarshi, R., Tripathi, H., Bhardwaj, A., Thakur, A., Thakur, A.: Performance metric analysis of modified LEACH routing protocol in wireless sensor network. Int. J. Eng. Technol. **7**(1–5), 196 (2017)
19. Akkaya, K., Younis, M.: A survey on routing protocols for wireless sensor networks (2005)
20. Akkaya, K., Younis, M.: A survey on routing protocols for wireless sensor networks. Ad Hoc Netw. **3**, 1–38 (2010)

21. Jizan, Z.: Congestion avoidance and control mechanism for multi-paths routing in WSN. In: Proceedings—International Conference on Computer Science and Software Engineering, CSSE 2008, pp. 1318–1322 (2008)
22. Hull, B., Jamieson, K., Balakrishnan, H.: Mitigating congestion in wireless sensor networks. In: Proceedings of the 2nd International Conference on Embedded Networked Sensor Systems—SenSys '04, p. 134 (2004)
23. Aguirre-Guerrero, D., Marcelín-Jiménez, R., Rodriguez-Colina, E., Pascoe-Chalke, M.: Congestion control for a fair packet delivery in WSN: from a complex system perspective. Sci. World J. 2014 (2014)
24. Kasi, M.K., Hinze, A., Jones, S., Legg, C.: Energy-efficient context-aware routing in heterogeneous WSN. In: Proceedings of the 8th ACM International Conference on Distributed Event-Based Systems—DEBS '14, pp. 166–176 (2014)
25. Attea, B., Khalil, E.: A new evolutionary based routing protocol for clustered heterogeneous wireless sensor networks. Appl. Soft Comput. **12**, 1950–1957 (2012)
26. Wang, N., Zhu, H.: An energy efficient algorithm based on LEACH protocol. In: Proceedings—2012 International Conference on Computer Science and Electronics Engineering, ICCSEE 2012, pp. 339–342 (2012)
27. Wu, X., Wang, S.: Performance comparison of LEACH and LEACH-C protocols by NS2. In: Proceedings—9th International Symposium on Distributed Computing and Applications to Business, Engineering and Science, DCABES 2010, pp. 254–258 (2010)
28. Xu, J., Jin, N., Lou, X., Peng, T., Zhou, Q., Chen, Y.: Improvement of LEACH protocol for WSN. In: Proceedings—2012 9th International Conference on Fuzzy Systems and Knowledge Discovery, FSKD 2012. pp. 2174–2177 (2012)
29. Xiangning, F., Yulin, S.: Improvement on LEACH protocol of wireless sensor network. In: Sensor Technologies and Applications, 2007. SensorComm 2007. International Conference, pp. 260–264 (2007)
30. Zhao, F., Xu, Y., Li, R., Zhang, W.: Improved leach communication protocol for WSN. In: Proceedings—2012 International Conference on Control Engineering and Communication Technology, ICCECT 2012, pp. 700–702 (2012)
31. Priyadarshi, R., Singh, L., Singh, A., Thakur, A.: SEEN: Stable energy efficient network for wireless sensor network. In: 2018 5th International Conference on Signal Processing and Integrated Networks (SPIN), pp. 338–342. Noida, Delhi-NCR, India (2018)
32. Priyadarshi, R., Singh, L., Randheer, Singh, A.: A novel HEED protocol for wireless sensor networks. In: 2018 5th International Conference on Signal Processing and Integrated Networks (SPIN), pp. 296–300. Noida, Delhi-NCR, India (2018)

Part IV
Information Systems

Leveraging Big Data Analytics Utilizing Hadoop Framework in Sports Science

Gagandeep Jagdev and Sarabjeet Kaur

Abstract The first ever utilization of statistics in professional sports has been made possible to make better personal decisions with the assistance of Big Data. Each day, a number of matches are played under different categories of sports and each day, new records are set up and old records are broken and all the concerned data, statistics, and records undergo major changes. With the introduction of innovative sensor enabled technologies and wearable devices, the data generated from different sources can be collecting easily and accurately and analysts can make most of it. This helps in taking decisions like when to substitute the player. A team can predict the policies and tactics to be adopted by the opposition prior to the next scheduled encounter with the assistance of Big Data. The same can be applied on the team itself to check out the shortcomings and flaws in the game plan of the team. The fundamental purpose of the research work is to investigate how sports have profited with the utilization of Big Data and how further enhancement can be made possible in this field. The major challenge in sports science is to gain the competitive advantage over opposition using big data and it can be accomplished via appropriately mining the collected data. The research work focuses on the comparison of conventional Apriori data mining algorithm with the Hadoop-based MapReduce algorithm capable of handling the enormous amount of data. With the use of the Apache Hadoop framework, all this generated data can be collected in huge servers and can be mined when and as required with much ease.

Keywords Apriori algorithm · Big data · MapReduce

G. Jagdev (✉)
Department of Computer Science, Punjabi University Guru Kashi College, Damdama Sahib, Punjab, India
e-mail: drgagan137@pbi.ac.in

S. Kaur
Research Scholar, Punjabi University, Patiala, Punjab, India
e-mail: sarabjeet1412@gmail.com

A. K. Luhach et al. (eds.), *Smart Computational Strategies: Theoretical and Practical Aspects*, https://doi.org/10.1007/978-981-13-6295-8_22

1 Introduction

With the advent of innovative technologies and knowledge, there has been tremendous growth in the pool of data and as much one tries to get to the surface, the level rises more above. It requires mining, and appropriate mining yields accurate results [1]. The data has grown to such an extent that new terms like Exabyte, Zettabyte, and Yottabyte have been introduced to measure the amount of accessible data [2]. Big Data has influenced the majority of public and private sectors where strategic decisions are of utmost importance, comprising banks, retail sector, medical science, stock market, elections, and sports science. There is no exact size which can be used as an indicator to declare data as a big data. It depends upon the planned purposes and economic sector.

1.1 Issues Related with Big Data

Major concerns relevant to big data are mentioned as under [3–5].

- Data degree—Degree refers to the size of an average data set which could be as excessive as a couple of petabytes. The degree is higher in the case of unstructured statistics, which includes video and audio.
- Data pace—The pace at which the records are being generated is beyond the handling power of traditional structures. Social media and e-commerce had unexpectedly increased the pace [6].
- Data range—Information might be stored in diverse codecs, including text, audio, video, and photos. Over 80% of the records generated today is unstructured and conventional methods are not capable of handling it.
- Data charge—The information that is being generated in large quantity is analyzed and used by extraordinary organizations for statistics analytics. It is witnessed that there is a large gap between enterprise leaders and the IT specialist. The industrial leaders aim is to maximize their income and are least concerned about the technicalities of the storage and processing [7].
- Data volume—Enormous amount of data is being generated regularly with each passing second. This data is been measured using big terms like Exabyte, Zettabyte, and Yottabyte [7].
- Data veracity—The amount of trust that data carries with itself is referred to as data veracity. With the availability of different forms of data, the preciseness of data is always doubtful. For example, a male can pretend himself to be a female on social networking platform and indulge in antisocial practices.

1.2 Advancements in the World of Sports

In modern times, the entire sports world is bound to concentrate on the collection, analysis, and use of data. With the help of measurable equipment, one can gather millions of data which can range from the swinging of a tennis racket to spin of a baseball. The below-mentioned facts emphasize on the change and advancements witnessed by the sports science [8, 9].

- *Wearable devices for live data collection*

Wearable devices have assisted people in fitness awareness. In the past, coaches have always relied on manual data collection. But today technology permits for live data collection as the data is being created at very high-speed. These devices are usually attached to player's body to have precise readings of player's heartbeat, acceleration, and speed. This allows coaches to create effective training plan for the player.

- *Support better judgment*

As the frequency of actions is too fast, often it becomes a challenge for referees to reach a correct decision. It was observed that due to one wrong decision of referee, the game changes entirely as balance shifts towards either of the two teams. But with the assistance of smart devices, such situations can be avoided as referees can refer to the data collected by on field devices for error-free judgment.

- *Predicting a fan preference*

In modern sports, even ticket vendors are making the best use of technology to understand what fans want. Today, the organizing authorities and bodies are constantly involved in deciding the venue and date of the match, keeping in mind to make it possible for maximum fans to have access to that particular match either in stadium live or via any other electronics communication medium.

- *Influencing coach decision*

The attributes and considerations relevant to player's real-time performance help coach to make better decisions in favor of the team. While standing outside the playing field, with the help of wearable devices, coach can study about player's health, playing orders, acceleration, and decide on substitutions to be made and plan out the formation of team for upcoming events.

- *Meeting business goals*

The companies investing in sports keep strict vigil on player's performance from commercial angle and target high performers to be their brand ambassadors to promote their products.

2 Hadoop-Based MapReduce Programming Model

2.1 Functioning of MapReduce

The Hadoop is a Java framework developed by Dough Cutting and sponsored by the Apache Software Foundation with the capability of handling enormous, complex, and variety of data sets in a distributed environment. The system is constituted of thousands of nodes having number of applications running on it with the objective of processing Petabytes of data [10]. Despite the failure of any particular node or few nodes, the Hadoop keeps on facilitating the fast transfer of data and system remains functional as always. It is because of this that even if multiple nodes becomes nonoperational, the risk concerned with devastating system failure is minimum. The motivational algorithm behind the working of the Hadoop is the Google MapReduce algorithm [11].

The following steps detail the functioning of MapReduce [12, 13] as shown in Fig. 1.

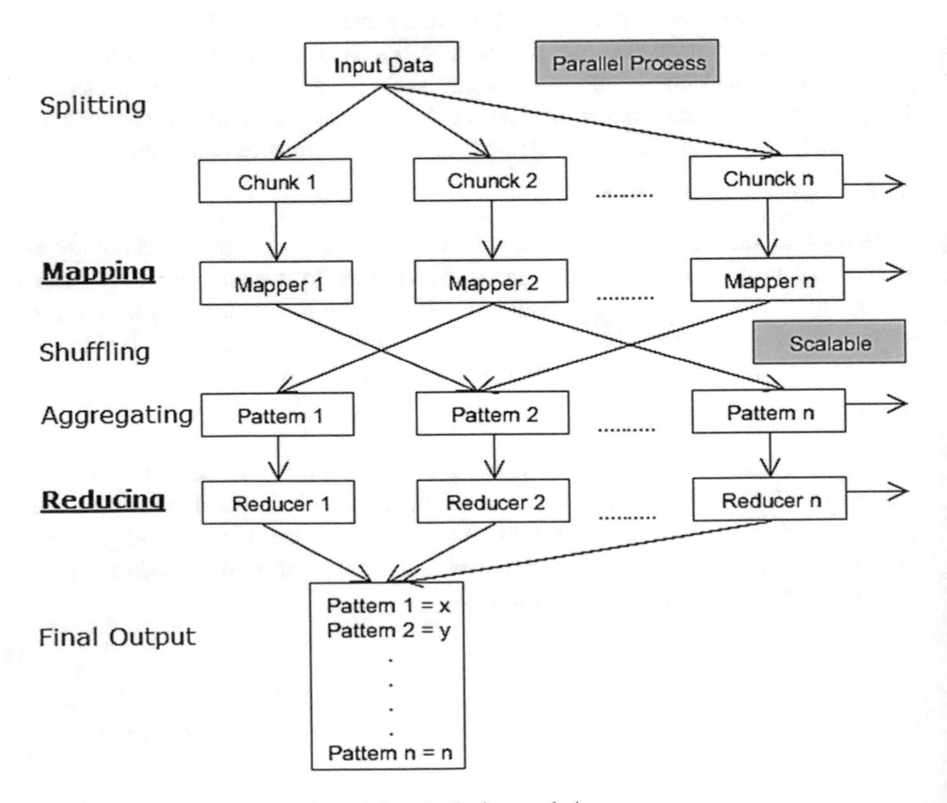

Fig. 1 Figure illustrates working of the MapReduce technique

- A single map task is generated for each split which is responsible for executing the concerned map function for every record involved in the split.
- It is a conventionally known fact that multiple splits are always beneficial as the time taken in processing split is comparatively much less than the time taken for processing whole input. Further, in the case of small splits, the load balancing concerned with the processing is well maintained because of engrossment of parallel processing.
- On the other hand, it should be taken care of that even too small splits are not preferred. The burden concerned with the maintenance of splits and creation of different map task are enough to overpower the total time involved in job execution.
- Usually, the split size should be approximately equal to the size of HDFS block which is 64 Megabytes by default.
- The map tasks which have been successfully executed write the output to a local disk present in the respective node and avoids writing it on HDFS. This practice reduces the practice of replication which often occurs in HDFS store operation.
- The output obtained from mapping is an intermediate output and needs to be further processed by reduce tasks to obtain the final output.
- On completion of the job, the output obtained from mapping is of no use. So, there is no need to save it in HDFS and give birth to overindulgence.
- On the failure of the node before the consumption of map output by reducing task, the Hadoop is supposed to make use of another node and reruns the map task and create new map output.
- The obtained output of each map task is provided as input to reduce task. The location of reduce task is traced and the required map output is transported there.
- The output is merged with this machine and later on passed to the user-defined reduce function. The output obtained from reducing task is stored in HDFS. The local node receives the first copy and the remaining copies find their place on off-rack nodes.

2.2 *Example Illustrating Stepwise Execution of MapReduce*

The example mentioned below depicts the working of MapReduce technique. Suppose we have a record of bowling average of the Indian bowlers in test matches from March 1980 to July 2017. There are four files. Each file further consists of four different bowlers records. The objective of studying the example is to find out the most efficient bowler in terms of his bowling average.

File A:	A Kumble, 29.65	Kapil Dev, 30.54	J Srinath, 30.49	Z Khan, 32.94
File B:	B Kumar, 29.88	J Srinath, 30.49	M Prabhakar, 37.3	Z Khan, 32.94
File C:	Kapil Dev, 30.54	Z Khan, 32.94	R Ashwin, 25.22	A Kumble, 29.65
File D:	J Srinath, 30.49	Kapil Dev, 30.54	M Prabhakar, 37.3	Z Khan, 32.94

The Map phase—The first and foremost need is to form key-value pairs which engages the closure of each record in the form <k, v>:<bowler, average>.

File A:	<A Kumble, 29.65>	<Kapil Dev, 30.54>	<J Srinath, 30.49>	<Z Khan, 32.94>
File B:	<B Kumar, 29.88>	<J Srinath, 30.49>	<M Prabhakar, 37.3>	<Z Khan, 32.94>
File C:	<Kapil Dev, 30.54>	<Z Khan, 32.94>	<R Ashwin, 25.22>	<A Kumble, 29.65>
File D:	<J Srinath, 30.49>	<Kapil Dev, 30.54>	<M Prabhakar, 37.3>	<Z Khan, 32.94>

The combiner phase (searching technique)—The code is mentioned as under to find out which bowler has the least (best) bowling average from each file as follows:

<k: bowler, v: average>
Min = the least bowling average.
if(v(second bowler).average < Min)
{Min = v(average);}
else{proceed with checking;}
The expected result is as follows:
A—<A Kumble, 29.65> B—<B Kumar, 29.88> C—<R Ashwin, 25.22> D—<J Srinath, 30.49>.

Reducer phase—Each file outputs the name of the bowler with least bowling average in the combiner phase. The same code is executed once more with four outputs of combiner phase as input of reducer phase in one file. The final output obtained is <R Ashwin, 25.22>.

3 MapReduce Versus Apriori Algorithm

Among different conventional data mining algorithms, the Apriori algorithm dominates the list when it comes to association rules [14–16]. This section of the research paper compares the efficiency of Apriori algorithm with the MapReduce algorithm in terms of time consumed. The source code for Apriori algorithm has been constructed using C language. A small excel file titled "numbers.csv" shown in Fig. 2 is provided as input to both the algorithms.

By executing the source code of Apriori algorithm and setting the minimum acceptance level to 3, the following output was obtained as shown in Figs. 3 and 4.

The result displays the occurrence of 1, 2, and 3 three times together in the input file and the amount of time engaged in conducting this practice is 13.516711 s.

	A	B	C	D	E	F
1	A1	A2	A3	A4	A5	Sum
2	1	5	2	0	0	8
3	2	3	0	1	0	6
4	3	4	0	0	0	7
5	2	1	3	0	0	6
6	1	2	3	0	0	6

Fig. 2 Input data file “numbers.csv”

1 2 3 3

```
Initial Input:

Trasaction        Items
1:        1       5       2       0       0
2:        2       3       0       1       0
3:        3       4       0       0       0
4:        2       1       3       0       0
5:        1       2       3       0       0
start time.....62.692308
```

Fig. 3 The snapshot depicts the digits in different sets in tabular form along with the start time of the operation

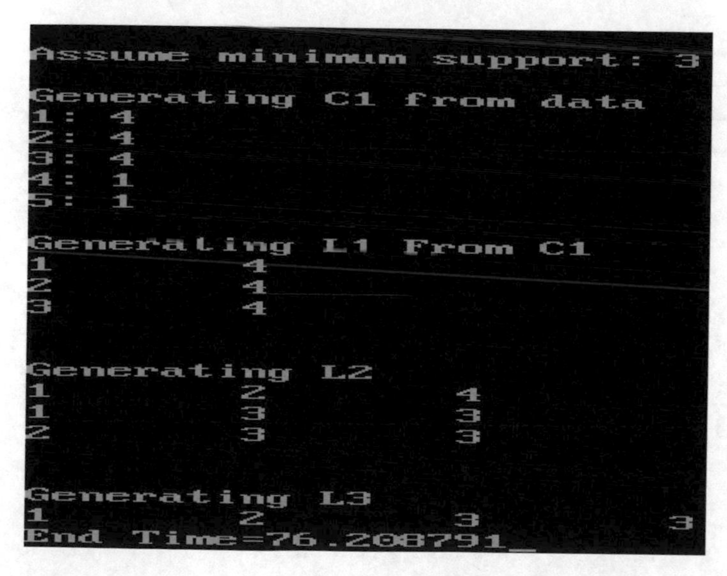

Fig. 4 The snapshot shows the digits grouped in three levels along with the end time of the operation

$$\text{Start time} = 62.692308\text{ s} \quad \text{End time} = 76.208791\text{ s}$$

$$\begin{aligned}\text{Time taken} &= \text{Start time} - \text{End time} \\ &= 76.208791 - 62.692308 \\ &= 13.516711\text{ s}\end{aligned}$$

When the same input was given to Hadoop Framework as shown in Tables 1 and 2, the result was obtained in 0.86 s (860 ms) as shown in Fig. 5.

Table 1 The table depicts the input file "numbers" being uploaded in Hadoop framework and construction of table titled "MapReduce"

	a1	a2	a3	a4	a5
0	1	5	2	0	0
1	2	3	0	1	0
2	3	4	0	0	0
3	2	1	3	0	0
4	1	2	3	0	0

Table 2 The table depicts the result obtained after mining the table "MapReduce"

	a1	a2	a3	a4	a5
0	2	3	0	1	0
1	2	1	3	0	0
2	1	2	3	0	0

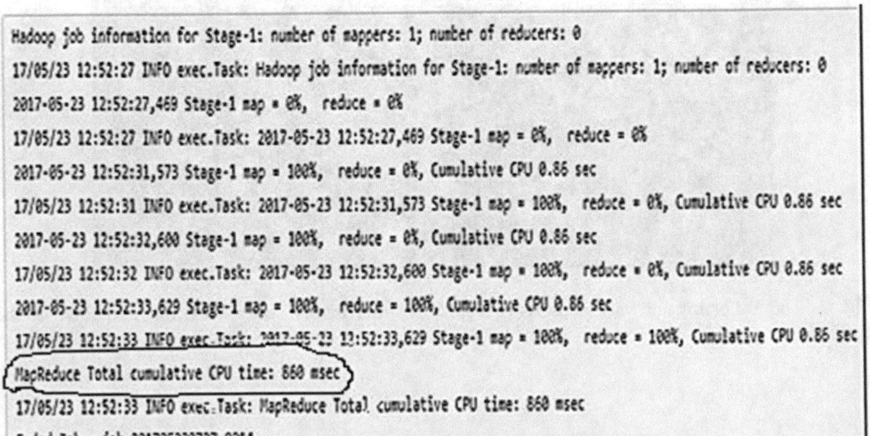

```
Hadoop job information for Stage-1: number of mappers: 1; number of reducers: 0
17/05/23 12:52:27 INFO exec.Task: Hadoop job information for Stage-1: number of mappers: 1; number of reducers: 0
2017-05-23 12:52:27,469 Stage-1 map = 0%,  reduce = 0%
17/05/23 12:52:27 INFO exec.Task: 2017-05-23 12:52:27,469 Stage-1 map = 0%,  reduce = 0%
2017-05-23 12:52:31,573 Stage-1 map = 100%,  reduce = 0%, Cumulative CPU 0.86 sec
17/05/23 12:52:31 INFO exec.Task: 2017-05-23 12:52:31,573 Stage-1 map = 100%,  reduce = 0%, Cumulative CPU 0.86 sec
2017-05-23 12:52:32,600 Stage-1 map = 100%,  reduce = 0%, Cumulative CPU 0.86 sec
17/05/23 12:52:32 INFO exec.Task: 2017-05-23 12:52:32,600 Stage-1 map = 100%,  reduce = 0%, Cumulative CPU 0.86 sec
2017-05-23 12:52:33,629 Stage-1 map = 100%,  reduce = 100%, Cumulative CPU 0.86 sec
17/05/23 12:52:33 INFO exec.Task: 2017-05-23 12:52:33,629 Stage-1 map = 100%,  reduce = 100%, Cumulative CPU 0.86 sec
MapReduce Total cumulative CPU time: 860 msec
17/05/23 12:52:33 INFO exec.Task: MapReduce Total cumulative CPU time: 860 msec
Ended Job = job_201705220727_0014
```

Fig. 5 The figure shows the log details depicting the total time (860 ms) taken to obtain the result

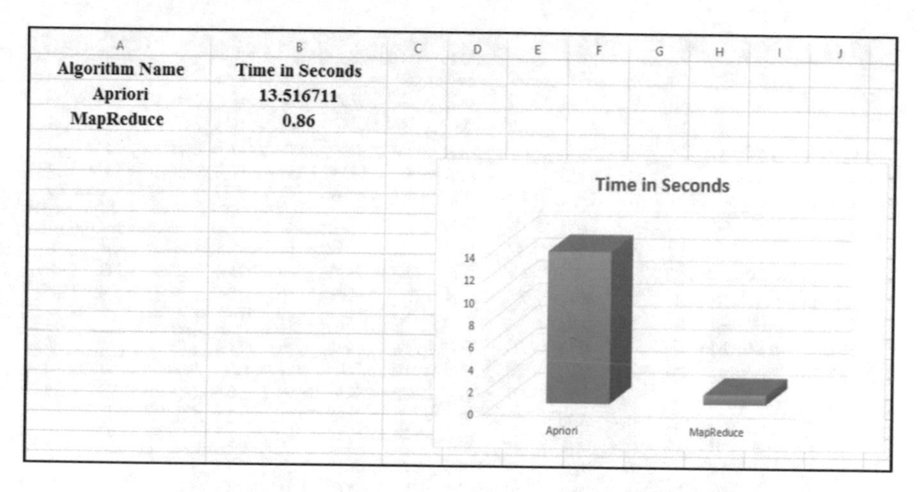

Fig. 6 Comparison of Apriori and MapReduce algorithm in terms of time in seconds

The time taken for obtaining the desired result via Hadoop-based MapReduce algorithm is recorded in log as 0.86 s or 860 ms as shown in Fig. 5.

The result obtained after execution of Apriori and MapReduce algorithms is shown in Fig. 6. The result shows that MapReduce algorithm is much more speedy and efficient in mining as compared to Apriori algorithm.

4 Implementation and Contribution

The research work concentrates on the mining the statistics of bowlers in Indian cricket from March 1980 to till date. A primary structured database has been generated involving 15 attributes mentioned below depicting player's record in international cricket irrespective of the format (Test matches, ODIs, T20s).

- *Player*—Name of the cricketer (bowler).
- *Span*—Time period during which played international cricket irrespective of format.
- *Mat*—Number of matches played.
- *Inns*—Number of innings played.
- *Overs*—Number of overs bowled by player in international cricket irrespective of format.
- *Mdns*—Number of maiden overs bowled by the player.
- *Runs*—Runs given by the bowler to opposition.
- *Wkts*—Number of wickets taken by the player.
- *Ave*—Bowling average of the player.
- *Econ*—Economy rate of the player.
- *SR*—Strike rate of the player.

Player	Span	Mat	Inns	Overs	Mdns	Runs	Wkts	Ave	Econ	SR	MWI	MWM	Fivewkts	Tenwkts
A Kumble	1990-2008	401	499	9204.2	1685	28655	953	30.06	3.11	57.9	10	14	37	8
Harbhajan Singh	1998-2016	365	442	6925.1	959	23042	707	32.59	3.32	58.7	8	15	28	5
Z Khan	2000-2014	303	373	4825.2	736	18797	597	31.48	3.89	48.4	7	10	12	1
N Kapil Dev	1980-1994	324	396	5516.5	1086	16810	578	29.08	3.04	57.2	9	10	18	1
J Srinath	1991-2003	296	348	4506.3	736	16043	551	29.11	3.55	49	8	13	13	1
R Ashwin	2010-2017	206	247	3581.1	527	13067	477	27.39	3.64	45	7	13	25	7
AB Agarkar	1998-2007	221	237	2400.4	269	10851	349	31.09	4.51	41.2	6	8	3	0
I Sharma	2007-2017	171	228	3131	503	12014	341	35.23	3.83	55	7	10	7	1
RA Jadeja	2009-2017	206	229	2727.5	441	9828	328	29.96	3.6	49.8	7	10	9	1
IK Pathan	2003-2012	173	195	2033.3	266	8986	301	29.85	4.41	40.5	7	12	9	2
BKV Prasad	1994-2001	194	218	2528.2	354	9692	292	33.19	3.83	51.9	6	10	8	1
RJ Shastri	1981-1992	230	261	3727.2	713	10835	280	38.69	2.9	79.8	5	8	3	0
M Prabhakar	1984-1996	169	195	2305.5	350	8115	253	32.07	3.51	54.6	6	6	5	0
A Nehra	1999-2017	160	172	1608	177	7494	233	32.16	4.66	41.4	6	6	2	0
SR Tendulkar	1989-2013	664	416	2051.3	107	9354	201	46.53	4.55	61.2	5	5	2	0
UT Yadav	2010-2017	102	129	1422	176	6406	187	34.25	4.5	45.6	5	7	1	0
Mohammed Shami	2013-2017	78	97	1152.1	142	4975	175	28.42	4.31	39.5	5	9	2	0
S Sreesanth	2005-2011	90	111	1349.5	180	6067	169	35.89	4.49	47.9	6	8	4	0
A Mishra	2003-2017	68	84	1208	143	4466	156	28.62	3.69	46.4	6	7	3	0
SLV Raju	1990-2001	81	103	1728.4	378	4871	156	31.22	2.81	66.4	6	11	5	1
Maninder Singh	1982-1993	94	109	1891.5	392	5354	154	34.76	2.83	73.7	7	10	3	2

Fig. 7 A self-constructed database depicting players bowling record

- *MWI*—Maximum wickets taken by the player in any particular innings.
- *MWM*—Maximum wickets taken by the player in any particular match.
- *Fivewkts*—Number of matches in which player claimed 5 or more than 5 wickets but less than 10 wickets.
- *Tenwkts*—Number of matches in which player claimed 10 or more than 10 wickets.

The glimpse of constructing database has been depicted in Fig. 7.

The research work is carried out on Hadoop-based framework, Hortonworks Sandbox 2.2.0 in collaboration with VMware. It comprises of components like HDP-1.3, Hadoop 1.2.0.1.3.0.0-107, HCatalog 0.11.0.1.3.0.0-107, Pig 0.11.1.1.3.0.0-107, Hive 0.11.0.1.0.0.0-107, and Oozie 3.3.2.1.3.0.0-107 [17].

The primary focus concerned with the research work is to conduct mining of constructed database and get fruitful results. Appropriately constructed queries and scripts need to be written to extract information like which bowler has the best average, which bowler has the best economy rate, which bowler has the maximum wickets to his credit, and much more. Further by writing appropriate scripts, sub-tables can be created from the enormous database based on any particular condition as per requirement. The results can be acquired in tabular form, CSV files, and graphical format. The steps followed in extracting fruitful result from the large database using the Hortonworks Sandbox 2.2.0 with detailed working are illustrated as under.

First, the self-constructed database "indian_bowlers_record.csv" is uploaded into the Sandbox using *File Browser* and handed over to *HCatalog* for constructing table titled "indian_bowlers".

After table construction, one can write appropriate scripts and queries using *Query Editor* of *Apache Hive* as shown in Fig. 8.

The execution of the script written in Fig. 8 results in construction of sub-tables "matches" and "bowlerperformance" from main table "indian_bowlers". The result of the table "bowlerperformance" is shown in Table 3.

Figure 9 shows the results in the form of bar graphs. The player name forms the *x*-axis and the values of inns (red color), mdns (green color), wkts (light blue color), and econ (dark blue color) forms the bar graphs of *y*-axis.

The matter concerning the second table "matches" is shown in Table 4.

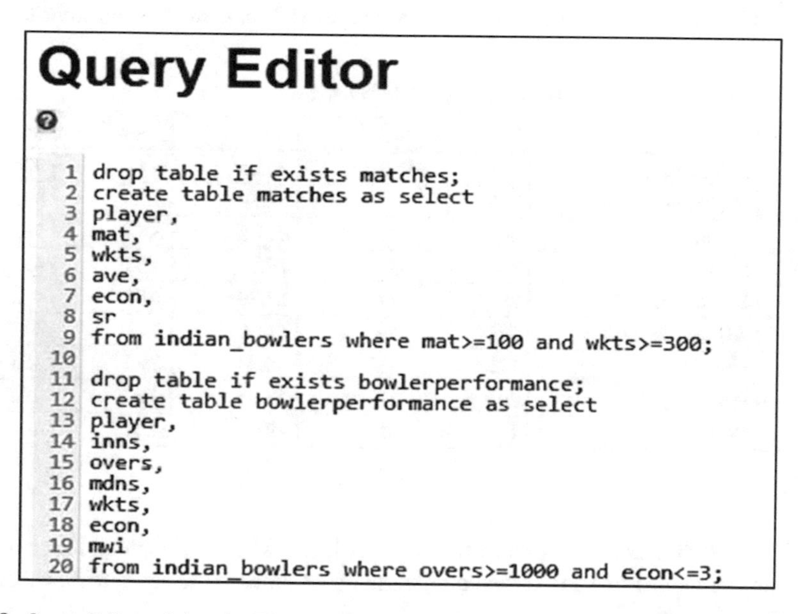

Fig. 8 Query Editor of Apache Hive displays the written query under study

Table 3 The figure displays the contents of table "bowlerperformance"

	Player	Inns	Overs	Mdns	Wkts	Econ	Mwi
0	RJ Shastri	261	3727.2	713	280	2.9	5
1	SLV Raju	103	1728.4	378	156	2.81	6
2	Maninder Singh	109	1891.5	392	154	2.83	7
3	PP Ojha	71	1429.1	303	144	2.92	6
4	DR Doshi	45	1099.5	292	90	2.56	6
5	NS Yadav	47	1093.1	250	78	2.59	5
6	RK Chauhan	68	1063.5	250	76	2.88	4

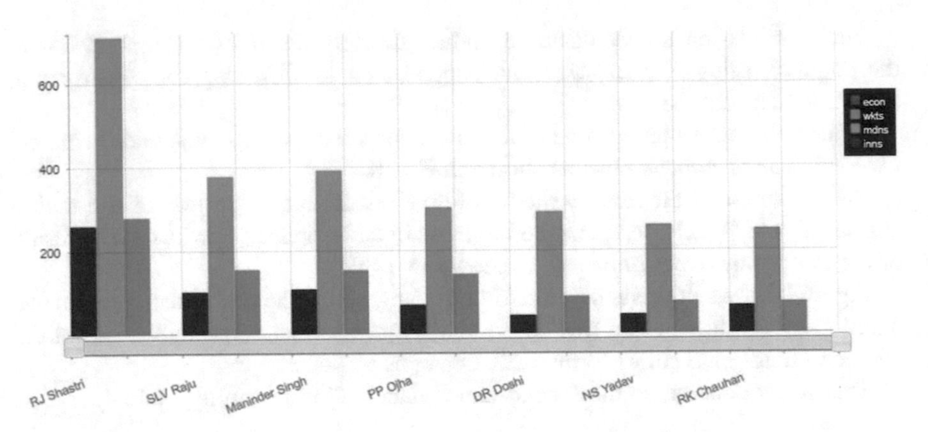

Fig. 9 The figure displays the graphical outcomes acquired from table "bowlerperformance"

Table 4 The figure displays the contents of table "matches"

	Player	Mat	Wkts	Ave	Econ	Sr
0	A Kumble	401	953	30.05	3.11	57.9
1	Harbhajan Singh	365	707	32.59	3.32	58.7
3	Z Khan	303	597	31.48	3.89	48.4
4	J Srinath	296	551	29.11	3.55	49.0
5	R Ashwin	206	477	27.39	3.64	45.0
6	AB Agarkar	221	349	31.09	4.51	41.2
7	I Sharma	171	341	35.23	3.83	55.0
8	RA Jadeja	206	328	29.95	3.6	49.8
9	IK Pathan	173	301	29.85	4.41	40.5

Figure 10 shows the results in the form of lines. The player name forms the *x*-axis and the values of wkts (green color), mat (red color), and sr (blue color) forms the lines of *y*-axis.

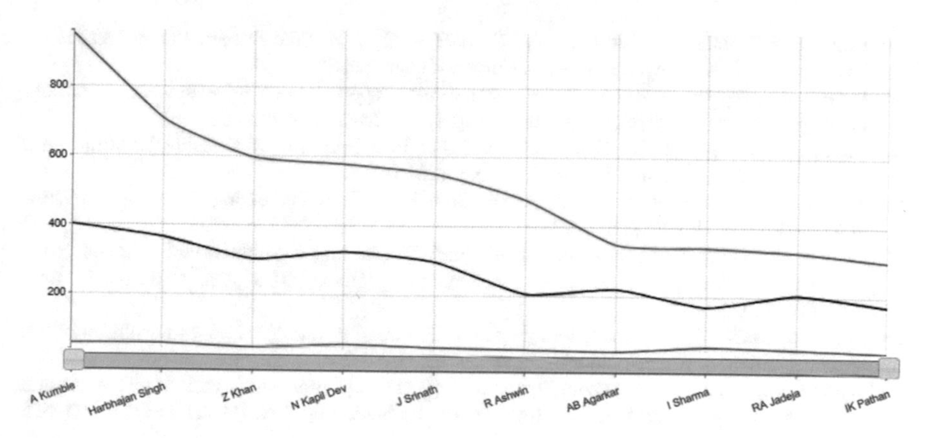

Fig. 10 The figure shows the graphical results obtained from table "matches"

5 Conclusion

The research paper elaborated the role played by big data in enhancing a team performance along with the working of MapReduce algorithm and Hortonworks Sandbox 2.2.0. The world of sports has witnessed a major change with the advent of practices and tools concerned with Big Data, but it should also be remembered that all this should be for informative purpose. There is no second thought in the fact that if a team stops having faith on its own players, coaches and most importantly on its skills, the enthusiasm related with the game and its fans would come to a tragic end. So, the Big Data should be preferred as assistance to improve the overall quality of the game and should not be regarded as a game itself.

References

1. Smolan, R., Erwitt, J.: The Human Face of Big Data, 1st edn. Sterling Publishing Company Incorporated (2012)
2. Han, J., Kamber, M., Pei, J.: Data Mining Concepts and Techniques, 3rd edn. Elsevier Morgan Kaufmann, USA (2012)
3. Katal, A., Wazi, M., Goudar, R.H.: Big data: issues, challenges, tools and good practices. In: Proceedings of IEEE, pp. 404–409 (2013)
4. Kaisler, S., Armour, F., Espinosa, J.A., Money, W.: Big data: issues and challenges moving forward. In: Proceedings of 46th International Conference on System Sciences, pp. 995–1004. IEEE Computer Society, Hawaii (2013)
5. Big data—insights and challenges. http://www.slideshare.net/rupenmomaya/big-data-insights-challenges. Accessed: 03/05/2017
6. Suthakar, U., Magnono, L., Smith, D.R., Khan, A., Andreeva, J.: An efficient strategy for collection and storage of large volumes of data for computation. J. Big Data 3–21 (2016)
7. Rajaraman, V.: Big Data Analytics, pp. 695–716. Resonance, India (2016)

8. Rein, R., Memmert, D.: Big Data and Tactical Analysis in Elite Soccer: Future Challenges and Opportunities for Sports Science. SpringerOpen (2016)
9. Leveraging big data analytics to revolutionize sports. http://www.tatvasoft.com/blog/leveraging-big-data-analytics-revolutionize-sports/. Accessed: 20/07/2017
10. Marz, N., Warren, J.: Big Data: Principles and Best Practices of Scalable Realtime Data Systems, 1st edn. Manning Publications, NY (2013)
11. Jeffery, D., Ghemawat, S.: MapReduce: Simplified Data Processing on Large Clusters. Google Research Publication (2004)
12. Jagdev, G., Kaur, A.: Comparing conventional data mining algorithms with Hadoop based Map-Reduce algorithm considering elections perspective. Int. J. Innov. Res. Sci. Eng. (IJIRSE) **3**(3), 57–68 (2017)
13. Basics of MapReduce algorithm explained with a simple example. http://www.thegeekstuff.com/2014/05/Map-Reduce-algorithm/. Accessed: 03/05/2017
14. Jagdev, G., Kaur, S.: Analyzing maneuver of Hadoop framework and MapR algorithm proficient in supervising big data. Int. J. Adv. Technol. Eng. Sci. (IJATES) **5**(5), 505–515 (2017)
15. Tao, Y., Lin, W., Xiao, X.: Minimal MapReduce algorithms. In: Proceedings of SIGMOD'13, New York (2013)
16. Gandomi, A., Haider, M.: Beyond the hype: big data concepts, methods and analytics. Int. J. Inf. Manage. **35**(2), 137–144 (2015)
17. Jagdev, G., Kaur, A., Kaur, A.: Excavating big data associated to Indian election scenario via Apache Hadoop. Int. J. Adv. Res. Comput. Sci. **7**(6), 117–123 (2016)

PEFT-Based Hybrid PSO for Scheduling Complex Applications in IoT

Komal Middha and Amandeep Verma

Abstract Internet of Things (IoT) is one of the buzzwords of the recent era and the most attractive field for researchers. It is defined as a system of connected physical objects which are approachable through the Internet and are capable of exchanging data using immerse technologies such as sensors, actuators. With the continuous evolution in IoT, number of issues arises such as confined storage space as well as limited processing capabilities. These issues can be resolved by merging IoT with cloud computing, as cloud has the immeasurable storage space as well as processing ability. This combination has proved as a boon for Internet and this combination can also be used to solve workflow scheduling problem as well. Large complex applications are often represented as workflows. Workflow scheduling is one of the eminent obstacles in both IoT and cloud computing. Several approaches have been proposed for workflow scheduling such as heuristic and meta-heuristic approaches. Commonly meta-heuristics approaches include Genetic Algorithm (GA), Simulated Annealing (SA), Particle Swarm Optimization (PSO) and heuristic approaches include Critical Path on Processor (CPOP), Heterogeneous Earliest Finish Time (HEFT), and Predict Earliest Finish Time (PEFT). But, mostly these approaches fail due to increasing of tasks, unable to execute tasks within specified budget, time, cost, and many more reasons. To overcome these above mention issues, this paper presents a hybrid PSO algorithm that uses a combine approach of both heuristic and meta-heuristic techniques namely PEFT and PSO, respectively.

Keywords Cloud computing · Internet of Things · Workflow scheduling · Predict earliest finish time · Particle swarm optimization

---

K. Middha (✉) · A. Verma
U.I.E.T., Panjab University, Chandigarh, India
e-mail: mail2komalmiddha@gmail.com

A. Verma
e-mail: amandeepverma@pu.ac.in

A. K. Luhach et al. (eds.), *Smart Computational Strategies: Theoretical and Practical Aspects*, https://doi.org/10.1007/978-981-13-6295-8_23

1 Introduction

Cloud Computing (CC) has emerged as an irresistible platform to solve multiple scientific applications. It provides on-demand network access to various measures such as storage, server, and services that can be rapidly purveyed and released with minimum efforts [1]. It has revolutionized the traditional business models by offering multiple services over the internet with the help of virtualization. It employs "pay-per-use" model, that means users are charged according to resource consumption [2]. It uses Internet as well as remote servers to provide numerous services to the users.

Cloud has three different forms namely private, public, and hybrid [3]. Each one of them provides different aspects regarding security. Public cloud is nothing but the Internet and can be accessed by anyone while private cloud is managed and owned by single company but it is more expensive than public cloud. Hybrid cloud relies both on private and public cloud and uses features of both these clouds in combination. Nowadays, community cloud is also in trend. It provides a framework for sharing data between organizations. Including the different types of cloud, we have different services offered by cloud namely Software as a Service (SaaS), Platform as a Service (PaaS), and Infrastructure as a Service (IaaS). SaaS provides multiple services to the host which can be accessed with the help of different clients such as web browser. Examples of SaaS are Google Mail, Google Docs [4]. PaaS provides a platform to the users to develop different applications and follows complete "software lifecycle" process. The most common example of PaaS is Google App Engine [5]. With the help of IaaS, the user can directly access various resources such as network and storage, for example, Amazon's EC2 [6].

Virtualization is the key concept used in CC. It allows a number of virtual machines to be centralized into a single machine and hence it helps in saving energy, reducing the cost which will help companies to maximize the profit [7]. CC offers a number of benefits such as scalability, flexibility, cost efficient [8]. Besides these benefits, it faces certain challenges also. Security [8] is one of the major challenges, as data stored on cloud can be accessed by anyone. Other challenges include lack of standards [8], technical issues [9], prone to attack [9].

Also, IoT is another new wave in the field of global industry. It is defined as a network of physical devices having sensing and network capabilities that enable the devices to store and exchange data but with limited storage capacity. As cloud offers unlimited storage and processing, IoT and cloud computing are amalgamated together. This combination is known as Cloud of Things [10]. Both of these applications can take the advantages of each other, such as IoT can overcome its storage, processing, resource allocation issues, and cloud computing which can expand its capabilities from virtual to real world. The mingling of these applications has given origin to multiple new services as shown in Fig. 1.

This new concept can also be used to solve the workflow scheduling issue also. It affects the performance of both IoT and cloud. It is NP-complete problem [12]. A workflow is defined as a set of multiple tasks having dependencies among them.

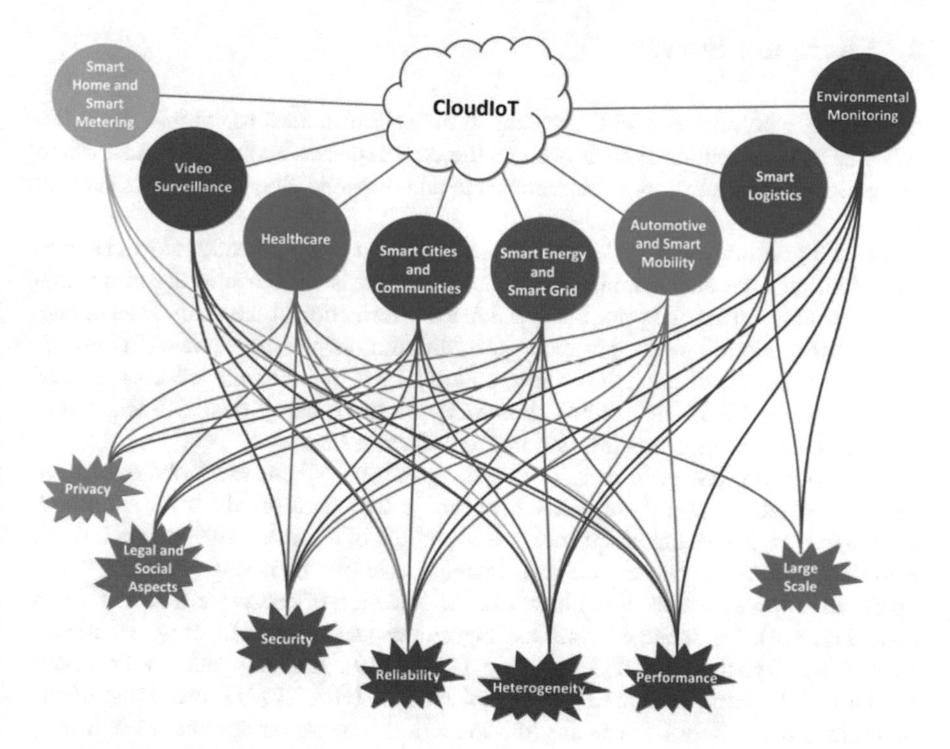

Fig. 1 New services offered [11]

Its main aim is to complete the execution of task before the deadline as well as within the budget constraints. But its complexity increases with increase in size of the task. There are multiple techniques to solve this problem such as heuristic algorithm, hybrid algorithm and many more. There are different categories under heuristic approach and among them list scheduling heuristics yield the best outcome in terms of quality as well as with efficient work schedule. Predict Earliest Finish Time (PFET) is also one of the heuristic techniques categorized under list heuristics and has better performance as compared to other list heuristics [13]. Meta-heuristic techniques also include multiple approaches like Genetic Algorithm (GA), Simulated Annealing (SA) but Particle Swarm Optimization (PSO) being an optimization technique finds the best solution among all prevailing solutions [14]. So, to solve the abovementioned problem more efficiently and effectively, this paper presents a hybrid combination PSO with PEFT algorithm, using PEFT as the inceptive basis.

The rest of the paper is divided into different sections. Section 2 reviews the literature work. Section 3 defines scheduling problem formulation followed by Sect. 4 defining the PEFT algorithm. Section 5 defines PSO algorithm and also discuss the proposed hybrid PEFT-PSO algorithm. The last section concludes the whole work along with the future scope.

2 Literature Survey

Scheduling with reference to CC means to allocate resources to the task in such a way that their execution time as well as the cost expenses should minimize. Main categories of scheduling algorithm are static and dynamic algorithm [15], shown in Fig. 2.

Static algorithm is one in which information about task is already known such as its execution time and mapping of resources whereas in dynamic algorithm this information is not known prior but it estimates information at the ready state of task [16]. Static algorithm is grouped into different types as shown in Fig. 3. Heuristic-based algorithm finds optimal solution by using a sample space of random solutions [17]. These are further classified into three categories: list scheduling heuristics, clustering heuristics, and task duplication heuristics.

A lot of work has been done in this field and many heuristic techniques have been proposed. Among these, list scheduling is frequently used for the workflow scheduling. In this technique, it makes a sorted list of all tasks with respect to their priorities. It has two phases: the prioritizing phase (for assigning priorities to the task) and a processor selection phase (for the selection of processor that minimizes cost function). It includes various algorithms such as Mapping Heuristics (MP) [18], Dynamic Level Scheduling (DLS) [19], Critical Path on Processor (CPOP) [20], Heterogeneous Earliest Finish Time (HEFT) [21], etc. Most of the algorithms focus only on reducing the execution time while ignores other factors such as cost, energy, and many more.

El-Rewini and Lewis [18] proposed MP scheduling algorithm. In this, the task is allocated to the processor which is in ready state or which has just recently completed its task execution. It does not consider other processors that are executing the task as these can complete the execution of the same task in shorter time span. Hence, scheduling done is not much efficient. Sih and Lee [19] proposed a compile time scheduling heuristic known as DLS. The outcomes of DLS are much better than the MH because it considers the allocated processors as well while scheduling tasks. It also has certain shortcomings as it does not consider the idle time between the two tasks which are scheduled on the same processor. For the better scheduling results, Hariri et al. [20, 21] have presented two algorithms namely CPOP and

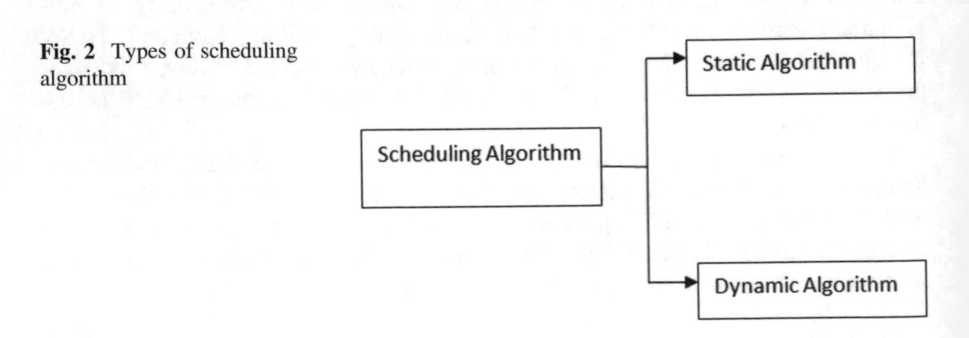

Fig. 2 Types of scheduling algorithm

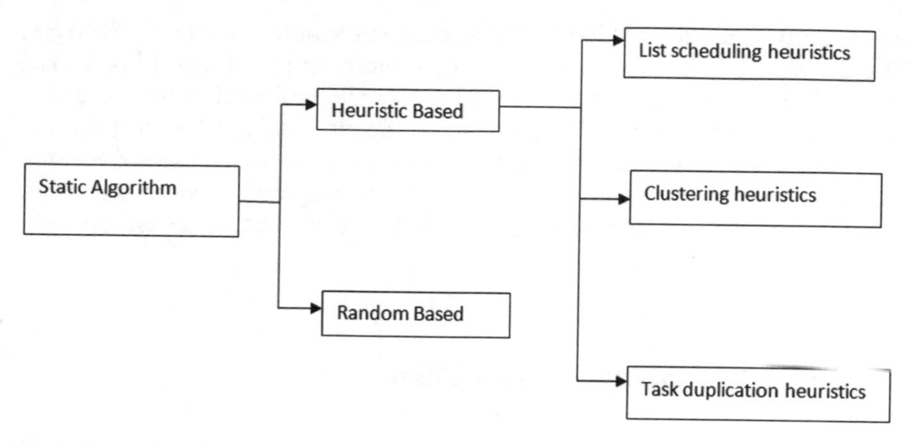

Fig. 3 Categories of static task scheduling algorithm

HEFT, which are less complex and more efficient than above-defined algorithms. HEFT has mainly two stages, first, it assigns priorities to all the tasks and after that, it assigns the topmost task to the processor that yields the earliest finish time for the task.

Based on HEFT, Sakellario and Zaho [22] gave two algorithms LOSS and GAIN. It schedules the task in DAG in such a way that the execution is within the budget constraint along with optimization of either cost or time. For the optimization of cost, Bossche et al. [23] proposed a set of algorithms for both public and private cloud. Including cost it also considers network bandwidth, but still it is applicable to applications having independent tasks. Lookahead algorithm given by Bittencourt et al. [24] is also based on HEFT. Its main feature is that its make span is lowest because of its processor selection technique. But it makes the whole process complicated and hence increases the complexity also. Rodriguez and Buyya [25] present an algorithm using Particle Swarm Optimization (PSO) meta-heuristic technique for scheduling workflows on IaaS with the aim to minimize cost within the specified deadline. Zhang et al. [26] proposed another algorithm that also aims to minimize cost under the deadline constraint and it is based on genetic algorithm called Dynamic Objective Genetic Algorithm (DOGA). It is suitable for business organizations as well. Verma et al. [27] proposed Multi-Objective Particle Swarm Optimization (MOPSO) for scheduling tasks within the time constraint as well as with minimum energy consumption on mobile cores or offloaded to cloud. It provides value-added services to the end user as well as it reduces overall cost.

Recently, many more algorithms have also been proposed for workflow scheduling. PEFT is one of the list heuristic techniques [13]. It has a distinct feature from proposed algorithms that, it can foresee the impact of each child of parent task using Optimistic Cost Table in quadratic time complexity. It yields better results than CPOP [20], HEFT [21], and other list scheduling heuristic in terms of efficiency and schedule length ratio. Kaur and Kalra [28] have recently presented a new hybrid PEFT-based genetic algorithm (PGA) using PEFT as initial seed and merged it with genetic algorithm, with the aim to minimize execution time and cost. But, in many previous works [27, 29–31], it has been proved that PSO is much more efficient than genetic algorithm in terms of multiple factors such as simplicity, convergence speed, high optimization speed. So, for further improving the performance of PEFT in terms of cost, time and energy, we have proposed a hybrid algorithm using optimization technique PSO. The proposed hybrid heuristic will provide a trade-off solution between execution time and cost under user-specified deadline and budget constraint.

3 Scheduling Problem Formulation

3.1 Workflow Model

A workflow model is often presented as Directed Acyclic Graph (DAG), with $G = (T, E)$ where T denotes the set of tasks (t_1, t_2, t_3 …, t_n) and E denotes the set of edges. In general terms, we can define workflow tasks as ready tasks once all its predecessor tasks are completed by virtual machines [32]. An example of DAG is shown in Fig. 4. Any task t_i can start its execution only when all its predecessors have finished their execution and necessary data and resources are also available at virtual machine which is going to execute it [21]. The positive value associated with each edge, i.e., the weight denotes the transmission cost between the tasks t_i and t_j.

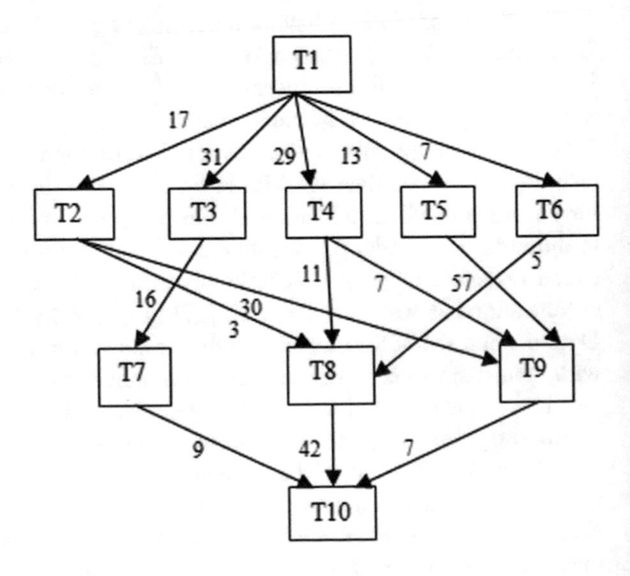

Fig. 4 Sample DAG

3.2 Application Model

Infrastructure as a Service (IaaS) provides many services such as security, networking, etc, and along with services it provides wide range of VM types. The performance of VM depends on its processing speed and on the cost per unit time. To measure execution time (ET) and transfer cost (TT), we have the following formulas:

$$\mathrm{ET}(t_{i,}, p_j) = \frac{(\mathrm{length})t_j}{\mathrm{PS}(r_j)^*(1 - (\mathrm{PD}(p_j))} \quad (1)$$

$$\mathrm{TT}\ (t_i, t_j) = \frac{\mathrm{Data}(t_i, \mathrm{out})}{\beta} \quad (2)$$

where, Data(t_i, out) is the output data size, β is the average bandwidth. When t_i, t_j are executed on the same resources than TT(t_i, t_j) becomes equal to zero.

3.3 Objective

The main objective of the workflow scheduling is to reduce the execution cost (TEC) along with keeping total execution time (TET) in the specified deadline. To calculate TEC and TET, we have the following formulas:

$$\mathrm{TEC} = \sum_{j=1}^{P} C(p_j)^* \left(\frac{\mathrm{LFT}(p_j) - \mathrm{LST}(p_j)}{T}\right) \quad (3)$$

$$\mathrm{TFT} = \max\{\mathrm{FT}(t_i) : t_i \in T\} \quad (4)$$

where LST(p_j) is the least start time, i.e., when the resource is enrolled by the processor, LFT(p_j) is the least end time, i.e., when the resource is released by the processor, T is the unit of time, and FT(t_i) is the finish time of task t_i.

4 Predict Earliest Finish Time Algorithm (PEFT)

PEFT [13] is one of the algorithms that comes under list scheduling algorithm. Mainly, it has two main stages: task prioritizing stage and processor selection stage. The most attractive feature about this algorithm is that it makes the use of lookahead feature which helps in measuring the impact of the assignments made by the current task for its children tasks. For this, it makes use of Optimistic Cost Table (OCT) and maintains the complexity in the quadratic order.

4.1 OCT

Both, the abovementioned phases of PEFT are dependent on OCT. Usually, OCT is represented as a matrix, where rows and columns depict the tasks and processors respectively. It is calculated as follows:

$$\mathrm{OCT}(t_i, p_j) = \max_{t_j \in \mathrm{succ}(t_i)} \left[\min_{r_w \in R}\left\{\mathrm{OCT}(t_j, p_w) + w(t_j, p_w) + c(t_i, t_j)\right\}\right] \quad (5)$$

$$c(t_i,\ t_j) = 0, \quad \text{if } p_w = p_j$$

where $c(t_i,\ t_j)$ is the average transfer cost of task t_i to task t_j and $w(t_j, p_w)$ is the execution time of task t_i on processor p_w. For the computation, it uses backward approach and for the exit node $\mathrm{OCT}(t_i,\ p_j) = 0$.

4.2 Task Prioritizing Stage

The average OCT is calculated for every task which depicts the rank of each one of them. The computed ranks are arranged in descending order and formula for the rank computation is:

$$\mathrm{rank}_{\mathrm{OCT}}(t_j) = \frac{\sum_{j=1}^{P} \mathrm{OCT}(t_i, r_j)}{P} \quad (6)$$

4.3 Processor Selection Stage

For the selection of processor, we calculate the optimistic EFT ($\mathrm{O_{EFT}}$) using (7). The main objective of this stage is to fortify that the task that is leading will finish their execution earlier.

$$\mathrm{O_{EFT}}(t_i, p_j) = \mathrm{EFT}(t_i,\ p_j) + \mathrm{OCT}(t_i, p_j) \quad (7)$$

4.4 PEFT Algorithm

(1) Compute the OCT matrix as defined in Eq. (5).
(2) Calculate $\mathrm{rank_{OCT}}$ for each matrix as given in Eq. (6) and arrange them in descending order.

(3) Repeat Step 4–6, until the list is empty.
(4) for all the processor (p_i) do
(5) Calculate EFT(n_i, p_j) value using insertion based scheduling policy for the topmost task [17].
(6) Using Eq. (7), calculate $O_{EFT}(n_i, p_j)$.
(7) Allocate task to the processor with minimum O_{EFT}.
(8) Return the optimal schedule.

5 Hybrid PSO Algorithm

This section gives an overview of PSO algorithm along with that it includes the proposed algorithm. Each step followed in the algorithm is also explained.

5.1 *PSO Algorithm*

PSO is a heuristic optimization method proposed by Dr. Kennedy and E. Berhart in 1995 [33]. It is based on Swarm Intelligence mechanism and has many advantages over Genetic Algorithm (GA) such as, it has high convergence rate as compared to GA [34], also it is more algebraically proficient than GA [29]. PSO mainly has three main steps namely: evaluating the fitness value, calculating the velocity and position of particles and the last step is to update the particle position and velocity, shown in Fig. 5.

Initially, a particle represents a swarm of individuals and is represented by two variables namely, its current position and current velocity [35]. The position depends on two parameters, the first is p_{best}, i.e., the best position traversed by the particle so far and other is global best position g_{best}. The values of position and velocity are updated using the formulas defined below:

For updating velocity:

$$v_i(t+1) = w.\, v_i(t) + c_1 r_1 (x_i{}^*(t) - x_i(t)) + c_2 r_2 (x^*(t) - x_i(t))) \tag{8}$$

For updating position:

$$x_i(t+1) = x_i(t) + v_i(t) \tag{9}$$

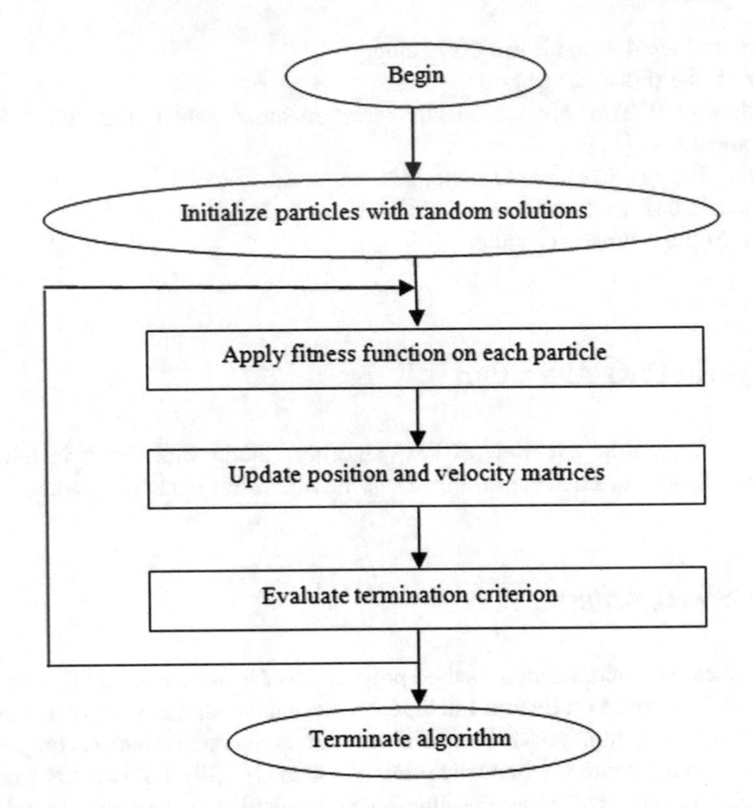

Fig. 5 Flowchart of PSO algorithm

where w represents inertia, c_i represents acceleration coefficient, and r_i is random number where r_i € [0,1], x_i* is the best position of particle, i.e., p_{best}, x^* represents the best particle in the population, i.e., g_{best} and x_i is the current position of particle i.

Both these values depend on the fitness function. To get a trade-off solution between execution time and execution cost, the fitness function for proposed heuristic is calculated as [36]:

$$\text{Fitness} = \alpha * \text{Time} + (1 - \alpha) * \text{Cost} \tag{10}$$

where time and cost denote the total execution time and cost respectively required for producing the workflow schedule and α represents the cost-time balance element in the range [0–1].

The whole process of PSO iterates until it finds an optimal solution or until termination criterion is not met as shown in Fig. 5.

5.2 Proposed Algorithm

Initially, PSO takes a random set of swarm of particles, but here we propose a new hybrid PSO algorithm, which amalgamates heuristic and meta-heuristic approaches namely PEFT and PSO respectively with PEFT being an initial input. In this, it will initialize first set of swarm using PEFT schedule and rest are initialized randomly. The whole description of each step is explained below:

Initialization. This is the most crucial step in any algorithm as, the whole performance and results both depend on input. Here, initialization is done in such a way that the first set of swarm is obtained using schedule of PEFT algorithm and the rest of the swarm are chosen randomly.

Fitness Value Computation. The computation of fitness value is directly dependent on the objective function. In this algorithm, the main aim is to minimize the execution cost, time and energy. The swarm with minimum cost as well as the one with minimum energy consumed will be opted else the swarm with minimum execution time and with minimum energy consumption will be opted. Mathematically, it can be stated as [37]:

$$\text{Min}(\text{Time}, \text{Cost}, \text{Energy})$$
$$\text{Subject to constraint are: Time} < D \text{ and Cost} < B$$

where D represents the deadline and B represents the estimated budget which is initially decided.

Parameters. The two main parameters used are p_{best} and g_{best}, which are responsible for optimizing the performance of the algorithm. The p_{best} defines the best solution attained till now and g_{best} value determines the global best value achieved by the swarm. After selecting the parameters, the position and velocity for each swarm are calculated using formulas (10) and (8) defined in Sect. 5.1.

Proposed PEFT-PSO Algorithm:

(1) Initialize the first set of swarm using schedule obtained from PEFT algorithm and the remaining are initialized randomly.
(2) Calculate the fitness value for each particle position as stated in Eq. (10).
(3) Compare the calculated fitness value with p_{best}, if the current value is better than p_{best} then update the value of particle with the current value.
(4) Also compare the fitness value with g_{best}, if the current value is better than g_{best} then update the value of particle with the current value.
(5) Assign best of all p_{best} as g_{best}.
(6) Update particle position and velocity using Eqs. (8) and (9).
(7) Repeat steps 2–5 until optimal solution is obtained.

The proposed heuristic will be evaluated using the scientific applications.

6 Conclusion and Future Scope

Although, a lot of scheduling algorithms exists but each one has certain issues associated with them. This paper presents a new methodology to solve the abovementioned issue. It employs a combination of two meta-heuristic approaches, i.e., PEFT and PSO. The main objective is to optimize scheduling along with certain constraints such as minimum cost, time and energy. In the future, we would like to implement our proposed technique and many more new initiatives can also be put in this field.

References

1. Srivinas, J., Reddy, K.V.S., Qyser, A.M.: Cloud computing basics. Int. J. Adv. Res. Comput. Commun. Eng. **1**(5), 343–347 (2012)
2. Al-Roomi, M., Al-Ebrahim, S., Buqrais, S., Ahmad, I.: Cloud computing pricing models: a survey. Int. J. Grid Distrib. Comp. **6**(5), 93–106 (2013). https://doi.org/10.14257/ijgdc.2013.6.5.09
3. AlZain, M., et al.: Cloud computing security: from single to multi-clouds. In: Proceedings of the 45th Hawaii International Conference on System Science (HICSS), pp. 5490–5499. IEEE Press, California (2012). https://doi.org/10.1109/hicss.2012.153
4. Abrishami, S., Naghibzadeh, M.: Deadline-constrained workflow scheduling in software as a service cloud. Sci. Iran. **19**(13), 680–689 (2012). https://doi.org/10.1016/j.scient.2011.11.047
5. Kumar, S., Goudar, R.H.: Cloud computing-research issues, challenges, architecture, platforms and services: a survey. Int. J. Fut. Comp. and Comm., vol. 1, no. 4, pp. 356–360 (2012). https://doi.org/10.7763/ijfcc.2012.v1.95
6. Abrishami, S., Naghibzadeh, M., Epema, D.H.J.: Deadline-constrained workflow scheduling algorithms for infrastructure as a service clouds. Fut. Gener. Comput. Syst., **29**, 158–169 (2013). https://doi.org/10.1016/j.future.2012.05.004
7. Jain, N., Choudhary, S.: Overview of virtualization in cloud computing. In: Symposium on Colossal Data Analysis and Networking, Indore (2016)
8. Kavitha, K.: Study on cloud computing models and its benefits, challenges. Int. J. Innov. Res. Comput. Commun. Eng. (IJIRCCE) **2**(1), 2423–2431 (2014)
9. Apostu, A., Puican, F., Ularu, G., Suciu, G., Todoran, G.: New classes of applications in the cloud evaluating advantage and disadvantage of cloud computing for telementary applications. Database Syst. J. **5**(1),3–14 (2014)
10. Aazam, M., Khan, I., Alsaffar, A.A.: Cloud of things: integrating internet of things and cloud computing and the issues involved. In: Proceeding of 11th International Bhurban Conference on Applied Sciences and Technology (IBCAST), Pakistan (2014). https://doi.org/10.1109/ibcast.2014.6778179
11. Boota, A., Donato, W., Pescape, A.: Integration of cloud computing and internet of thing: a survey. Fut. Gener. Comput. Syst. (FGCS) **56**, 684–700 (2016). http://dx.doi.org/10.1016/j.future.2015.09.021
12. Ullaman, J.D.: NP-complete scheduling problems. J. Comput. Syst. Sci. **10**(3), 384–393 (1975). https://doi.org/10.1016/s0022-0000(75)80008-0
13. Arabnejad, H., Barbosa, J.G.: List scheduling algorithm for heterogeneous systems by an optimistic cost table. IEEE Trans. Parallel Distrib. Syst. **25**(3), 682–694 (2014)

14. Abdi, S., Motamedi, S.A., Shorfian, S.: Task scheduling using modified particle swarm optimization algorithm in cloud environment. In: International Conference on Machine Learning, Dubai, pp. 37–41 (8–9 Jan). https://doi.org/10.15242/iie.e0114078
15. Arya, L.K., Verma, A.: Workflow scheduling algorithm in cloud environment—a survey. Eng. Comput. Sci. (RAECS), Chandigarh, India (2014). https://doi.org/10.1109/races.2014.6799514
16. Kwok, Y.K., Ahmad, I.: Dynamic critical path scheduling: an effective technique for allocating task graph to multiprocessors. IEEE Trans. Parallel Distrib. Syst. **7**(5), 506–521 (1996)
17. Sharma, N., Tyagi, S., Atri, S.: A survey on heuristic approach on task scheduling in cloud computing. Int. J. Adv. Res. Comput. Sci. **8**(3), 260–274 (2002)
18. El-Rewini, H., Lewis, T.G.: Scheduling parallel program tasks onto arbitrary target machines. J. Parallel Distrib. Comput. **9**(2), 138–153 (1990). https://doi.org/10.16/0743-7315(90)90042-n
19. Sih, G.C., Lee, E.A.: A compile-time scheduling heuristic for interconnection-constrained heterogeneous processor architecture. IEEE Trans. Parallel Distrib. Syst. **4**(2), 175–187 (1993). https://doi.org/10.1109/71.207593
20. Topcuoglu, H., Hariri, S., Wu, M.: Task scheduling algorithms for heterogeneous processors. In: Proceeding of 8th Heterogeneous Computing Workshop (HCS), USA, pp. 3–14 (1999). https://doi.org/10.1109/hcw.1999.765092
21. Topcuoglu, H., Hariri, S., Wu, M.Y.: Performance effective and low complexity task scheduling for heterogeneous computing. IEEE Trans. Parallel Distrib. Syst. **13**(3), 260–274 (2002). https://doi.org/10.1109/71.80160
22. Sakellarion, R., Zhao, H., Tsiakkouri, E., Dikaiakos, M. D.: Scheduling workflows with budget constraint. In: Proceeding of Integrated Research in GRID Computing, pp. 189–202 (2007)
23. Bossche, R.V., Vanmechelen, K., Brockhone, J.: Online cost efficient scheduling of deadline constrained workloads on hybrid clouds. Fut. Gener. Comput. Syst. **29**(4), 973–985 (2013). https://doi.org/10.1016/j.future.2012.12.012
24. Bittencourt, L.F., Sakellariou, R., Madeira, E.R.M.: DAG scheduling using a lookahead variant of the heterogeneous earliest finish time algorithm. In: Proceeding of 18th Euromicro International Conference on Parallel, Distributed and Network-Based Processing, Washington, USA, pp. 27–34 (2010). https://doi.org/10.1109/pdp.2010.56
25. Rodriguiz, M.A., Buyya, R.: Deadline based resource provisioning and scheduling algorithm for scientific workflows on clouds. IEEE Trans. Cloud Comput. **2**(2), 222–235 (2014)
26. Chen, Z.G., Du, K.J., Zhan, Z.H., Zhang, J.: Deadline constrained cloud computing resources scheduling for cost optimization based on dynamic objective genetic algorithm. In: IEEE Congress Evolutionary Computation (CEC), pp. 708–714 (2015). https://doi.org/10.1109/cccri.2015.14
27. Verma, A., Kaushal, S., Sangaiah, A.K.: Computational intelligence based heuristic approach for maximizing energy efficiency in internet of things. In: Intelligent Decision Support Systems for Sustainable Computing. Studies in Computational Intelligence, vol. 705, pp. 53–76. Springer, Berlin (2017). https://doi.org/10.1007/978-3-319-53153-3_4
28. Kaur, G., Kalra, M.: Deadline constrained scheduling of scientific workflows on cloud using hybrid genetic algorithm. In: IEEE Conference, Noida (2017). https://doi.org/10.1109/confluence.2017.7943162
29. Hassan, R., Cohanim, B., Weck, O.: A comparison of particle swarm optimization and genetic algorithm. In: 46th AIAA/ASME/ASCE/AHS/ASC structures, Structural Dynamics and Material Conference, pp. 1–13 (2005). https://doi.org/10.2514/6.2005-1897
30. Li, Z., Liu, X., Duan, X.: Comparative research on particle swarm optimization and genetic algorithm. Comput. Inf. Sci. (CCSE) **3**(1), 120–127 (2010)
31. Elbetagi, E., Hegazy, T., Grierson, D.: Comparison among five evolutionary based optimization algorithm. Adv. Eng. Inf. (AEI), pp. 43–53 (2005). https://doi.org/10.1016/j.aci.2005.01.004

32. Smanchat, S., Viriyapant, K.: Taxonomies of workflow scheduling problem and techniques in the cloud. Fut. Gener. Comput. Syst. **52**, 1–12 (2015). https://doi.org/10.1016/j.future.2015.04.019
33. Bai, Q.: Analysis of particle swarm optimization algorithm. Comput. Inf. Sci. **3**(1), 180–184 (2010). https://doi.org/10.5539/cis.v31p180
34. Chavan, S.D., Adgokar, N.P.: An overview on particle swarm optimization: basic concepts and variants. Int. J. Sci. Res. (IJSR) **4**(5), 255–260 (2015)
35. Verma, A., Singh, M.: Particle swarm optimization techniques for workflow scheduling in cloud: a survey. Int. J. Inf. Commun. Technol. **6**(1–2), 385–390
36. Verma, A., Kaushal, S.: Cost minimization PSO based workflow scheduling plan for cloud computing. Int. J. Inf. Technol. Comput. Sci. **8**, 37–43 (2015). https://doi.org/10.5815/ijites.2015.08.06
37. Verma, A., Kaushal, S.: A hybrid multi-objective PSO for scientific workflow scheduling. Parallel Comput. **62**, 1–9 (2017). https://doi.org/10.1016/j.parco.2017.01.002

Printed in the United States
By Bookmasters